P9-DIE-081

THE MODEL THINKER

THE MODEL THINKER

What You Need to Know to Make Data Work for You

SCOTT E. PAGE

BASIC BOOKS

New York

Copyright © 2018 by Scott E. Page
Cover Design by Chin-Yee Lai
Cover © 2018 Hachette Book Group, Inc.

Hachette Book Group supports the right to free expression and the value
of copyright. The purpose of copyright is to encourage writers and artists to
produce the creative works that enrich our culture.

The scanning, uploading, and distribution of this book without permission
is a theft of the author's intellectual property. If you would like permission to
use material from the book (other than for review purposes), please contact
permissions@hbgusa.com. Thank you for your support of the author's rights.

Basic Books
Hachette Book Group
290 Avenue of the Americas, New York, NY 10104
www.basicbooks.com

Printed in the United States of America

First Edition: November 2018

Published by Basic Books, an imprint of Perseus Books, LLC, a subsidiary of
Hachette Book Group, Inc. The Basic Books name and logo is a trademark of
the Hachette Book Group.

The Hachette Speakers Bureau provides a wide range of authors for speaking
events. To find out more, go to www.hachettespeakersbureau.com or call
(866) 376-6591.

The publisher is not responsible for websites (or their content) that are not
owned by the publisher.

Library of Congress Control Number: 2018942802

ISBNs: 978-0-465-09462-2 (hardcover); 978-0-465-09463-9 (ebook)

LSC-C

10 9 8 7 6 5 4 3 2 1

To Michael D. Cohen

(1945–2013)

CONTENTS

CONTENTS

It can scarcely be denied that the supreme goal of all theory is to make the irreducible basic elements as simple and as few as possible without having to surrender the adequate representation of a single datum of experience.

—ALBERT EINSTEIN

Prologue

To me success means effectiveness in the world, that I am able to carry my ideas and values into the world—that I am able to change it in positive ways.

—Maxine Hong Kingston

This book began as the result of a chance meeting with Michael Cohen in 2005 near the flower garden in the mall adjacent to the University of Michigan's West Hall. Michael, a scholar known for his generosity, made a comment that altered my teaching career. With a twinkle in his eyes, Michael said, "Scottie, I once taught a course called Introduction to Modeling for Social Scientists, based on a book written by Charles Lave and James March. You should resurrect the course. It needs you."

It needed me? I returned to my office a little confused, so I chased down an old course syllabus. I discovered that Michael had misled me. The course did not need me. I needed it. I had been wanting to develop a course that would introduce students to the core ideas of complex systems—networks, diversity, learning, large events, path dependence, tipping points—that would be relevant to their daily lives and future careers. By teaching modeling, I could make students better thinkers while introducing them to complexity. I could teach them tools that would improve their abilities to reason, explain, predict, design, communicate, act, and explore

The course's motivating idea would be that we must confront the complexity of the modern world with multiple models. At semester's end, rather than see the world from a particular angle, students would see the world through many lenses. They would be standing in houses with many windows, able to look in multiple directions. My students would be better prepared for the complex challenges before them—improving education, reducing poverty, creating sustainable growth, finding meaningful work in an age of artificial intelligence, managing resources, and designing robust financial, economic, and political systems.

The next fall, I resurrected the course. I contemplated rebranding it as Thirty-Two Models That Will Turn You into a Genius, but the culture at Michigan frowns on hyperbole, so I stuck with Michael's title: An Introduction to Modeling. Lave and March's book proved to be a brilliant introduction. However, modeling had made huge advances in the intervening decades. I needed an updated version that included models of long-tailed distributions, networks, rugged landscapes, and random walks. I needed a book that discussed complexity.

So I began to write. For two years, the ground proved rocky. My plow moved at a slow place. One spring day, I again ran into Michael, this time in the archway underneath West Hall. I had been questioning the course, which was now drawing twenty students. Were models too abstract for undergraduates? Should I teach a different course on a specific issue or policy domain? Michael offered up a smile, noting that any endeavor worth pursuing merited questioning. As we parted, Michael commented on the importance and value of helping people think clearly. He told me not to give up, that he took joy in my challenges.

In the fall of 2012, the ground under the course shifted. Vice Provost Martha Pollack asked me to teach an online version—what is now called a MOOC. With a tablet computer, a $29 camera, and a $90 microphone, Model Thinking was born. With assistance from too many people at Michigan, Coursera, and Stanford University to thank properly (a quick shout-out to Tom Hickey, who did yeoman's work), I reorganized my lectures into a form suitable for an online course, dividing each subject into modules and removing all copyrighted material. With my dog Bounder as an audience, I taped and retaped lectures.

The first offering of Model Thinking drew 60,000 students. That number now approaches a million. The popularity of the online course led me to abandon the book. I thought the project unnecessary, but, over the next two years, my email inbox began to fill with requests for a book to complement the online lectures. Then Michael Cohen lost his battle with cancer, and I felt that I needed to finish the book. I reopened the manuscript folder.

Writing a book requires large blocks of time and spaces that allow for clear thought. The poet Wallace Stevens wrote, "Perhaps the truth depends on a walk around the lake." I relied on a close analog: mind-clearing swims across Winans Lake, where my family spends our summers. Throughout the writing process, the continuous life I share with the love of my life, Jenna Bednar, our sons, Orrie and Cooper, and our enormous dogs, Bounder, Oda, and Hildy, has brought laughter, comfort, and opportunities—among them Orrie having one week to correct the penultimate draft's mathematical errors and Jenna having two weeks to identify instances of unclear writing, logical flaws, and muddled thinking. As has been true of most of my written work, this manuscript might be best described as an original draft by Scott Page with substantial revision by Jenna Bednar.

During the seven-year period of writing this book, my children have transitioned from pre-teens to young adults. Orrie is now off to college. Cooper follows next year. In the interval between sketching the initial outline and submitting the final version, my family has consumed copious amounts of bibimbap, pasta carbonara, and oatmeal chocolate chip cookies, taken the saws and loppers to scores of fallen branches and limbs, repaired dozens of breaks in the backyard fence, embarked on numerous failed initiatives to reduce the entropy in the basement and garage, and wished and hoped for the ice on the lake to be suitable for skating. We have also had to accept loss. Midway through the project, my mother, Marilyn Tamboer Page, died from a sudden heart attack while enjoying the bliss of her routine daily walk with her dog. Not a day goes by when I do not reflect on the love she showered on her family and the support she gave to others.

The book before you is as complete as it can be at this moment in time. Doubtless, new models will be created, and old models will find new uses creating gaps in this current offering. As I humbly send the manuscript out into the world, I feel that my efforts will have been repaid if you, the reader, find the models and ideas within to be useful and generative, and that you are able to carry them out into the world and change it in positive ways.

If one day, when sitting in some professor's or graduate student's office, preferably at a college or university in my beloved Midwest, I scan the bookshelves and find this book leaning, as it has during its writing, on a well-worn copy of Lave and March, then my efforts will have been all the sweeter.

1. The Many-Model Thinker

To become wise you've got to have models in your head. And you've got to array your experience—both vicarious and direct—on this latticework of models.

—Charlie Munger

This is a book about models. It describes dozens of models in straightforward language and explains how to apply them. Models are formal structures represented in mathematics and diagrams that help us to understand the world. Mastery of models improves your ability to reason, explain, design, communicate, act, predict, and explore.

This book promotes a *many-model thinking* approach: the application of ensembles of models to make sense of complex phenomena. The core idea is that many-model thinking produces wisdom through a diverse ensemble of logical frames. The various models accentuate different causal forces. Their insights and implications overlap and interweave. By engaging many models as frames, we develop nuanced, deep understandings. The book includes formal arguments to make the case for multiple models along with myriad real-world examples.

The book has a pragmatic focus. Many-model thinking has tremendous practical value. Practice it, and you will better understand complex phenomena. You will reason better. You exhibit fewer gaps in your reasoning and make more robust decisions in your career, community activities, and personal life. You may even become wise.

Twenty-five years ago, a book of models would have been intended for professors and graduate students studying business, policy, and the social sciences along with financial analysts, actuaries, and members of the intelligence community. These were the people who applied models and, not coincidentally, they were also the people most engaged with large data sets. Today, a book of models has a much larger audience: the vast universe of knowledge workers, who, owing to the rise of big data, now find working with models a part of their daily lives.

1

Organizing and interpreting data with models has become a core competency for business strategists, urban planners, economists, medical professionals, engineers, actuaries, and environmental scientists among others. Anyone who analyzes data, formulates business strategies, allocates resources, designs products and protocols, or makes hiring decisions encounters models. It follows that mastering the material in this book—particularly the models covering innovation, forecasting, data binning, learning, and market entry timing—will be of practical value to many.

Thinking with models will do more than improve your performance at work. It will make you a better citizen and a more thoughtful contributor to civic life. It will make you more adept at evaluating economic and political events. You will be able to identify flaws in your logic and in that of others. You will learn to identify when you are allowing ideology to supplant reason and have richer, more layered insights into the implications of policy initiatives, whether they be proposed greenbelts or mandatory drug tests.

These benefits will accrue from an engagement with a variety of models—not hundreds, but a few dozen. The models in this book offer a good starting collection. They come from multiple disciplines and include the Prisoners' Dilemma, the Race to the Bottom, and the SIR model of disease transmission. All of these models share a common form: they assume a set of entities—often people or organizations—and describe how they interact.

The models we cover fall into three classes: simplifications of the world, mathematical analogies, and exploratory, artificial constructs. In whatever form, a model must be tractable. It must be simple enough that within it we can apply logic. For example, we cover a model of communicable diseases that consists of infected, susceptible, and recovered people that assumes a rate of contagion. Using the model we can derive a contagion threshold, a tipping point, above which the disease spreads. We can also determine the proportion of people we must vaccinate to stop the disease from spreading.

As powerful as single models can be, a collection of models accomplishes even more. With many models, we avoid the narrowness inherent in each individual model. A many-models approach illuminates each component model's blind spots. Policy choices made based on single models may ignore important features of the world such as income disparity, identity diversity, and interdependencies with other systems.[1] With many models, we build logical understandings of multiple processes. We see how causal processes overlap and interact. We create the possibility of making sense of the complexity that characterizes our economic, political, and social worlds. And, we do so without abandoning rigor—model thinking ensures logical coherence. That logic can be then be grounded in evidence by taking models to data to test, refine, and improve them. In sum, when our thinking is informed by diverse logically consistent, empirically validated frames, we are more likely to make wise choices.

Models in the Age of Data

The appearance of a book on models may seem out of place in the era of big data. Today, data exists in unprecedented dimensionality and granularity. Customer purchase data, which used to arrive in monthly aggregates on printed paper, now streams instantaneously with geospatial, temporal, and consumer tags. Student academic performance data now includes scores on every homework, paper, quiz, and exam, as opposed to semester-end summary grades. In the past, a farmer might mention dry ground at a monthly Grange meeting. Now, tractors transmit instantaneous data on soil conditions and moisture levels in square-foot increments. Investment firms track dozens of ratios and trends for thousands of stocks and use natural-language processing tools to parse documents. Doctors can pull up page upon page of individual patient records that can include relevant genetic markers.

A mere twenty-five years ago, most of us had access to little more than a few bookshelves' worth of knowledge. Perhaps your place of work had a small reference library, or at home you had a collection of encyclopedias and a few dozen reference books. Academics and government and private-sector researchers had access to large library collections, but even they had to physically visit the material. As late as the turn of the millennium, academics could be found shuttling back and forth between card catalog rooms, microfiche collections, library stacks, and special collections in search of information.

That has all changed. Content that had been paper-bound for centuries now flows in tiny packets through the air. So too does the information about the here and now. News that arrived on our doorsteps on newsprint once a day now flows in a continuous digital stream into our personal devices. Stock prices, sports scores, and news of political events and cultural happenings can all be accessed with a swipe or query.

As impressive as the data may be, it is no panacea. We now know what has happened and is happening, but, owing to the complexity of the modern world, we may be less capable of understanding why it happened. Empirical findings may be misleading. Data on piece-rate work often shows that the more people are paid per unit of output, the less they produce. A model in which pay depends on work conditions can explain those data. If conditions are poor so that producing output is difficult, per unit pay may be high. If conditions are good, per unit pay may be low. Thus, higher pay does not lead to less productivity. Instead, more difficult work conditions require higher per unit pay.[2]

In addition, most of our social data—that is, data about our economic, social, and political phenomena—documents only moments or intervals in time. It rarely tells us universal truths. Our economic, social, and political worlds are not stationary. Boys may outscore girls on standardized tests in one decade and girls may outscore boys the next. The reasons people vote today may differ from the reasons they vote in coming decades.

We need models to make sense of the fire-hose-like streams of data that cross our computer screens. Thus, it is because we have so much data that this might also be called the age of many models. Look across the academy, government, the business world, and the nonprofit sector, and you struggle to find a domain of inquiry or decision not informed by models. Consulting giants McKinsey and Deloitte build models to formulate business strategies. Financial firms such as BlackRock and JPMorgan Chase apply models to select investments. Actuaries at State Farm and Allstate use models to calibrate risk when pricing insurance policies. The people team at Google builds predictive analytic models to evaluate its more than three million job applicants. College and university admissions officers construct predictive models to select from among tens of thousands of applicants.

The Office of Management and Budget constructs economic models to predict the effects of tax policies. Warner Brothers applies data analytics to create models of audience responses. Amazon develops machine learning models to make product recommendations. Researchers funded by the National Institutes of Health build mathematical models of human genomics to search for and evaluate potential cures for cancer. The Gates Foundation uses epidemiological models to design vaccination strategies. Even sports teams use models to evaluate draft prospects and trade opportunities and to formulate within-game strategies. By relying on models to select players and strategies, the Chicago Cubs won a World Series championship after more than a century of failures.

To people who use models, the rise of model thinking has an even simpler explanation: *models make us smarter*. Without models, people suffer from a laundry list of cognitive shortcomings: we overweight recent events, we assign probabilities based on reasonableness, and we ignore base rates. Without models, we have limited capacity to include data. With models, we clarify assumptions and think logically. And, we can leverage big data to fit, calibrate, and test causal and correlative claims. With models, we think better. In head-to-head competitions between models and people, models win.[3]

Why We Need Many Models

In this book we advocate using not just one model in a given situation but many models. The logic behind the many-model approach builds on the age-old idea that we achieve wisdom through a multiplicity of lenses. This idea traces back to Aristotle, who wrote of the value of combining the excellences of many. A diversity of perspectives was also a motivation for the great-books movement, which collected 102 important transferable ideas in *The Great Ideas: A Syntopicon of Great Books of the Western World*. The approach finds a modern voice in the work of Maxine Hong Kingston, who wrote in *The Woman Warrior*, "I learned to make my mind large, as the universe is large, so that there is room for paradoxes." It is also the basis for pragmatic actions in the world of business and policy. Recent books argue that if we want to understand of international relations, we should not model the world exclusively as a group of self-interested nations with well-defined objectives, or only as an evolving nexus of multinational corporations and intergovernmental organizations. We should do both.[4]

As commonsensical as the many-model approach may seem, keep in mind that it runs counter to how we teach models and the practice of modeling. The traditional approach—the one taught in high school—relies on a one-to-one logic: one problem requires one model. For example: now we apply Newton's first law; now we apply the second; now the third. Or: here we use the replicator equation to show the size of the rabbit population in the next period. In this traditional approach, the objective is to (a) identify the one proper model and (b) apply it correctly. Many-model thinking challenges that approach. It advocates trying many models. Had you used many-model thinking in ninth grade, you might have been held back. Use it now, and you will move forward.

Academic papers, for the most part, follow the one-to-one approach as well, even though they use those single models to explain complex phenomena: Trump voters in the 2016 election were those who had been left behind economically. Or: the quality of a child's second-grade teacher determines how economically successful that child will be as an adult.[5] A stream of best-selling nonfiction titles present cures for our ills based on single-model thinking: Educational success depends on grit. Inequality results from concentrations of capital. Our nation's poor health is due to sugar consumption. Each of these models may be true, but none is comprehensive. To confront the complexity of these challenges, to create a world of broader educational achievement, will require lattices of models.

By learning the models in this book, you can begin to build your own lattice. The models originate from a broad spectrum of disciplines, addressing phenomena as varied as the causes of income inequality, the distribution of power, the spread of diseases and fads, the conditions that precede social uprisings, the evolution of cooperation, the emergence of order in cities, and the structure of the internet. The models vary in their assumptions and their structure. Some describe small numbers of rational, self-interested actors. Others describe large populations of rule-following altruists. Some describe equilibrium processes. Others produce path dependence and complexity. The models also differ in their uses. Some help predict and explain. Others guide actions, inform designs, or facilitate communication. Still others create artificial worlds for our minds to explore.

The models share three common characteristics: First, they simplify, stripping away unnecessary details, abstracting from reality, or creating anew from whole cloth. Second, they formalize, making precise definitions. Models use mathematics, not words. A model might represent beliefs as probability distributions over states of the world or preferences as rankings of alternatives. By simplifying and making precise, they create tractable spaces within which we can work through logic, generate hypotheses, design solutions, and fit data. Models create structures within which we can think logically. As Wittgenstein wrote in his *Tractatus Logico-Philosophicus*, "Logic takes care of itself; all we have to do is to look and see how it does it." The logic will help to explain, predict, communicate, and design. But the logic comes at a cost, which leads to their third characteristic: *all models are wrong*, as George Box noted.[6] That is true of all models; even the sublime creations of Newton that we refer to as laws hold only at certain scales. Models are wrong because they simplify. They omit details. By considering many models, we can overcome the narrowing of rigor by crisscrossing the landscape of the possible.

To rely on a single model is hubris. It invites disaster. To believe that a single equation can explain or predict complex real-world phenomena is to fall prey to the charisma of clean, spare mathematical forms. We should not expect any one model to produce exact numerical predictions of sea levels in 10,000 years or of unemployment rates in 10 months. We need many models to make sense of complex systems. Complex systems like politics, the economy, international relations, or the brain exhibit ever-changing emergent structures and patterns that lie between ordered and random. By definition, complex phenomena are difficult to explain, evolve, or predict.[7]

Thus, we confront a disconnect. On the one hand, we need models to think coherently. On the other hand, any single model with a few moving parts cannot make sense of high-dimensional, complex phenomena such as patterns in international trade policy, trends in the consumer products industry, or adaptive responses within the brain. No Newton can write a three-variable equation that explains monthly employment, election outcomes, or reductions in crime. If we hope to understand the spread of diseases, variation in educational performance, the variety of flora and fauna, the effect of artificial intelligence on job markets, the impact of humans on the earth's climate, or the likelihood of social uprisings, we must come at them with machine learning models, systems dynamics models, game theory models, and agent-based models.

The Wisdom Hierarchy

To sketch the argument for many-model thinking, we begin with a query from poet and dramatist T. S. Eliot: "Where is the wisdom we have lost in knowledge? Where is the knowledge we have lost in information?" To that we might add, where is the information we have lost in all this data?

Eliot's questioning can be formalized as the *wisdom hierarchy*. At the bottom of the hierarchy lie *data*: raw, uncoded events, experiences, and phenomena. Births, deaths, market transactions, votes, music downloads, rainfall, soccer matches, and speciation events. Data can be long strings of zeros and ones, time stamps, and linkages between pages. Data lack meaning, organization, or structure.

Information names and partitions data into categories. Examples clarify the distinction between data and information. Rain falling on your head is data. Total rainfall for the month of July in Burlington, Vermont, and Lake Ontario's water level are information. The bright red peppers and yellow corn on farmers' stands surrounding the capitol in Madison, Wisconsin, on market Saturdays are data. The farmers' total sales are information.

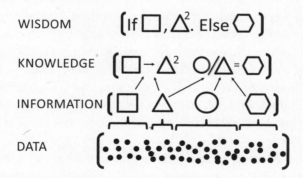

Figure 1.1: How Models Transform Data into Wisdom

We live in an age of abundant information. A century and a half ago, knowing information brought great economic and social status. Jane Austen's Emma asks if Frank Churchill is "a young man of information." Today she would not care. Churchill, like everyone else, would have a smartphone. The question is whether he could put that information to use. As Fyodor Dostoyevsky writes in *Crime and Punishment*, "We've got facts, they say. But facts aren't everything; at least half the battle consists in how one makes use of them!"

Plato defined *knowledge* as justified true belief. More modern definitions refer to it as understandings of correlative, causal, and logical relationships. Knowledge organizes information. Knowledge often takes model form. Economic models of market competition, sociological models of networks, geological models of earthquakes, ecological models of niche formation, and psychological models of learning all embed knowledge. Those models explain and predict. Models of chemical bonds explain why metallic bonds prevent us from putting our hands through steel doors while hydrogen bonds yield to our weight when we dive into a lake.[8]

Atop the hierarchy lies *wisdom*, the ability to identify and apply relevant knowledge. Wisdom requires many-model thinking. Sometimes, wisdom consists of selecting the best model, as if drawing from a quiver of arrows. Other times, wisdom can be achieved by averaging models; this is common when making predictions. (We discuss the value of model averaging in the next section.) When taking actions, wise people apply multiple models like a doctor's set of diagnostic tests. They use models to rule out some actions and privilege others. Wise people and teams construct a dialogue across models, exploring their overlaps and differences.

Wisdom can consist of selecting the correct knowledge or model; consider the following physics problem: A small stuffed cheetah falls from an airplane's hold at 20,000 feet. How much damage will it do upon landing? A student might know a gravity model and a terminal velocity model. The two models give different insights. The gravity model predicts that the stuffed animal would tear through a car's roof. The terminal velocity model predicts that the toy cheetah's speed tops out at around 10 mph.[9] Wisdom consists of knowing to apply the terminal velocity model. A person could stand on the ground and catch the soft cheetah in her hands. To quote the evolutionary biologist J. B. S. Haldane, "You can drop a mouse down a thousand-yard mine shaft; and, on arriving at the bottom, it gets a slight shock and walks away, provided that the ground is fairly soft. A rat is killed, a man is broken, a horse splashes."

In the stuffed-cheetah problem, arriving at the correct solution requires information (the weight of the toy), knowledge (the terminal velocity model), and wisdom (selecting the correct model). Business and policy leaders also rely on information and knowledge to make wise choices. On October 9, 2008, the value of Iceland's currency, the króna, began a free fall. Eric Ball, then treasurer of software giant Oracle, was faced with a decision. A few weeks prior he had dealt with the domestic repercussions of the home mortgage crisis. Iceland's situations posed an international concern. Oracle held billions of dollars in overseas assets. Ball considered network contagion models of financial collapse. He also thought of economic models of supply and demand in which the magnitude of a price change correlates with the size of the market shock. In 2008, Iceland had a GDP of $12 billion, or less than six months' revenues for McDonald's Corporation. Ball recollected thinking, "Iceland is smaller than Fresno. Go back to work."[10] The key to understanding this event, and many-model thinking generally, lies in recognizing that Ball did not search among many models to find one that supported an action that he had already decided to take. He did not use many models to find one that justified his action. Instead, he evaluated two models as possibly useful and then chose the better one. Ball had the right information (Iceland is small), chose the right model (supply and demand), and made a wise choice.

We next show how to create a dialogue among multiple models by reconsidering two historical events: the 2008 global financial market collapse, which reduced total wealth (or what had been thought to be wealth) by trillions of dollars, resulting in a four-year global recession, and the 1961 Cuban missile crisis, which nearly resulted in nuclear war.

The 2008 financial collapse has multiple explanations: too much foreign investment, over-leveraged investment banks, lack of oversight in the mortgage approval process, blissful optimism among home-flipping consumers, the complexity of financial instruments, a misunderstanding of risk, and greedy bankers who knew the bubble existed and expected a bailout. Superficial evidence aligns with each of these accounts: money flowed in from China, loan originators wrote toxic mortgages, investment banks had high leverage ratios, financial instruments were too complex for most to understand, and some banks expected a bailout. With models we can adjudicate between these accounts and check the internal consistency of these accounts: Do they make logical sense? We can also calibrate the models and test the magnitude of the effects.

The economist Andrew Lo, exercising many-model thinking, evaluates twenty-one accounts of the crisis. He finds each to be lacking. It does not make sense that investors would contribute to a bubble that they knew would lead to a global crisis. Hence, the extent of the bubble must have been a surprise to many. Financial firms may well have assumed the other firms had done due diligence when in fact they had not. Second, what were, in retrospect, clearly toxic (low-quality) bundles of mortgages found buyers. Had global collapse been a foregone conclusion, the buyers would not have existed. And while leverage ratios had increased since 2002, they were not much higher than they had been in 1998. And as for the notion that the government would bail out the banks, Lehman Brothers collapsed on September 15, 2008; with over $600 billion in holdings, it was the largest bankruptcy in US history. The government did not intervene.

Lo finds that each account contains a logical gap. The data, such as it is, privileges no single explanation. As Lo summarizes: "We should strive at the outset to entertain as many interpretations of the same set of objective facts as we can, and hope that a more nuanced and internally consistent understanding of the crisis emerges in the fullness of time." He goes on to say, "Only by collecting a diverse and often mutually contradictory set of narratives can we eventually develop a more complete understanding of the crisis." No single model suffices.[11]

In *Essence of Decision*, Graham Allison undertakes a many-model approach to explain the Cuban missile crisis. On April 17, 1961, a CIA-trained paramilitary group landed on the shores of Cuba in a failed attempt to overthrow Fidel Castro's communist regime, increasing tensions between the United States and the Soviet Union, Cuba's ally. In response, Soviet premier Nikita Khrushchev moved short-range nuclear missiles to Cuba. President John F. Kennedy responded by blockading Cuba. The Soviet Union backed down, and the crisis ended.

Allison interprets events with three models. He applies a rational-actor model to show that Kennedy had three possible actions: start a nuclear war, invade Cuba, or impose a blockade. He chose the blockade. The rational-actor model assumes that Kennedy draws a game tree with each action followed by the possible responses by the Soviets. Kennedy then thinks through the Soviets' optimal response. If, for example, Kennedy launched a nuclear attack, the Soviets would strike back, resulting in millions dead. If Kennedy imposed a blockade, he would starve the Cubans. The Soviet Union could either back down or launch missiles. Given that choice, the Soviet Union should back down. The model reveals the central strategic logic at play and provides a rationale for Kennedy's bold choice to blockade Cuba.

Like all models, though, it is wrong. It ignores relevant details, allowing it to initially appear a better explanation than it really is. The model neglects to add a stage in which the Soviets put the missiles in Cuba. If the Soviets had been rational, they should have drawn the same tree as Kennedy and realized that they would have to remove the missiles. The rational-actor model also fails to explain why the Soviets did not hide the missiles.

Allison applies an organizational process model to explain these inconsistencies. A lack of organizational capacity explains the Soviets' failure to hide the missiles. The same model can explain Kennedy's choice to blockade. At the time, the United States Air Force lacked the capacity to wipe out the missiles in a single strike. If even a single missile remained, it could kill millions of Americans. Allison deftly combines the two models. An insight from the organizational model changes the payoffs in the rational-choice model.

Allison adds a governmental process model. The other two models reduce countries to their leaders: Kennedy acts for the United States and Khrushchev for the Soviet Union. The government process model recognizes that Kennedy had to contend with Congress and that Khrushchev needed to maintain a political base of support. Thus, Khrushchev's placing of the missiles in Cuba signaled strength.

Allison's book shows the power of models alone and in dialogue. Each model clarifies our thinking. The rational-actor model identifies possible actions once the missiles have arrived and allows us to see the implications of those actions. The organizational model draws our attention to the fact that organizations, not individuals, carry out those actions. The governmental process model highlights the political cost of invasion. By evaluating events through all three lenses, we gain a broader and deeper understanding. All models are wrong; many are useful.

In both examples, the different models explicate distinct causal forces. Multiple models can also focus on different scales. In an oft-repeated tale, a child claims that the Earth rests on the back of a giant elephant. A scientist asks the child what the elephant stands on, to which the child replies, "A giant turtle." Anticipating what's about to come next, the child quickly adds, "Don't even ask. It's turtles all the way down."[12] If the world were turtles all the way—if the world were self-similar— then a model of the top level would apply at every level. But the economy, the political world, and society are not turtles all the way down, nor is the brain. At the sub-micron level, the brain is made up of molecules that form synapses, which in turn form neurons. The neurons combine in networks. The networks overlap in elaborate ways that can be studied with brain imaging. These neuronal networks exist on a scale below that of functional systems such as the cerebellum. Given that the brain differs at each level, we need multiple models, and those models differ. The models that characterize the robustness of neuronal networks bear little resemblance to the molecular biology models used to explain brain cell function, which in turn differ from the psychological models used to explain cognitive biases.

The success of many-model thinking depends on a degree of separability. In analyzing the 2008 financial crisis, we rely on separate models of foreign purchases of assets, of the bundling of assets, and of increased leverage ratios. Allison drew implications from the game theoretic model without considering the organizational model. In studying the human body, doctors separate the skeletal, muscular, limbic, and nervous systems. That said, many-model thinking does not require that these distinct models divide the system into independent parts. Confronted with a complex system, we cannot, to paraphrase Plato, carve the world at its joints. We can partially isolate the major causal threads and then explore how they are interwoven. In doing so, we will find that the data produced by our economic, political, and social systems exhibits coherence. Social data is more than sequences of incomprehensible hairballs that might have been spit up by the family cat.

Summary and Outline of the Book

To summarize, we live in a time awash in information and data. The same technological advances generating those data shrink time and distance. They make economic, political, and social actors more agile, capable of responding to economic and political events in an instant. They also increase connectedness, and therefore complexity. We face a technologically induced paradox: we know more about the world, but that world is more complex. In light of that complexity, any single model will be more likely to fail. We should not though abandon models. To the contrary, we should privilege logical coherence over intuition and double, triple, and even quadruple down on models and become many-model thinkers.

Becoming a many-model thinker requires learning multiple models of which we gain a working knowledge; we need to understand the formal descriptions of the models and know how to apply them. We need not be experts. Hence, this book balances accessibility and depth. It can function both as a resource and as a guide. The formal descriptions are isolated in stand-alone boxes. It avoids line after line of equations, which overwhelm even the most dedicated readers. The formalism that remains should be engaged and absorbed. Modeling is a craft, mastered through engagement; it is not a spectator sport. It requires deliberate practice. In modeling, mathematics and logic play the role of an expert coach. They correct our flaws.

The remainder of the book is organized as follows: Chapters 2 and 3 motivate the many-model approach. Chapter 4 discusses the challenges of modeling people. The next twenty or so chapters cover individual models or classes of models. By considering one type of model at a time, we can better wrap our heads around its assumptions, implications, and applications. This structure also means that we can pull the book from our bookshelves or open it in our browsers and find self-contained analyses of linear models, prediction models, network models, contagion models, and models of long-tailed distributions, learning, spatial competition, consumer preferences, path dependence, innovation, and economic growth. Interspersed throughout the chapters are applications of many-model thinking to a variety of problems and issues. The book concludes with two deeper dives into the opioid epidemic and income inequality.

2. *Why Model?*

Knowing reality means constructing systems of transformations that correspond, more or less adequately, to reality.

—Jean Piaget

In this chapter, we define types of models. Models are often described as simplifications of the world. They can be, but models can also take the form of analogies or be fictional worlds mined for ideas and insights. We also describe the uses of models. In school, we apply models to explain data. In practice, we can also use models to predict, design, and take actions. We can use models to explore ideas and possibilities. And we can use models to communicate ideas and understandings.

The value of models also resides in their ability to reveal conditions under which results hold. Most of what we know holds only in some cases: the square of the longest side of a triangle equals the sum of the squares of the other sides only if the longest side is opposite a right angle. Models reveal similar conditions for our intuitions. With models we can parse out when diseases spread, when markets work, when voting leads to good outcomes, and when crowds make accurate predictions. None of those is a sure thing.

This chapter consists of two parts. In the first, we describe the three types of models. In the second, we cover the uses of models: to reason, explain, design, communicate, act, predict, and explore. These form the acronym REDCAPE, a not-so-subtle reminder that many-model thinking endows us with superpowers.[1]

Types of Models

When constructing a model, we take one of three approaches. We can aim for realism and follow an *embodiment approach*. Such models include the important parts and either strip away unnecessary dimensions and attributes or lump them

13

together. Models of ecological glades, legislatures, and traffic systems take this approach, as do climate models and models of the brain. Or we can take an *analogy approach* and abstract from reality. We can model crime spreading like a disease and the taking of political positions as choices on a left-right continuum. The spherical cow is a favorite classroom example of the analogy approach: to make an estimate of the amount of leather in a cowhide, we assume a spherical cow. We do so because the integral tables in the back of calculus textbooks include tan(x) and cos(x) but not cow(x).[2]

While the embodiment approach stresses realism, the analogy approach tries to capture the essence of a process, system, or phenomenon. When a physicist assumes away friction but otherwise makes realistic assumptions, she takes the embodiment approach. When an economist represents competing firms as different species and defines product niches, she makes an analogy. She does so using a model developed to embody a different system. No bright line differentiates the embodiment approach from the analogy approach. Psychological models of learning that assign weights to alternatives lump together dopamine responses and other factors; they also invoke the analogy of a scale on which we balance alternatives.

A third approach, the *alternative reality approach,* purposely does not represent or capture reality. These models function as analytic and computational playgrounds in which we can explore possibilities. This approach allows us to discover general insights that apply outside our physical and social world. They help us to understand the implications of real-world constraints: What if energy could be sent safely and efficiently through the air? And they allow us to run impossible experiments: What if we tried to evolve a brain? This book contains a few such models, notably the Game of Life, which consists of a checkerboard whose squares are classified as either alive (black) or dead (white) that switch between alive and dead according to fixed rules. Though unrealistic, the model produces insights into self-organization, complexity, and, some argue, even life itself.

Whether embodying a more complex reality, creating an analogy, or building a made-up world for exploring ideas, a model must be *communicable* and *tractable.* We should be able to write the model in a formal language such as mathematics or computer code. When describing a model, we cannot toss out terms like *beliefs* or *preferences* without providing a formal description. Beliefs can be represented as a probability distribution over a set of events or priors. Preferences can be represented in several ways such as a ranking over a set of alternatives or as a mathematical function.

How tractable something is means how amenable it is to analysis. In the past, analysis relied on mathematical or logical reasoning. A modeler had to be able to prove each step in an argument. This constraint led to an aesthetic that valued stark models. English friar and theologian William of Ockham (1287–1347) wrote, "Plurality must never be posited without necessity." Einstein summed up this principle, known as *Ockham's Razor,* as follows: *everything should be made as simple as possible, but not simpler.* Today, when we run up against the constraint of analytic tractability, we can turn to computation. We can build elaborate models with many moving parts without concern for analytic tractability. Scientists take this approach when constructing models of the global climate, the brain, forest fires, and traffic. They still pay heed to Ockham's advice, but recognize that "as simple as possible" might require a lot of moving parts.

The Seven Uses of Models

The academic literature describes dozens of uses of models. Here, we focus on seven categories of uses: to *reason, explain, design, communicate, act, predict,* and *explore.*

The Uses of Models (REDCAPE)

Reason: To identify conditions and deduce logical implications.

Explain: To provide (testable) explanations for empirical phenomena.

Design: To choose features of institutions, policies, and rules.

Communicate: To relate knowledge and understandings.

Act: To guide policy choices and strategic actions.

Predict: To make numerical and categorical predictions of future and unknown phenomena.

Explore: To investigate possibilities and hypotheticals.

REDCAPE: Reason

When constructing a model, we identify the most important actors and entities along with relevant characteristics. We then describe how those parts interact and aggregate, enabling us to derive what follows from what, and why. In doing so, we improve our reasoning. While what we can derive depends upon what we assume, we uncover more than tautologies. Rarely can we infer the full range of implications of our assumptions from inspection alone. We need formal logic. Logic also reveals impossibilities and possibilities. With it, we can derive precise and sometimes unexpected relationships. We can discover the conditionality of our intuitions.

Arrow's theorem provides an example of how logic reveals impossibilities. The model addresses the question of whether individual preferences aggregate to form a collective preference. This model represents preferences as ordinal rankings over alternatives. If applied to five Italian restaurants, denoted by the letters A through E, the model allows any of the 120 orderings. Arrow required that the collective ordering be *monotonic* (if everyone ranks A above B, then so does the collective), *independent of irrelevant alternatives* (if no person's relative rankings of A and B are unchanged but rankings of other alternatives change, then the order of A and B in the collective ranking does not change), and *nondictatorial* (no single person should decide the collective ordering). Arrow then proved that if any preferences are allowed, then no collective ordering necessarily exists.[3]

Logic can also reveal paradoxes. Using models we can show the possibility of each subpopulation containing a larger percentage of women than men but the total population containing a larger percentage of men, a phenomenon (*Simpson's paradox*). This actually happened: 1973, the University of California, Berkeley, accepted a larger percentage of women in most departments. Overall, it accepted men at a higher rate. Models also show that it is possible for two losing bets, when played alternately, to produce a positive expected return (*Parrondo's paradox*). With models, we can show that it is possible to add a node to a network and reduce the total length of the edges needed to connect all the nodes.[4]

We should not dismiss these examples as mathematical novelties. Each has practical applications: efforts to increase the population of women could backfire, combinations of losing investments could win, and the total length of a network of electric lines, pipelines, ethernet lines, or roads could be be reduced by adding more nodes.

Logic also uncovers mathematical relationships. Given Euclid's axioms, a triangle can be uniquely determined by any two angles and a side, or by any two sides and an angle. With standard assumptions about consumer and firm behaviors, in markets with a large number of competing firms, price equals marginal cost. Some results are unexpected: among them the *friendship paradox*, which states that in any friendship network, on average, people's friends have more friends than they do.

The paradox arises because highly popular people have more friends. Figure 2.1 shows Zachary's Karate Network. The person represented by the dark circle has six friends, denoted by gray circles. His friends have nine friends on average. These

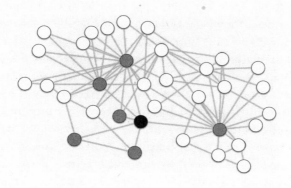

Figure 2.1: The Friendship Paradox: A Person's Friends Have More Friends

people are represented by white circles. Over the entire network, twenty-nine of the thirty-four people have friends who are more popular than they are.[5] Later we show that if we make a few more assumptions, most people's friends will also be, on average, better-looking, kinder, richer, and smarter than they are.

Last, and most important of all, logic reveals the conditionality of truths. A politician may claim that lowering income taxes increases government revenue by spurring economic growth. A rudimentary model in which revenue equals the tax rate times the income level proves that revenue increases only if the percentage growth in income exceeds the percentage cut in taxes.[6] Thus, a 10% cut in income taxes increases revenue only if it causes income to grow by more than 10%. The politician's logic only holds given certain conditions. Models identify those conditions.

The power of conditionality becomes evident when we contrast claims derived from models with narrative claims, even when the latter have empirical support. Consider the management proverb *first things first*: the idea that when facing multiple tasks, you should do the most important task first. This rule is also known as *big rocks first*, because when filling a bucket with rocks of various sizes, you should put the big rocks in first—if you put the little rocks in first, the big rocks will not fit.

The rule *big rocks first,* inferred from expert observation, may be a good rule most of the time, but it is unconditional. A model-based approach would make specific assumptions about the task and then derive an optimal rule. In the *bin packing problem,* a set of objects of various sizes (or weights) must be allocated into bins of finite capacity. The objective is to use as few bins as possible. Imagine, for example, you are packing up your apartment and putting everything into

2-foot-by-2-foot boxes. Ordering your possessions by size and putting each object in the first box with sufficient space (known as the *first fit algorithm*) turns out to be quite effective. Big rocks first works well. However, suppose that we consider a more complex task: allocating space on the International Space Station for research projects. Each project has a payload weight, a size, and power requirements along with demands on the astronauts' time and cognitive abilities. Each also makes a potential scientific contribution. Even if we came up with some measure of bigness as a weighted average of these attributes, big rocks first would prove a poor rule given the dimensionality of interdependencies. More sophisticated algorithms and possibly market mechanisms would perform much better.[7] Thus, under some conditions, big rocks first is a good rule. Under other conditions, it is not. With models, we can trace the boundaries of when we should place the big rocks first and when we should not.

Critics of formalism claim that models repackage what we already know, that they pour old wine into shiny mathematical bottles, that we do not need a model to know that two heads are better than one or that he who hesitates is lost. We can learn the value of commitment from reading of Odysseus tying himself to the mast. That criticism fails to recognize that inferences drawn from models take conditional forms: if condition *A* holds, then result *B* follows (e.g., if you are packing bins and size is the only constraint, pack the biggest objects first). Lessons drawn from literature or proverbial advice from great thinkers often provide no conditions. If we try to lead our lives or manage others by unconditional rules, we find ourselves lost in a sea of *opposite proverbs*. Are two heads better than one? Or, do too many cooks spoil the broth?

Proverb	Opposite
Two heads are better than one	Too many cooks spoil the broth
He who hesitates is lost	A stitch in time saves nine
Tie yourself to the mast	Keep your options open
The perfect is the enemy of the good	Do it well or not at all
Actions speak louder than words	The pen is mightier than the sword

While opposite proverbs abound, opposite theorems cannot. Within models, we make assumptions and prove theorems. Two theorems that disagree on the optimal action, make different predictions, or offer distinct explanations must make different assumptions.

REDCAPE: Explain

Models provide clear logical explanations for empirical phenomena. Economic models explain price movements and market shares. Physics models explain the rate of falling objects and the shape of trajectories. Biological models explain the distributions of species. Epidemiological models explain the speed and patterns of disease spread. Geophysical models explain the size distribution of earthquakes.

Models can explain point values and changes in their values. A model can explain the current price of pork belly futures and why prices rose over the past six months. A model can explain why a president appoints a moderate Supreme Court justice and why a candidate moves to the left or right. Models also explain shape: models of the diffusion of ideas, technologies, and diseases produce an S-shaped curve of adoption (or contagion).

The models we learn in physics, such as Boyle's Law (a model stating that the pressure of oxygen times the volume equals a constant ($PV = k$)), explain phenomena unreasonably well.[8] If we know the volume, we can estimate the constant k, and then explain or predict pressure P as a function of V and k. The model owes its accuracy to the fact that gases consist of simple parts that exist in large numbers and follow fixed rules: any two oxygen molecules placed in the identical situation follow the same physical laws. They exist in such large numbers that statistical averaging cancels out any randomness. Most social phenomena share none of these three attributes: social actors are heterogeneous, interact in small groups, and do not follow fixed rules. People also think. Even more problematic, people respond to social influences, meaning that behavioral variations may not cancel out. As a result, social phenomena are much less predictable than physical phenomena.[9]

The most effective models explain both straightforward outcomes and puzzling ones. Textbook models of markets can explain why an unanticipated increase in the demand for a normal good like shoes or potato chips increases the price in the short run, an intuitive result. These same models explain why in the long run, demand increases have less of an effect on price than the marginal cost of producing the good. Increases in demand can even produce reductions in price that result from increased returns to scale in production, a more surprising result. The same models can explain paradoxes such as why diamonds, which have little practical value, have high prices, but water, a necessity for survival, costs little.

As for the claim that models can explain anything: it is true, they can. However, a model-based explanation includes formal assumptions and explicit causal chains. Those assumptions and causal chains can be taken to data. A model that claims that high levels of criminal behavior can be explained by low probabilities of being caught can be tested.

REDCAPE: Design

Models aid in design by providing frameworks within which we can contemplate the implications of choices. Engineers use models to design supply chains. Computer scientists use models to design web protocols. Social scientists used models to design institutions.

In July 1993, a group of economists met at Caltech in Pasadena, California, to design an auction to allocate the electronic spectrum for cellular phones. In the past, the government had allocated spectrum rights to large companies for modest fees. A provision within the Omnibus Budget Reconciliation Act of 1993 allowed for auctioning the spectrum to raise money.

The radio signal from a tower covers a geographic range. Therefore, the government sought to sell licenses for specific regions: Western Oklahoma, Northern California, Massachusetts, Eastern Texas, and so on. This created a design challenge. The value of any given license for a company depended on the other licenses that company won. The license for Southern California would be worth more to a company that also owned the license for Northern California, for example. Economists refer to these interdependent valuations as *externalities*. The externalities had two main sources: construction and advertising. Holding neighboring licenses meant lower construction costs and the potential to exploit overlapping media markets.

The externalities created a problem with holding simultaneous auctions. A company trying to win a bundle of licenses might lose one license to another bidder and therefore lose the externalities. That company might then want to back out of its bids on other licenses. Sequential auctions had a different shortcoming. Bidders would underbid in early auctions to hedge against losing subsequent licenses.

A successful auction design had to be immune to strategic manipulation, generate efficient outcomes, and be comprehensible to participants. The economists used game theory models to analyze whether features could be exploited by strategic bidders, computer simulation models to compare the efficiency of various designs, and statistical models to choose parameters for experiments with real people. The final design, a multiple-round auction that allowed participants to back out of bids and prohibited sitting out early periods to mask intentions, proved successful. Over the past thirty years, the FCC has raised nearly $60 billion using this type of auction.[10]

REDCAPE: Communicate

By creating a common representation, models improve communication. Models require formal definitions of the relevant features and their relationships that we can then communicate with precision. The model $F = MA$ relates three measurable quantities, force, mass, and acceleration, and does so in equation form. Each term is expressed in measurable units that can be communicated without fear of misinterpretation. By comparison, the claim that "bigger, faster things generate more power" offers far less precision. Much can get lost in translation. Does bigger mean weight or size? Does faster mean velocity or acceleration? Does power mean energy or force? And how do bigger and faster combine to produce power? Attempts to formalize the claim could result in any of several forms: power could be written incorrectly as weight plus velocity ($P = W + V$), weight times velocity ($P = WV$), or weight plus acceleration ($P = W + A$).

When we formally define an abstract concept like political ideology using a reproducible methodology, those concepts take on some of the same features as physical qualities such as mass and acceleration. We can use a model to say that one politician is more liberal than another based on their voting records. We can then communicate that claim with precision. Liberalness is well defined and measurable. Someone can use the same method to compare other politicians. Of course, voting records may not be the only measure of liberalness. We might construct a second model that assigns ideologies based on textual analysis of speeches. With that model as well, we can communicate with clarity what we mean by more liberal.

Many underappreciate the impact of communication on progress. An idea that cannot be communicated is like a tree falling in a forest with no one around to notice it. The remarkable economic growth in the Age of Enlightenment was due in no small part to the transferability of knowledge, often in model form. In fact, the evidence suggests that the transferability of ideas may have contributed more to economic growth during that time than did levels of education: city-level growth in eighteenth-century France correlates more strongly with the number of subscriptions to Diderot's *Encyclopédie* than with literacy rates.[11]

REDCAPE: Act

Francis Bacon wrote, "The great end of life is not knowledge but action." Good actions require good models. Governments, corporations, and nonprofits all use models to guide actions. Whether it be raising or lowering prices, opening a new location, acquiring a company, offering universal health care, or funding an afterschool program, decision-makers rely on models. On the most important actions, decision-makers use sophisticated models. Models are linked to data.

In 2008, as part of the Troubled Asset Relief Program (TARP), the Federal Reserve gave $182 billion in financial assistance to bail out the multinational insurance company American International Group (AIG). According to the US Department of the Treasury, the government chose to stabilize AIG "because its failure during the financial crisis would have had a devastating impact on our financial system and the economy."[12] The purpose of the bailout was not to save AIG but to prop up the entire financial system. Businesses fail every day, and the government does not intervene.[13]

The particular choices made within TARP were based on models. Figure 2.2 shows a version of a network model produced by the International Monetary Fund. The nodes (circles) represent financial institutions. The edges (the lines between the circles) represent correlations between the values of the holdings of those institutions. The color and width of an edge corresponds to the strength of the correlation between the institutions, with darker and thicker lines implying greater correlation.[14]

AIG occupies a central position in the network because it sold insurance to other firms. AIG held promises to pay other firms if those firms' assets lost value. If prices fell, then AIG owed those firms money. By implication, if AIG failed, so too would the firms connected to AIG. A cascade of failures might ensue. By stabilizing AIG's position, the government could prop up the market values of other firms in the network.[15]

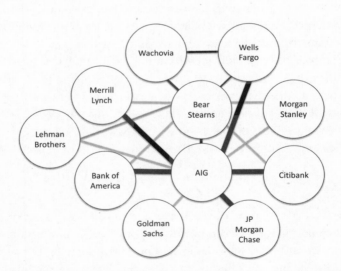

Figure 2.2: Correlation Graph Between Financial Institutions

Figure 2.2 also helps to explain why the government let Lehman Brothers fail. Lehman did not occupy a central position in the network. We cannot rerun history, so we cannot know if the Federal Reserve took the correct action. We do know that the financial industry did not collapse as a result of Lehman's failure. We also know that the government earned a $23 billion profit on its loan to AIG. So, we can infer that the policy choices—based on many-model thinking—were not a failure.

Models that guide action, such as policy models, often rely on data, but not all do. Most policy models also use mathematics, though that was not always true. In the past, policymakers built physical models as well. Phillips's hydraulic model of the British economy was used to think through policy choices in the mid-twentieth century, and a physical model of San Francisco Bay was instrumental in the decision not to dam the bay for fresh water.[16] The Mississippi River Basin Model Waterways Experiment Station, which covers nearly 200 acres near Clinton, Mississippi, is a miniature replica of the river's basin built on a horizontal scale of 1:100. The model can test the upstream and downstream effects of building new dams and reservoirs. The released water follows the laws of physics within the physical structure. In these physical models, the entities themselves are analogs of the real world. The models are logical because they follow the laws of physics.

Our examples so far have considered organizations using models to act. People can do the same. When taking important actions in our personal lives, we should also use models. In deciding to purchase a home, take a new job, return to graduate school, or buy or lease a car, we can use models to guide our thinking. Those models may be qualitative rather than tied to data. Even in those cases, the models will oblige us to ask relevant questions.

REDCAPE: Predict

Models have long been used to predict. Weather forecasters, consultants, sports handicappers, and central bankers all predict using models. Police agencies and the intelligence community use models to predict criminal behavior. Epidemiologists use models to predict which strain of flu will be most widespread in the upcoming flu season. As data has become more available and granular, this use of models has grown. Twitter feeds and internet searches are used to predict consumer trends and social uprisings.

Models can predict individual events as well as general trends. On June 1, 2009, Air France flight AF 477, en route from Rio de Janeiro to Paris, crashed over the Atlantic. In the days following, rescuers found floating debris but could not locate the fuselage. By July, the batteries in the plane's acoustic beacons were depleted, halting search efforts. A year later, a second search led by the Woods Hole Oceanographic Institution using US Navy side-scan sonar vessels and autonomous underwater vehicles also proved unsuccessful. The French Bureau d'Enquêtes et d'Analyses eventually turned to models. They applied probabilistic models to ocean currents and identified a small rectangular region as being most likely to contain the fuselage. Using the model's prediction, searchers found the wreckage within a week.[17]

In the past, explanation and prediction tended to go hand in hand. Electrical engineering models that explain voltage patterns can also predict voltages. Spatial models that explain politicians' past votes can also predict future votes. In perhaps the most famous example of applying an explanatory model to predict, the French mathematician Urbain Le Verrier applied the Newtonian laws created to explain planetary movements to evaluate the discrepancies in the orbit of Uranus. He discovered the orbits to be consistent with the presence of a large planet in the outer region of the solar system. On September 18, 1846, he sent his prediction to the Berlin Observatory. Five days later, astronomers located the planet Neptune exactly where Le Verrier had predicted it would be.

That said, prediction differs from explanation. A model can predict without explaining. Deep-learning algorithms can predict product sales, tomorrow's weather, price trends, and some health outcomes, but they offer little in the way of explanation. Such models resemble bomb-sniffing dogs. Even though a dog's olfactory system can determine whether a package contains explosives, we should not look to the dog for an explanation of why the bomb is there, how it works, or how to disarm it.

Note also that other models can explain but have little value as predictors. Plate tectonics models explain how earthquakes arise but do not predict when they occur. Dynamical systems models can explain hurricanes, but they cannot predict with much success when hurricanes will form or what paths they will take. And while ecology models can explain patterns of speciation, they cannot predict new types of species.[18]

REDCAP_E: Explore

Last, we use models to explore intuitions and possibilities. These explorations can be policy-related: What if we make all city buses free? What if we let students choose which assignments determine their course grades? What if we put signs on people's lawns showing their energy consumption? Each of these hypotheticals can be explored with models. We can also use models to explore unrealistic environments. What if Lamarck had been correct and acquired traits could be passed on to our offspring, so the children of parents with orthodontically corrected teeth would not need braces? What happens in such a world? Asking that question and exploring its implications can help to reveal the limits of evolutionary processes. Abandoning the constraints of reality can spur creativity. For this reason, advocates of the critical design movement engage in speculative fictions to generate new ideas.[19]

Exploration sometimes consists of comparing common assumptions across domains. To understand network effects, a modeler might begin a collection of stylized network structures and then ask whether and how network structure affects cooperation, disease spread, or social uprisings. Or a modeler might apply a collection of learning models to decisions, two-person games, and multiperson games. The purpose of these exercises is not to explain, predict, act, or design. It is to explore and learn.

When we apply a model in practice, we may use it in any of several ways. The same model may explain, predict, and guide action. As an example, on August 14, 2003, sagging trees leaning on power lines near Toledo, Ohio, created a localized power outage that spread when a software failure prevented an alarm from alerting technicians to redistribute power. Within a day, more than 50 million people in the northeastern United States and Canada had lost power. That same year, a storm knocked out a power line between Italy and Switzerland, leaving 60 million Europeans without power. Engineers and scientists turned to models that represent the power grid as a network. The models helped to explain how the failures occurred, offered predictions of regions where future failures might be likely, and also guided actions by identifying locations where new lines, transformers, and power supplies would enhance the robustness of the network. Putting one model to many uses will be a recurrent theme in this book. As we see next, one-to-many is a necessary complement to our central theme of applying many models to make sense of complex phenomena.

3. *The Science of Many Models*

Nothing is less real than realism. Details are confusing. It is only by selection, by elimination, by emphasis that we get to the real meaning of things.

—Georgia O'Keeffe

In this chapter, we take a scientific approach to motivate the many-model approach. We begin with the Condorcet jury theorem and the diversity prediction theorem, which make quantifiable cases for the value of many models in helping us act, predict, and explain. These theorems may overstate the case for many models. To show why, we introduce categorization models, which partition the world into boxes. Using categorization models shows us that constructing many models may be harder than we expect. We then apply this same class of model to discuss model granularity—how specific our models should be—and help us decide whether to use one big model or many small models. The choice will depend on the use. When predicting, we often want to go big. When explaining, smaller is better.

The conclusion addresses a lingering concern. Many-model thinking might seem to require learning a lot of models. While we must learn some models, we need not learn as many as you might think. We do not need to master a hundred models, or even fifty, because models possess a one-to-many property. We can apply any one model to many cases by reassigning names and identifiers and modifying assumptions. This property of models offers a counterpoise to the demands of many-model thinking. Applying a model in a new domain requires creativity, an openness of mind, and skepticism. We must recognize that not every model will appropriate to every task. If a model cannot explain, predict, or help us reason, we must set it aside.

The skills required to excel at one-to-many differ from the mathematical and analytic talents many people think of as necessary for being a good modeler. The process of one-to-many involves creativity. It is to ask: *How many uses can I think of for a random walk?* To provide a hint of the forms that creativity takes, at the end of the chapter we apply the geometric formula for area and volume as a model and use it to explain the size of supertankers, to criticize the body mass index, to predict the scaling of metabolisms, and to explain why we see so few women CEOs.

Many Models as Independent Lies

We now turn to formal models that help reveal the benefits of many-model thinking. Within those models, we describe two theorems: the Condorcet jury theorem and the diversity prediction theorem. The *Condorcet jury theorem* is derived from a model constructed to explain the advantages of majority rule. In the model, jurors make binary decisions of guilt or innocence. Each juror is correct more often than not. In order to apply the theorem to collections of models instead of jurors, we interpret each juror's decision as a classification by a model. These classifications could be actions (buy or sell) or predictions (Democratic or Republican winner). The theorem then tells us that by constructing multiple models and using majority rule we will be more accurate than if we used one of the constituent models. The model relies on the concept of a *state of the world,* a full description of all relevant information. For a jury, the state of the world consists of the evidence presented at trial. For models that measure the social contribution of a charitable project, the state of the world might correspond to the project's team, the organizational structure, the operational plan, and the characteristics of the problem or situation the project would address.

Condorcet Jury Theorem

Each of an odd number of people (models) classifies an unknown state of the world as either true or false. Each person (model) classifies correctly with a probability $p > \frac{1}{2}$, and the probability that any person (model) classifies correctly is statistically independent of the correctness of any other person (model).

Condorcet jury theorem: A majority vote classifies correctly with higher probability than any person (model), and as the number of people (models) becomes large, the accuracy of the majority vote approaches 100%.

Ecologist Richard Levins elaborates on how the logic of the theorem applies to the many-model approach: "Therefore, we attempt to treat the same problem with several alternative models each with different simplifications but with a common biological assumption. Then, if these models, despite their different assumptions, lead to similar results, we have what we can call a robust theorem, which is relatively free of the details of the model. Hence our truth is the intersection of independent lies."[1] Note that here he aspires to a unanimity of classification. When many models make a common classification, our confidence should soar.

Our next theorem, the *diversity prediction theorem,* applies to models that make numerical predictions or valuations. It quantifies the contributions of model accuracy and model diversity to the accuracy of the average of those models.[2]

Diversity Prediction Theorem

Many-Model Error = Average-Model Error − Diversity of Model Predictions

$$\left(\bar{M}-V\right)^2 = \sum_{i=1}^{N}\frac{\left(M_i-V\right)^2}{N} - \sum_{i=1}^{N}\frac{\left(M_i-\bar{M}\right)^2}{N}$$

where M_i equals model i's prediction, $\bar{M}$ equals the average of the model's values, and V equals the true value.

The diversity prediction theorem describes a mathematical identity. We need not test it. It always holds. Here is an example. Two models predict the number of Oscars a film will be awarded. One model predicts two Oscars, and the other predicts eight. The average of the two models' predictions—the many-model prediction—equals five. If, as it turns out, the film wins four Oscars, the first model's error equals 4 (2 squared), the second model's error equals 16 (4 squared), and the many-model error equals 1. The diversity of the models' predictions equals 9 because each differs from the mean prediction by 3. The diversity prediction theorem can then be expressed as follows: 1 (the many-model error) = 10 (the average-model error) − 9 (the diversity of the predictive models).

The logic of the theorem relies on opposite types of errors (pluses and minuses) canceling each other out. If one model predicts a value that is too high and another model predicts a value that is too low, then the models exhibit predictive diversity. The two errors cancel, and the average of the models will be more accurate than either model by itself. Even if both predict values that are too high, the error of the average of those predictions will still not be worse than the average error of the two high predictions.

The theorem does not imply that any collection of diverse models will be accurate. If all of the models share a common bias, their average will also contain that bias. The theorem does imply that any collection of diverse models (or people) will be more accurate than its average member, a phenomenon referred to as the *wisdom of crowds*. That mathematical fact explains the success of ensemble methods in computer science that average multiple classifications as well as evidence that individuals who think using multiple models and frameworks predict with higher accuracy than people who use single models. Any single way of looking at the world leaves out details and makes us prone to blind spots. Single-model thinkers are less likely to anticipate large events, such as market collapses or the Arab Spring of 2011.[3]

These two theorems make a compelling case for using many models, at least in the context of prediction. The case may be too compelling, however. The Condorcet jury theorem implies that with enough models, we would almost never make a mistake. The diversity prediction theorem implies that if we could construct a diverse set of moderately accurate predictive models, we can reduce our many-model error to near zero. As we see next, our ability to construct many diverse models has limits.

Categorization Models

To demonstrate why the two theorems may overstate the case, we rely on *categorization models*. These models provide micro-foundations for the Condorcet jury theorem. Categorization models partition the states of the world into disjoint boxes. Such models date to antiquity. In *The Categories,* Aristotle defined ten attributes that could be used to partition the world. These included *substance, quantity, location,* and *positioning*. Each combination of attributes would create a distinct category.

We use categories any time we use a common noun. "Pants" is a category; so are "dogs," "spoons," "fireplaces," and "summer vacations." We use categories to guide actions. We categorize restaurants by ethnicity—Italian, French, Turkish, or Korean—to decide where to have lunch. We categorize stocks by their price-to-earnings ratios and sell stocks with low price-to-earnings ratios. We use categories to explain, as when we claim that Arizona's population has grown because the state has good weather. We also use categories to predict: we might forecast that a candidate for political office with military experience has an increased chance of winning.

We can interpret the contributions of categorization models within the wisdom hierarchy. The objects constitute the data. Binning the objects into categories creates information. The assigning of valuations to categories requires knowledge. To critique the Condorcet jury theorem, we rely on a *binary categorization model* that partitions the objects or states into two categories, one labeled "guilty" and one "innocent." The key insight will be that the number of relevant attributes constrains the number of distinct categorizations, and therefore the number of useful models.

Categorization Models

There exists a set of objects or states of the world, each defined by a set of attributes and each with a value. A **categorization model**, M, partitions these objects or states into a finite set of categories $\{S_1, S_2, \ldots, S_n\}$ based on the object's attributes and assigns **valuations** $\{M_1, M_2, \ldots, M_n\}$ for each category.

Imagine we have one hundred student loan applications, half of which were paid back and half of which were defaulted. We know two pieces of information for each loan: whether the loan amount exceeded $50,000, and whether the recipient majored in engineering or the liberal arts. These are the two attributes. With two attributes we can distinguish between four types of loans: large loans to engineers, small loans to engineers, large loans to liberal arts majors, and small loans to liberal arts majors.

A binary categorization model classifies each of these four types as either repaid or defaulted. One model might classify small loans as repaid and large loans as defaulted. Another model might classify loans to engineers as repaid and loans to liberal arts majors as defaulted. It seems plausible that each of these models could be correct more than half the time, and that the two models might be approximately independent of each other. A problem arises when we try to construct more models. There exist only sixteen unique models that map four categories into two outcomes. Two of those models classify all loans as repaid or defaulted. Each of the remaining fourteen has an exact opposite. Whenever the model classifies correctly, its opposite model classifies incorrectly. Thus, of the fourteen possible models, at most seven can be correct more than half the time. And if any model happens to be correct exactly half of the time, then so must its opposite.

The dimensionality of our data limits the number of models we can produce. At most we can have seven models. We cannot construct eleven independent models, much less seventy-seven. Even if we had higher-dimensional data—say, if we knew the recipient's age, grade point average, income, marital status, and address—the categorizations that relied on those attributes must yield accurate predictions. Each subset of attributes would have to be relevant to whether the loan was repaid and be uncorrelated with the other attributes. Both are strong assumptions. For example, if address, marital status, and income are correlated, then models that swap those attributes will be correlated as well.[4] In the stark probabilistic model, independence seemed reasonable: different models make independent mistakes. When we unpack that logic with categorization models, we see the difficulty of constructing multiple independent models.

Attempts to construct a collection of diverse, accurate models encounter a similar problem. Suppose that we want to build an ensemble of categorization models that predict unemployment rates across five hundred mid-size cities. An accurate model must partition cities into categories such that within a category the cities have similar unemployment rates. The model must also predict unemployment accurately for each category. For two models to make diverse predictions, they must categorize cities differently, predict differently, or do both. Those two criteria, though not in contradiction, can be difficult to satisfy. If one categorization relies on average education level and a second relies on average income, they may categorize similarly. If so, the two models will be accurate but not diverse. Creating twenty-six categories using the first letter of each city's name will create a diverse categorization but probably not an accurate model. Here as well, the takeaway is that in practice "many" may be closer to five than fifty.

Empirical studies of prediction align with that inference. While adding models improves accuracy (they have to, given the theorems), the marginal contribution of each model falls off after a handful of models. Google found that using one interviewer to evaluate job candidates (instead of picking at random) increases the probability of an above-average hire from 50% to 74%, adding a second interviewer increases the probability to 81%, adding a third raises it to 84%, and using a fourth lifts it to 86%. Using twenty interviewers only increases the probability to a little over 90%. That evidence suggests a limit to the number of relevant ways of looking at a potential hire.

A similar finding holds for an evaluation of tens of thousands of forecasts by economists regarding unemployment, growth, and inflation. In this case, we should think of the economists as models. Adding a second economist improves the accuracy of the prediction by about 8%, two more increase it by 12%, and three more by 15%. Ten economists improve the accuracy by about 19%. Incidentally, the best economist is only about 9% better than average—assuming you knew which economist was best. So three random economists perform better than the best one.[5] Another reason for averaging many and not relying on the economist who has been best historically is that the world changes. The economist who performs at the top today may be middling tomorrow. That same logic explains why the US Federal Reserve relies on an ensemble of economic models rather than just one: the average of many models will typically be better than the best model.

The lesson should be clear: if we can construct multiple diverse, accurate models, then we can make very accurate predictions and valuations and choose good actions. The theorems validate the logic of many-model thinking. What the theorems do not do, and cannot do, is construct the many models that meet their assumptions. In practice, we may find that we can construct three or maybe five good models. If so, that would be great. We need only read back one paragraph: adding a second model yields an 8% improvement, while adding a third gets us to 15%. Keep in mind, these second and third models need not be better than the first model. They could be worse. If they are a little less accurate, but categorically (in the literal sense) different, they should be added to the mix.

One Big Model and the Granularity Question

Many models work in theory and in practice. That does not mean that they are always the correct approach. Sometimes we are better off constructing a single large model. In this section, we put some thought into when we should use each approach and along the way take up the *granularity question* of how finely we should partition our data.

To take on the first question, of whether to use one big model or many small ones, recall the uses of models: to *reason, explain, design, communicate, act, predict,* and *explore*. Four of these uses—to reason, explain, communicate, and explore—require simplification. By simplifying, we can apply logic allowing us to explain phenomena, communicate our ideas, and explore possibilities.

Think back to the Condorcet jury theorem. Within it, we could unpack logic, explain why an approach that uses many models was more likely to produce a correct result, and communicate our findings. Had we constructed a model of jurors with personality types and described the evidence as vectors of words, we would have been lost in a mangle of detail. Borges elaborates on this point in an essay on science. He describes mapmakers who make ever more elaborate maps: "The Cartographers Guilds struck a Map of the Empire whose size was that of the Empire, and which coincided point for point with it. The following Generations, who were not so fond of the Study of Cartography as their Forebears had been, saw that this vast Map was useless."

The three other uses of models—to *predict, design,* and *act*—can benefit from high-fidelity models. If we have BIG data, we should use it. As a rule of thumb, the more data we have, the more granular we should make our model. This can be shown by using categorization models to structure our thinking. Suppose first that we want to construct a model to explain variation in a data set. To provide context, suppose that we have an enormous data set from a chain of grocery stores detailing monthly spending on food for several million households. These households differ in the amount they spend, which we measure as variation: the sum of the squared differences between what each family spends and average spending across all households. If average spending is $500 a month and a given family spends $520, that family contributes 400, or 20 squared, to the total variation. Statisticians call the proportion of the variation that a model explains the model's R^2.

If the data had a total variation of 1 billion and a model explains 800 million of that variation, then the model has an R^2 of 0.8. The amount of variation explained corresponds to how much the model improves on the mean estimate. If the model estimates that a household will spend $600 and the household in fact spent $600, then the model explains all 10,000 that the household contributes to total variation. If the household spent $800 and the model says $700, then what had been a contribution of 90,000 to total variation $((800-500)^2)$ is now only a 10,000 contribution $((800-700)^2)$. The model explains $\frac{8}{9}$ of the variation.

R^2: Percentage of Variance Explained

$$R^2 = \frac{\sum_{x \in X} \left(V(x) - \bar{V}\right)^2 - \sum_{x \in X} (M(x) - V(x))^2}{\sum_{x \in X} \left(V(x) - \bar{V}\right)^2}$$

where $V(x)$ equals the value of x in X, $\bar{V}$ equals the average value, and $M(x)$ equals the model's valuation.

In this context, a categorization model would partition the households into categories and estimate a value for each category. A more granular model would create more categories. This may require considering more attributes of the households to create those categories. As we add more categories, we can explain more of the variation, but we can go too far. If we follow the example of Borges's mapmakers and place each household in its own category, we can explain all of the variation. That explanation, like the life-sized map, would not be of much use.

Creating too many categories overfits the data, overfitting undermines prediction of future events. Suppose that we want to use last month's data on grocery purchases to predict this month's data. Households vary in their monthly spending. A model that places each household in its own category would predict that each household spends the same as in the previous month. That would not be a good predictor given monthly fluctuations in spending. By placing the household into a category with other similar households, we can use the average spending on groceries for similar households to create a more accurate predictor.

To do this, we think of each household's monthly purchases as a draw from a distribution (we will cover distributions in Chapter 5). That distribution has a mean and a variance. The objective in creating a categorization model is to construct categories based on attributes so that the households within the same category have similar means. If we can do that, one household's spending in the first month tells us about the other households' spending in the second month. No categorization will be perfect. The means of households within each category will differ by a little. We call this *categorization error*.

As we make larger categories, we increase categorization error, as we are more likely to clump households with different means into the same category. However, these larger categories rely on more data, so our estimates of the means in each category will be more accurate (see the square root rules in Chapter 5). The error from misestimating the mean is called the *valuation error*. Valuation error decreases as we make categories larger. One or even ten houses per category will not give an accurate estimate of the mean if households vary substantially in their monthly spending. A thousand households will.

We now have the key intuition: increasing the number of categories decreases the categorization error from binning households with different means into the same category. Statisticians call this *model bias*. However, making more categories increases the error from estimating the mean within each category. Statisticians refer to this as increasing the *variance* of the mean. The trade-off in how many categories to create can be expressed formally in the *model error decomposition theorem*. Statisticians refer to the result as the bias-variance trade-off.

Model Error Decomposition Theorem
The Bias-Variance Trade-off

Model Error = Categorization Error + Valuation Error

$$\sum_{x \in X}(M(x)-V(x))^2 = \sum_{i=1}^{n}\sum_{x \in S_i}(V(x)-V_i)^2 + \sum_{i=1}^{n}(M_i-V_i)^2$$

where $M(x)$ and M_i denote the model's values for data point x and category S_i and $V(x)$ and V_i denote their true values.[6]

One-to-Many

Learning models takes time, effort, and breadth. To reduce those demands, we take a *one-to-many* approach. We advocate mastering a modest number of flexible models and applying them creatively. We use a model from epidemiology to understand the diffusion of seed corn, Facebook, crime, and pop stars. We apply a model of signaling to advertising, marriage, peacock feathers, and insurance premiums. And we apply a rugged-landscape model of evolutionary adaption to explain why humans lack blowholes. Of course, we cannot take any model and apply it to any context, but most models are flexible. We gain even when we fail because attempts at creative uses of models reveal their limits. And it is fun.

The one-to-many approach is relatively new. In the past, models belonged to specific disciplines. Economists had models of supply and demand, monopolistic competition, and economic growth; political scientists had models of electoral competition; ecologists had models of speciation and replication; and physicists had models describing laws of motion. All of these models were developed with specific purposes in mind. One would not apply a model from physics to the economy or a model from economics to the brain any more than one would use a sewing machine to repair a leaky pipe.

Taking models out of their disciplinary silos and practicing one-to-many has produced notable successes. Paul Samuelson reinterpreted models from physics to explain how markets attain equilibria. Anthony Downs applied a model of ice cream vendors competing on a beach to explain the positioning of political candidates competing in ideological space. Social scientists have applied models of interacting particles to explain poverty traps, variation in crime rates, and even economic growth across countries. And economists have taken models of self-control based on economic principles to understand the functioning of the brain.[7]

One-to-Many: Higher Powers (X^N)

Creatively applying models requires practice. To provide a preview of the potential of the *many-to-one* principle, we take the familiar formula of a variable raised to a power, X^N, and apply it as a model. When the power equals 2, the formula gives the area of a square, when the power equals 3, it gives the volume of a cube. When raised to higher powers, it captures geometric expansion or decay.

Supertankers: Our first application considers a cubic supertanker whose length is eight times its depth and width, which we denote by S. As shown in figure 3.1, the supertanker has a surface area of $34S^2$ and a volume of $8S^3$. The cost of building a supertanker depends primarily on its surface area, which determines the amount of steel used. The amount of revenue a supertanker generates depends on its volume. Computing the ratio of volume to surface area, $\frac{8S^3}{34S^2} \approx \frac{S}{4}$, reveals a linear gain in profitability from increasing size.

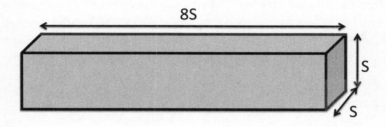

Figure 3.1: A Cubic Supertanker: Surface Area $= 34S^2$, Volume $= 8S^3$

Shipping magnate Stavros Niarchos, who knew this ratio, built the first modern supertankers and made billions during the period of rebuilding that followed World War II. To give some sense of scale: the T2 oil tanker used during World War II measured 500 feet long, 25 feet deep, and 50 feet wide. Modern supertankers such as the *Knock Nevis* measure 1,500 feet long, 80 feet deep, and 180 feet wide. Imagine tipping the Willis (Sears) Tower in Chicago on its side and floating it in Lake Michigan. The *Knock Nevis* resembles a T2 oil tanker scaled up by a factor of a little over three. The *Knock Nevis* has about ten times the surface area as a T2 oil tanker and over thirty times the volume. A question arises as to why supertankers are not even larger. The short answer is that tankers must pass through the Suez Canal; the *Knock Nevis* squeezes through with a gap of a few feet on each side.[8]

Body mass index: Body mass index (BMI) is used by the medical profession to define weight categories. Developed in England, BMI equals the ratio of a person's weight (in kilograms) to her height in meters squared.[9] Holding height constant, BMI increases linearly with weight. If one person weighs 20% more than another person of the same height, the first person's BMI will be 20% higher.

We first apply our model to approximate a person as a perfect cube made up of some mixture of fat, muscle, and bone. Let M denote the weight of one cubic meter of our cubic person. The human cube's weight equals its volume times the weight per cubic meter, or $H^3 \cdot M$. Our cube's BMI equals $H \cdot M$. Our model reveals two flaws: BMI increases linearly with height, and given that muscle weighs more than fat, fit people have higher M and therefore higher BMIs. Height should be unrelated to obesity, and muscularity is the *opposite* of fatness. These flaws remain if we make the model more realistic. If we make a person's depth (thickness front to back) and width proportional to height using parameters d and w, then BMI can be written as follows: $BMI = \frac{H \cdot (dH) \cdot (wH) \cdot M}{H^2} = dwHM$. The BMIs of many NBA stars and other athletes place them in the overweight category (BMI > 25), along with many of the world's top male decathletes.[10] Given that even moderately tall, physically fit people will likely have high BMIs, we should not be surprised that a meta-analysis of nearly a hundred studies with a combined sample size in the millions found that slightly overweight people live longest.[11]

Metabolic rates: We now apply our model to predict an inverse relationship between an animal's size and its metabolic rate. Every living entity has a metabolism, a repeated sequence of chemical reactions that breaks down organic matter and transforms it into energy. An organism's metabolic rate, measured in calories, equals the amount of energy needed to remain alive. If we construct cubic models of a mouse and an elephant, figure 3.2 shows that the smaller cube has a much larger ratio of surface area to volume.

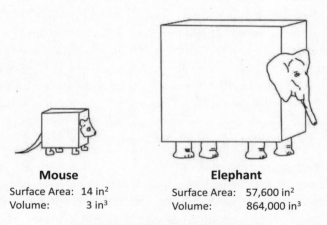

Mouse		**Elephant**	
Surface Area:	14 in²	Surface Area:	57,600 in²
Volume:	3 in³	Volume:	864,000 in³

Figure 3.2: The Exploding Elephant

We can model the mouse and the elephant as composed of cells 1 cubic inch in volume, each with a metabolism. Those metabolic reactions produce heat that must dissipate through the surface of the animal. Our mouse has a surface area of 14 square inches and a volume of 3 cubic inches, a surface-to-volume ratio of roughly 5:1.[12] For each cubic-inch cell in its volume, the mouse has five square inches of surface area through which it can dissipate heat. Each heat-producing cell in the elephant has only one-fifteenth of a square inch of surface area. The mouse can dissipate heat at seventy-five times the rate of the elephant.

For both animals to maintain the same internal temperature, the elephant must have a slower metabolism. It does. An elephant with a mouse's metabolism would require 15,000 pounds of food per day. The elephant's cells would also produce too much heat to be dissipated through its skin. As a result, elephants would smolder and then explode. The reason elephants do not blow up is that they have a metabolism roughly twenty times lower than that of mice. The model does not predict the rate at which metabolism scales with size, only the direction. More elaborate models can explain the scaling laws.[13]

Women CEOs: For our last application, we increase the exponent in the formula and use the model to explain why so few women become CEOs. In 2016, fewer than 5% of Fortune 500 companies had women CEOs. To become a CEO a person must receive multiple promotions. We can model those promotion opportunities as probabilistic events: a person has some probability of receiving a promotion. We further assume that to become CEO, a person must be promoted at each opportunity.

We assume fifteen promotion opportunities as a benchmark, as that corresponds to a promotion every two years on a thirty-year path to CEO. The weight of evidence reveals modest biases in favor of men, which we can model as men having a higher probability of being promoted.[14] We model this as a man's probability of promotion, P_M, being slightly larger than a woman's, P_W. If we benchmark these probabilities at 50% and 40%, respectively, then a man is nearly thirty times more likely than a woman to become CEO.[15] The model reveals how modest biases accumulate. A 10% difference in promotion rates becomes a 30-fold bias at the top. This same model provides a novel explanation for why a much larger percentage (about 25%) of college and university presidents are women. Colleges and universities have fewer administrative layers than Fortune 500 companies. A professor can become president in as few as three promotions: department chair, dean, and then president. Less bias accumulates over three levels. Thus, the larger proportion of women presidents need not imply that educational institutions are more egalitarian than corporations.

Summary

We began the chapter by laying logical foundations for the many-to-one approach using the Condorcet jury theorem and the diversity prediction theorem. We then used categorization models to show the limits of model diversity. We saw how many models can improve our abilities to predict, act, design, and so on. We also saw that it is not easy to come up with many diverse models. If we could, then we could predict with near perfect accuracy, which we know we cannot. Nevertheless, our goal will be to construct as many useful, diverse models as possible.

In the chapters that follow, we describe a core set of models. Those models make salient different parts of the world. They make different assumptions about causal interactions. Through their diversity they create the potential for productive many-model thinking. By emphasizing distinct parts of more complex wholes, each model contributes on its own. Each also can be part of an even more powerful ensemble of models.

As noted earlier, many-model thinking does require that we know more than one model. However, we need not know a huge number of models, so long as we can apply each model that we do know in multiple domains. That will not always be easy. Successful one-to-many thinking depends on creatively tweaking assumptions and constructing novel analogies in order to apply a model developed for one purpose in a new context. Thus, becoming a many-model thinker demands more than mathematical competence; it requires creativity as was evident in our many applications of our model of a cube.

Bagging and Many Models

Often we fit a model to a sample from an existing data set and then test that same model against the remainder of the data. Other times we fit a model to existing data and use that model to predict future data. This type of modeling creates a tension: the more parameters we include in our model, the better we can fit data and the more we risk overfitting. Good fit does not imply a good model. Physicist Freeman Dyson tells of Enrico Fermi's reaction to a piece of Dyson's research that had exceptional model fit. "In desperation I asked Fermi whether he was not impressed by the agreement between our calculated numbers and his measured numbers. He replied, 'How many arbitrary parameters did you use for your calculations?' I thought for a moment about our cut-off procedures and said, 'Four.' He said, 'I remember my friend Johnny von Neumann used to say, with four parameters I can fit an elephant, and with five I can make him wiggle his trunk.' With that, the conversation was over."[16]

The estimates used to "wiggle the trunk" often include higher-order terms: squares, cubes, and fourth powers. This introduces a risk of large errors, because higher-order terms amplify. While 10 is twice as large as 5, 10^4 is 16 times as large as 5^4. The figure below shows an example of overfitting.

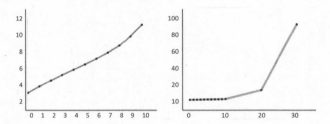

Overfitting and Out-of-Sample Error

The graph on the left shows (hypothetical) sales data from a company that manufactures industrial 3-D printers as a function of the number of site visits made (on average) per month by their sales team. The graph on the left shows a nonlinear best fit that includes nonlinear terms up to the fifth power. The graph on the right shows that the model predicts sales of 100 printers if sales visits reach 30. That cannot be correct if customers buy at most one 3-D printer. By overfitting, the model makes a huge error out of the sample.

To prevent overfitting, we could avoid higher-order terms. A more sophisticated solution known as *bootstrap aggregation* or *bagging* constructs many models. To bootstrap a data set, we create multiple data sets of equal size by randomly drawing data points from the original data. The points are drawn with replacement—after we draw a data point, we put it back in the "bag" so that we might draw it again. This technique produces a collection of data sets of equal size, each of which contains multiple copies of some data points and no copies of others.

We then fit (nonlinear) models to each data set, resulting in multiple models.[17] We can then plot all the models on the same set of axes, creating a *spaghetti graph* (see below). The dark line shows the average of the different models.

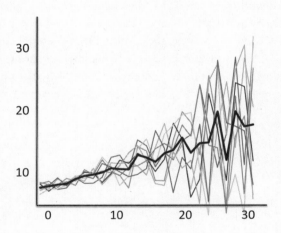

Bootstrapping and a Spaghetti Graph

Bagging will capture robust nonlinear effects, as they will be evident in multiple random samples of the data, while avoiding fitting idiosyncratic patterns in any single data set. By building diversity through random samples and then averaging the many models, bagging applies the logic that underpins the diversity prediction theorem. It creates diverse models, and as we know, the average of those models will be more accurate than the models themselves.

4. *Modeling Human Actors*

It is not possible yet to point to a single theory of human behavior that has been successfully formulated and tested in a variety of settings.

—Elinor Ostrom

In this chapter, we address a question that lies at the core of this book: How do we model people? In many of the models that follow, people will be the fundamental unit of analysis. We will construct models of people who vote, cooperate, start uprisings, participate in fads, invest in retirement accounts, and become addicted to drugs. Within each model, we will have to make assumptions about people. What are their objectives? Do they only care about themselves or are they altruistic? What are their potential actions? How do they choose what action to take, or do they not even have a choice?

We could make ad hoc assumptions for each model. But to do so would invite confusion and miss an opportunity. We would be left with an idiosyncratic set of constructions. Each new model would require new thinking about how people act. The resulting heterogeneity would limit our ability to think across and combine models. We could not be effective many-model thinkers.

The approach we follow stresses coherence along with variety. We will model people as either *rule-based actors* or *rational actors*. Within the set of rule-based actors, we consider those who act based on *simple fixed rules* and those who act based on *adaptive rules*. Someone acting based on an adaptive rule can change her behavior based upon information or past success or because she watches others. As we shall discuss, no bright lines distinguish these cases from one another; an adaptive rule can sometimes be interpreted as a fixed rule, and sometimes rational actions will take the form of simple rules.

How we choose to model people will depend on the context and on our goals. Are we predicting or explaining? Are we evaluating policy actions? Are we trying to design an institution? Or are we exploring? In low-stakes environments, such as modeling what color coat people buy or whether they stand for an ovation after a show, we will most often assume that people apply fixed rules. When people decide whether or not to cooperate in a venture or to trust another person, we will assume that people learn and adapt. Finally, in high-stakes environments, we will assume that informed, sophisticated people make optimal choices.

Before describing our approach in more detail, we first address some common misconceptions. Many people first encounter formal models of social phenomena in introductory economics courses. Those models often rely on a rudimentary rational-actor model in which everyone is self-interested and capable of optimizing. The model may also assume everyone has the same preferences and income level. Economists then solve for equilibria within these models, enabling them to evaluate the effects of shocks to a market or policy changes. These models, though based on incorrect assumptions, are useful. They help economists to communicate and students to understand.

Based on this experience, many people infer that formal modeling requires a narrow, unrealistic view of human nature, in which people are self-interested and never make mistakes. That is not the case. In fact, not even economists think that to be true. The frontiers of economics consist of models with imperfectly informed, heterogeneous actors who adapt in response to what they learn, and who sometimes, though not always, care about the payoffs to others. The extent to which people exhibit other-regarding preferences also depends on the situation; when donating to charity or volunteering, for example, a person may be more other-regarding than when buying a house.

Nevertheless, the unfortunate impression persists that modeling assumes selfish, unrealistically rational people. We must disabuse ourselves of that view. As an analogy, if you only wade a few steps into the ocean, you might infer that it is shallow. As you swim out farther, you begin to sense the depth. Here, we start near the shore. At times we venture further out and show how models can accommodate other-focused, boundedly rational people.

Whatever assumptions we do make, we cannot escape their implications. We are tied to the mast of logical coherence. We cannot manufacture implications. If we assume strong social influences in consumer choices, our model will produce a handful of products with large market shares. If we assume people obtain information through networks, then people who fill structural holes will possess power.

The remainder of the chapter begins with an overview of some challenges of modeling people: we are diverse, socially influenced, mistake-prone, purposive, adaptive, and possessed of agency. We cannot include all of these characteristics in a single model without creating a complicated mess, so we must pick and choose.

If heterogeneity matters little, perhaps we assume identical agents. If the problem is simple or the people sophisticated, perhaps we assume people do not make mistakes.

We next describe the rational-actor model and discuss its theoretical underpinnings and the justifications for its use despite its descriptive inaccuracy. We conclude that whether the rational-actor model functions as a gold standard, a straw man, or somewhere in between depends on our model's purpose. The rational actor will be less successful at predicting human behavior than as a tool for communicating, evaluating actions, and designing policies.

We then show how we can add psychological biases and altruistic preferences onto the standard rational-actor model. The choice of whether to include a bias or a concern for others rests again on what we are studying. Some human biases such as loss aversion and presentist bias—caring more about delays today than in the future—may be necessary to include in some instances. For example, those assumptions may be important for models of retirement savings or riots. The assumptions may be less important for models of driving behavior or disease transmission.

In the fourth section, we describe rule-based behavior. This category of models has the advantage of being both flexible—any behavior we can write down as a rule is fair game—and tractable. We need only encode that behavior in a computer program, an agent-based model, and watch what unfolds. That freedom comes with responsibility. As we can choose any behavioral rule, we must guard against ad hoc assumptions. In some cases, behavioral rules can be justified as optimal behavior given an objective function, though this will not always be the case.

The chapter concludes with a reconsideration of the value of rationality as a benchmark behavior. Even if people do not optimize, they do adapt to changing circumstances and new knowledge. That observation produces a conundrum of sorts. If we design an institution or policy based on the assumption that people have a bias or that they act in ways not in their self-interest, we run the risk that people will change their behaviors. People may be fooled once, but they may be harder to fool two or three times. We need not conclude that rationality is the only plausible assumption, but the logic does argue for rationality as a relevant benchmark. Logic also supports considering simple rules of behavior as well, as a lower bound on rationality. And, in modeling any given situation, we might apply any number of adaptive and psychological rules as a way for us to explore the giant space in between those extremes.

The Challenge of Modeling People

Modeling people presents a challenge because while models require low-dimensional representations, people defy simple characterizations. People are *diverse,* we are *socially influenced,* we are *error-prone,* we are *purposive,* and we *learn*. In addition, people *possess agency*—we have the capacity to act.

By way of contrast, physical objects, such as carbon atoms and billiard balls, exhibit none of these six properties. Carbon atoms lack diversity (though they can occupy heterogeneous positions within compounds, such as in propane). Carbon atoms never violate the laws of physics nor do the lead purposive lives. They do not change their behaviors based on past experiences. They lack agency; they do not decide to lead uprisings or switch careers. Hence the oft-repeated quip by social scientists: how difficult physics would be if electrons could think. Physics would be even harder if electrons could write models.

We can start with the problems created by diversity. People differ in our preferences, in our capacity to act, in the social networks we form, in our levels of altruism, and in the level of cognitive attention we allocate to actions. Modeling would be easier if everyone were the same. Sometimes we rely on statistical logic and assume that the behavioral diversity cancels. For example, we might construct a model that predicts charitable donations as a function of income. For a given income level and tax rate, some people may be more altruistic than we assume and others may be less. If the deviations from the model average out (and in Chapter 5 we cover models of distributions that explain why they might), then our model may be accurate. This canceling out of diversity will not occur unless actions are independent. When behavior is socially influenced, extreme actions can create spillovers. This occurs when political activists energize voters. We will encounter this effect of diversity when we model riots.

Whether or not mistakes cancel in the aggregate depends on the context. Errors that result from a lack of cognitive attachment may be random and independent. Errors that arise from cognitive biases may be systematic and correlated. For example, people may overweight recent events and recall narratives better than statistics. A shared bias like this will not cancel out.

The next challenge relates to what people desire. A central challenge in writing models of people will be making an accurate assessment of their goals and objectives. Some people desire wealth and fame. Others want to contribute to the betterment of their communities and the world. In the rational-actor model, we represent a person's payoff directly in the form of a function. In rule-based models, purposes are more implicit. A behavioral rule in which people seek to live in an integrated neighborhood but move out of a neighborhood if the percentage of people who share their racial identity falls below 10% embeds certain beliefs about what people desire.

The final challenge to modeling people results from the fact that people have agency: the ability to take action, to change what we do, and to learn. That said, in some contexts, people may be better characterized as creatures of habit. Actions may be outside our control. Few people choose to be addicted to opioids or to be poor. Yet people take actions that produce those outcomes.

Often, when people take actions that produce bad outcomes, they adapt their behavior. We can capture this by including learning in our models. How people learn varies by context. When learning how many hours they need to study for an exam in order to get a good grade, or how many times a week they need to exercise, people may learn based on individual experiences and introspection. When learning what grocery store to visit or whether to contribute to a charity, people may learn through observing others. In Chapter 26, we show how in non-strategic contexts, learning generally works. People learn the best action. We also show that in strategic contexts, which we model as games, all bets are off. Neither individual nor social learning necessarily produces good outcomes.

Each of these six characteristics are potential model features. If we include a feature, we must decide how much of it to include. How diverse do we make our actors? How much social influence do we include? Do people learn from others? How do we define objectives? How much agency do people possess? We may possess less agency than we believe. Jonathan Haidt describes our lack of agency with his metaphor of the rider and the elephant. "The image I came up with for myself, as I marveled at my weakness, was that I was a rider on the back of an elephant. I'm holding the reins in my hands, and by pulling one way or the other I can tell the elephant to turn, to stop, or to go. I can direct things, but only when the elephant doesn't have desires of his own. When the elephant really wants to do something, I'm no match for him."[1] Sometimes we do ride the elephant. Sometimes we do not. No single approach to modeling humans will be appropriate in all settings, so we model humans in a variety of ways.

The Rational-Actor Model

The *rational-actor model* assumes that people make optimal choices given a payoff or utility function. These actions can be *decisions,* where the payoff depends only on the individual's own action, or they can take place within a *game,* where payoffs depend on what others do. In a game with simultaneous choices or with incomplete information, the rational-actor model also specifies *beliefs* about what the other actors will do.

Rational-Actor Model

An individual's preferences are represented by a mathematical utility or pay-off function defined over a set of possible actions. The individual chooses the action that maximizes the function's value. In a game, that choice may require beliefs about the actions of other players.

As an example, we construct a primitive rational-actor model of an individual's decision for how much income to allocate to housing. The model characterizes her utility as a function of housing and all other consumption, with the latter including food, clothing, and entertainment (see box). The model assumes a price for housing and a price for all other goods. The model is far from realistic. It treats all housing the same. And, it lumps all other goods into one category called consumption and prices them all the same. We can set those inaccuracies aside for the moment, as the purpose of the model is to explain the proportion of income spent on housing.

A Rational-Actor Model of Consumption

Assumption: An individual's utility from general consumption, C, and housing, H, can be written as follows:

$$U(C,H) = C^{\frac{2}{3}}H^{\frac{1}{3}}$$

Result: A utility-maximizing individual (a rational actor) spends exactly one-third of her income on housing.[2]

In the model, the proportion of income a person spends on housing does not depend on the price of housing or on income. Both results are reasonable approximations of the data.[3] Other than people at the extremes of the income distribution, most people spend about a third of their income on housing. The finding has policy implications: if housing prices fall by 10%, people will buy 10% more housing. The finding also provides a justification for assuming identical agents. If people spend a fixed percentage of income on housing, total spending on housing depends only on average income.

Using a utility function makes our models analyzable, testable, and tractable. We can estimate the functions with data, we can derive optimal actions, and we can ask "what if" questions by changing parameter values. In assuming a utility function, we imply a coherence to preferences that may not exist. For preferences to be representable by a utility function, they must satisfy certain axioms. Theorems that prove the existence of utility functions assume a set of alternatives along with a *preference ordering*. Imagine that we can list all possible bundles of goods a person might buy. A preference ordering ranks these bundles from most to least favored. A person might prefer coffee with milk to tea with lemon; if so, she ranks the bundle {coffee, milk} above the bundle {tea, lemon}.

A utility function represents preferences if it assigns a higher value to bundle A than bundle B if and only if the preference ordering ranks A above B. For preferences to be consistent with a utility function they must satisfy completeness, transitivity, independence, and continuity. *Completeness* requires that the preference ordering is defined over all pairs of alternatives. *Transitivity* rules out preference cycles. If someone prefers bundle A to bundle B and bundle B to bundle C, she must also prefer A to C. In other words, if a person prefers apples to bananas and bananas to cheese, then she must also prefer apples to cheese. This condition rules out inconsistent preferences.

Independence requires that people evaluate the outcomes in a lottery separately. A lottery is a probability distribution over alternatives, such as a 60% probability of A and a 40% probability of B. Preferences satisfy independence if when A ranks above B, then in any lottery that includes B as an outcome, the person prefers an alternative lottery in which we replace B with A. Independence rules out strong risk aversion. A risk-averse person might rank a trip to New Orleans over a trip to Disney World but prefers knowing for certain that he will be going to Disney World over entering a lottery that sends him to New Orleans with probability $\frac{1}{2}$ and to Disney World otherwise. The final condition, *continuity,* requires that if a person ranks A above B and B above C, then there exists a lottery in which she gets A with probability p and C with probability $(1-p)$ that she likes exactly as much as B. This assumption also rules out strong preferences for certain outcomes.[4]

The assumptions of independence and transitivity, which people violate, on top of the dubious claim that people optimize leads many to question the widespread use of the rational-actor model, particularly by economists. Yet there exist good reasons to assume rationality. First, people may act "as if" they optimize. They may apply rules that produce nearly optimal behavior. When people play pool, catch a Frisbee, or drive a car, they do not write down mathematical equations. The mathematics required to time a leap to catch a Frisbee would overwhelm almost anyone. Yet people catch Frisbees. So, by the way, do dogs. Thus, both people and dogs act as if we solve a difficult optimization problem.

This same logic extends to high-dimensional problems. An analysis of the actions of Harold Zurcher, the superintendent of maintenance for the Metropolitan Bus Company in Madison, Wisconsin, found that he made near optimal decisions about when and whether to replace bus engines.[5] Though Zurcher did not write down any mathematics, he relied on heuristics. Those heuristics, informed by experience, meant that he acted (almost) as if he were a rational actor.

Second, even if people do make mistakes, in cases where a situation is repeated, our capacity to learn should push us toward optimal actions. Third, in cases where the stakes are large, people should put in the time and energy to make near-optimal choices. People may overpay 30% for coffee or AAA batteries, but they do not overpay 30% for cars or houses. The claim that learning and higher stakes increase rationality has ample empirical and experimental support.[6]

A fourth reason for adopting the rational-actor model, paradoxically, is that it simplifies the analysis. Most utility functions will have a unique optimal action. A person can behave suboptimally in thousands of ways. Saying that people do not optimize opens an enormous box of possibilities. If we assume people make choices to maintain their identities or to enforce cultural norms, we may lack a single clear answer. Rational choice is not realistic, but realism comes at the cost of messiness. An answer, even if it is known to be wrong, can be more useful then having no answer at all, as it allows us to bring the model to data and to work through the effects of changes in variables.[7]

Arguments for Rational Choice

"As if": Intelligent rule-based behavior may be indistinguishable from optimal or near-optimal behavior.

Learning: In situations that are repeated, people should approach optimal behavior.

Large stakes: On important decisions, people gather information and think slowly.

Uniqueness: Optimal behavior is often unique, making the model testable.

Consistency: Optimal behavior creates a consistent model. If people learn the model, they will not change their behavior.

Benchmark: Optimal behavior provides a benchmark as an upper bound on people's cognitive abilities.

Fifth, the rational-actor assumption guarantees internal consistency. If a model assumes suboptimal behavior and the model is in the public domain, the model can be learned. People can change their behavior. They might not optimize, but any assumption other than optimality is subject to the criticism that it is not consistent. We return to this point in the discussion at the end of the chapter.

Last, and some would argue most important, rationality can function as a benchmark.[8] When designing a policy, making a prediction, or choosing an action, we should consider what would happen if people had rational preferences and optimized. That exercise may point to flaws in our thinking. We should also be open to the possibility that the exercise will lead us to conclude that the rational-actor model does not apply and that we should privilege other models instead. To this list we might add a seventh reason: many-model thinking. If people apply many models, they are less likely to make mistakes.

Psychological Biases

The rational-actor model has been challenged by psychologists, economists, and neuroscientists, who note that it does not match up with how humans behave. Empirical findings from laboratory and natural experiments show that people suffer a variety of biases, including a status quo bias. We ignore base rates when making probability calculations, we attach too much significance to sure things, and we are loss-averse.

As researchers begin to link behavior and beliefs to processes within the brain, evidence of hardwired biases becomes more compelling. For example, neuroeconomics uses brain imaging studies to study economically relevant behaviors such as attitudes toward risk, levels of confidence, and responses to information.[9] Kahneman argues that what we know so far supports making a distinction between two types of thinking: quick, intuitive rules (*fast thinking*) and deliberate contemplation (*slow thinking*). Fast thinking is more likely to be subject to the aforementioned biases.[10] In the long run, we may be able to infer some behavioral patterns from brain structures, but we should keep in mind that the brain has tremendous plasticity. It is capable of overcoming biases by thinking slowly.

Further, we should be cautious about accepting as universal any finding documented in just a handful of studies. Many psychological findings have not proven robust. A recent study failed to replicate half of one hundred findings published in leading psychology journals.[11] Furthermore, replicability need not imply universality. Subject pools for many studies lack economic and cultural diversity.[12] We might expect that more diverse subject pools would produce fewer behavioral regularities, providing even greater reason to avoid generalizations about behavior.

Last, in attempting to make more realistic models, we must keep tractability in mind. More realistic models may require more sophisticated mathematics.[13] None of these concerns is so persuasive to suggest abandoning models with psychologically realistic behaviors, but collectively they imply that we proceed with caution and emphasize well-documented behavioral regularities.

Two deviations that have been replicated many times are loss aversion and hyperbolic discounting. *Loss aversion* states that people are risk-averse over gains and risk-loving over losses. Kahneman and Tversky refer to this general theory of behavior as *prospect theory*.[14] Loss aversion does not at first appear irrational, but it implies that people choose different actions when an identical scenario is presented as a potential loss as opposed to a potential gain.

For example, people prefer winning $400 for certain rather than entering a lottery with an even chance of winning $1,000. Yet they will enter a lottery with an even chance of losing $1,000 rather than lose $600 for certain. This same inconsistency extends to nonmonetary domains. Doctors given choices framed as gains are risk-averse. When choices are presented as losses, doctors take more risks.[15]

Prospect Theory: Example

Gain Framing: You have two options:
Option A: Win $400 for certain.
Option B: Win $1,000 if a fair coin comes up heads, and $0 if tails.

Loss Framing: You are given $1,000 and have two options:
Option Â: Lose $600 for certain.
Option B̂: Lose $0 if a fair coin comes up heads, and lose $1,000 if tails.

A and $\hat{A}$ are equivalent as are B and $\hat{B}$. According to prospect theory, more people choose A and $\hat{B}$.

Hyperbolic discounting implies stronger discounting of the immediate future. Standard economic models assume *exponential discounting,* a constant rate at which people discount the future. A person with an annual discount rate of 10% values $1,000 next year as equal to $900 today. She would discount next year's money by 10% for every year into the future. Evidence shows that most people do not discount the future at a fixed discount rate. Instead, they suffer from an *immediacy bias*: they discount the near future far more than the later future.[16] For example, if you ask people whether they would prefer $9,500 twenty years from now or $10,000 twenty years and one day from now, almost everyone will wait one more day for the extra $500. If you ask those same people if they would prefer $9,500 today or $10,000 tomorrow, many will take the $9,500 now. This is an example of immediacy bias.[17]

That bias produces time-inconsistent behavior. One year from now, most people prefer waiting one more day and taking the $10,000. Such preferences are not logically consistent. Hyperbolic discounting has been put forward as a reason why people run up credit card debts, eat unhealthy foods, have unprotected sexual relations, and fail to save for retirement.

In summary, depending on how we will use our model, we may choose to assume loss aversion and hyperbolic discounting given that these assumptions appear to better match behavior for most people. The main reason not to do so would be if they complicate the model without qualitatively changing what we find, or if by assuming hyperbolic discounting our model produces unrealistic behavior.

Rule-Based Models

We now turn to rule-based models.[18] While optimization-based models assume an underlying utility or payoff function that people maximize, rule-based models assume specific behaviors. A rule-based model might assume that in an auction, a person will bid 10% less than her true value for an item, or that a person will copy a friend's action if that friend consistently receives higher payoffs. Many people equate optimization-based models with mathematics and rule-based models with computation. The distinction between optimization-based models and rule-based models is not as clean as might be thought. Think back to our model of housing consumption. Optimal behavior took the form of a simple rule: spend one-third of income on housing. The key difference between the two approaches lies in their foundational assumptions. In an optimization-based model, preferences or payoffs are fundamental. In a rule-based model, the behavior is fundamental.

Behavioral rules can be fixed or adapt. A fixed rule applies the same algorithm at all times. Just as rational-choice models provide an upper bound on people's cognitive abilities, fixed-rule models provide a lower bound. A common fixed rule in markets, the *zero intelligence rule,* accepts any offer that produces a higher payoff. It never takes a stupid (i.e., utility-reducing) action. Suppose we want to gauge the efficiency of a one-sided market design in which sellers post take-it-or-leave-it offers for some good. A seller following a zero intelligence rule would randomly pick a price above her value. A buyer would purchase any good with a price below her value. When we encode those behaviors in a computer model, we find that in markets zero-intelligence traders produce nearly efficient outcomes. Thus, exchange markets do not need rational buyers and sellers to function well.[19]

An adaptive rule switches among a set of behaviors, evolves new behaviors, or copies the behaviors of others. It takes these actions in order to improve the payoff. Therefore, unlike fixed rules, adaptive rules require a utility or payoff function. Advocates of this approach argue that within any situation people tend toward simple and effective rules, and that if this is what people do, we should model them in that way.[20] Though rule-based models make no explicit assumption about rationality, adaptive-rule models exhibit ecological rationality—better rules come to predominate.[21]

To explain how adaptive-rule models operate, we describe the *El Farol model* of self-organized coordination.[22] El Farol is a nightclub in Santa Fe, New Mexico, that features dancing every Tuesday night. Each week, a population of 100 potential dancers decide whether to go dance at El Farol or stay home. All 100 people like to dance, but they do not want to go if the club is too crowded. The model assumes a stark form of preferences. A person earns a payoff of zero from staying home, a payoff of 1 from attending if 60 or fewer people attend, and a payoff of -1 from attending when more than 60 people attend.

If we construct a fixed-rule model, anything might result. For example, if we assigned everyone the rule "go the first week; if more than 60 people attend, do not go the next week; and then go the following week," the El Farol would have 100 attendees the first week, zero attendees the second week, and then 100 attendees the third week. The El Farol model creates adaptive rules by endowing each person with an ensemble of rules. Each rule tells the individual whether or not to go to El Farol. The rules take several forms. Some are fixed rules: go every other week. Others are based on trends in the number of people who attended El Farol in recent weeks. One rule might predict that the number of people that show up this week will be the same as last week. If fewer than 60 people attended last week, that rule would tell the person to go this week.

An adaptive behavioral rule model might assign a score to each rule equal to the percentage of weeks for which it gave correct advice. Each individual might then use the model in her ensemble with the highest score. The best rule will vary over the course of the weeks. Simulations of this type of model find that if individuals possess a large ensemble of rules, then approximately 60 people attend each week: coordination emerges without any central planner. In other words, the system of adaptive rules self-organizes into nearly efficient outcomes.

El Farol Model: Adaptive Rules

Each of 100 individuals decides independently whether or not to go to El Farol every week for a year. An individual who goes to El Farol earns a payoff of 1 if 60 or fewer people attend and a payoff of -1 otherwise. An individual who does not go to El Farol earns a payoff of zero.

Each individual has an ensemble of rules to decide whether to attend. These rules can be fixed or contingent on recent past attendances. Each week, each individual follows the rule in his ensemble that, if followed, would have produced the highest payoff in the past.

We can interpret behavior within adaptive-rule models, like the El Farol model, within the *micro-macro loop* (see figure 4.1). At the micro-level, a set of individuals take actions (denoted by the a_i's) according to rules. These rules create macro-level phenomena (denoted by Macro$_1$ and Macro$_2$), as represented by the upward arrows. In the El Farol problem, the macro-level phenomena are the sequences of past attendances. The downward arrows represent how these macro-level phenomena feed back into the behaviors of the individuals. In the El Farol model, each person may be applying a different rule. If the rules people apply produce a crowded El Farol four weeks in a row, then rules that tell people to attend less often will produce higher payoffs. As people switch to those rules, fewer people will attend. The micro-level rules produce a macro-level phenomenon (over-attendance) that feeds back to the micro-level rules.

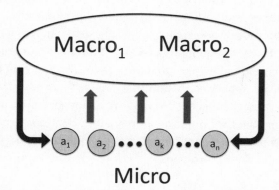

Figure 4.1: The Micro-Macro Loop

Cognitive Closure, a Big Question, and Many Models

The micro-macro loop elucidates a central tension as to how smart to make our agents. Should people infer all consequences of their actions? The loop also hints at a larger question that we encounter throughout the book as to what class of outcome a model produces: Does it go to equilibrium, produce randomness, create a cycle, or generate a complex series of outcomes?

We start with the question of how smart to make our agents. Suppose that we believe that individuals possess only modest cognitive abilities, so we build a model with *zero intelligence agents*. Their actions aggregate to produce aggregate macro-level phenomena. If the macro level produces efficient or nearly efficient outcomes, as we noted was the case with a one-sided market of buyers and sellers, then we may be justified in our assumption. An easy-to-follow fixed rule produces good outcomes. People would have little incentive to expend effort developing more sophisticated rules.

The tension arises when our model produces inefficient or even lousy macro-level outcomes. Such could be the case in the El Farol model, where a common fixed rule could lead to a cycle in which El Farol was overcrowded with dancers one week and empty the next. Confronted with an inefficient outcome, we might think that people would adapt. They might experiment. They might think through the logic of the situation to formulate a new action. If we follow that logic to its extreme and assume a low cost of thinking, then we find ourselves advocating the rational-actor model. Any person not behaving optimally could do better. While that is true, people also have to be able to formulate that better action.

This leads to a big question: What class of outcomes does the model produce? We have four options: equilibrium, cycles, randomness, or complexity. The class of outcome will matter for deciding how seriously we take the argument that people should learn their way to equilibrium. First, if the model produces randomness at the macro level, the individuals probably cannot learn anything. Our model is fine. A similar logic applies to models that produce complex patterns. In these cases, we would assume that people continue to adapt new rules, but we would not necessarily assume that they can choose optimally. To the contrary, the complexity of the macro-level phenomena makes optimal responses implausible. People would be more likely, as in the El Farol model, to confront complexity with an ensemble of simple rules.

The models that produce cycles or equilibria create a stationary environment. We therefore might expect that people can learn—that no one would continually take a suboptimal action. As an example, suppose we have a traffic model in which everyone chooses a route to work using a fixed rule. In our model, the traffic system settles into an equilibrium. In that equilibrium, one of the individuals, Layne, spends 75 minutes each morning traveling from Calabasas to downtown Los Angeles. Given the equilibrium, if Layne took side streets through Topanga Canyon, her trip would take only 45 minutes. Given the value of an extra 30 minutes per day and the frequency with which people in Los Angeles talk about traffic, Layne would likely find the shorter route. She has no shortage of methods for finding it. She might use a route recommender, talk to a neighbor, or experiment.

Thus, if our model produces an equilibrium (or a simple cycle) and that equilibrium is not consistent with optimizing behavior, then our model suffers a logical flaw. If people have a better action available to them, they should figure it out. They should learn. Notice that we need not assume optimal behavior in order to reach the equilibrium. People could follow simple rules and produce an equilibrium in which no one can benefit by changing her action. At that equilibrium, it would look "as if" people are optimizing, because they are. Again, that logic need not apply for complex or random outcomes. If traffic patterns in Los Angeles produce a complex sequence of traffic slowdowns and jams, we have little reason to believe that Layne selects an optimal route each day. She almost surely cannot.

If adaptive rules that can adopt any action produce an equilibrium, then the equilibrium must be consistent with behavior by optimizing agents. If those same adaptive rules produce complexity, the agents' behavior need not be optimal. We can restate this idea as follows: optimal behavior may be an unrealistic assumption, particularly in complex situations. On the other hand, if a system produces a stable outcome in which a person has better actions, she will probably figure out a better action to take.

An extension of this logic applies to policy interventions. Suppose that we use data to estimate people's behavioral rule—say, the likelihood a person shows up at a hospital's emergency room during lunch hour for minor health issues. If we assume a fixed rule, we might enlarge the size of our facility so that people do not have to wait. If people continue to follow that fixed rule, we have a new equilibrium with short midday wait times. However, with new, shorter wait times, people who had not been going to the emergency room for sprained ankles or chest colds may now decide to go. That equilibrium relies on people choosing suboptimal actions, such as not going to an emergency room even though they would not have to wait. If people learn, we cannot rely on past data to predict outcomes under a policy change. This insight, known as the *Lucas critique,* is a variant of *Campbell's law,* which states that people respond to any measure or standard in ways that render it less effective.[23]

The Lucas Critique

Changes in a policy or the environment likely produce behavioral responses by those affected. Models estimated with data on past human behaviors will therefore not be accurate. Models must take into account the fact that people respond to policy and environmental changes.

As should be clear at this point, there exists no best solution for how to model people. How rational we make them or how adaptive we make their rules depends on the circumstances. We should exercise our best judgment in each situation. Given the uncertainties, we should err on the side of more models rather than fewer.

Even if we are predisposed to dismiss rational-choice models as unrealistic, we must recognize their tractability, their capacity to reveal the directional forces of incentives, and their value as a benchmark. Simple rule-based behaviors, such as zero intelligence, are also unrealistic. Though wrong, they can be of use. They are easy to analyze and can reveal how much intelligence matters in a given setting.

Human behavior occurs within the extremes of zero intelligence and full rationality, so it makes sense to construct models in which individuals adapt using rules. Those rules should take into account the fact that people vary in their cognitive attachment and capabilities within a domain. We should therefore expect behavioral diversity. We might also expect some consistency within groups. This too can be included in models.[24]

In sum, given the complexities involved in modeling humans, we have abundant reasons to apply multiple diverse models. We may not be able to predict exactly what people will do, but we may be able to identify the set of possibilities. If we can, we have benefited from constructing models because we know what could happen.

We conclude with a plea for humility and empathy. In constructing models of people, a modeler must be humble. Given the challenges of diversity, social influence, cognitive errors, purpose, and adaptation, our models will inevitably be wrong, which is why we take a many-model approach. Austere models of behavior fit some situations well and allow us to focus on other aspects of the environment. Richer behavioral models will be more appropriate when we have better data. We must maintain modest expectations. People are diverse, purposive, adaptive, biased, and socially influenced, and we possess a degree of agency. How can we not expect any single model of human behavior to be wrong? It must be. Our aim is to construct many models that as an ensemble will be useful.

5. *Normal Distributions: The Bell Curve*

I couldn't claim that I was smarter than sixty-five other guys—but the average of sixty-five other guys, certainly.

—Richard Feynman

Distributions constitute part of the core knowledge base for any modeler. Later, we use distributions to construct and analyze models of path dependence, random walks, Markov processes, search, and learning. We also require a working knowledge of distributions to measure inequality in power, income, and wealth and to perform statistical tests. Our treatment of distributions unfolds over two short chapters—one each for normal and power law (long-tailed) distributions—in which we take the perspective of modelers rather than statisticians. As modelers, we are interested in two big questions: Why do we see the distributions we do, and why do distributions matter?

To address the first big question, we need to reacquaint ourselves with what distributions are. A *distribution* mathematically captures variation (differences within a type) and diversity (difference across types) by representing them as probability distributions defined over numerical values or categories. A *normal distribution* takes the familiar bell curve shape. Heights and weights of most species satisfy normal distributions. They are symmetric around their means and do not include particularly large or small events. We do not encounter many six-foot-long ants or four-pound elk. We can rely on the central limit theorem to explain the prevalence of normal distributions. It tells us that when we add up or average random variables, we can expect to obtain a normal distribution. Many empirical phenomena, in particular any aggregate like sales data or vote totals, can be written as sums of random events.

Not all event sizes are normal. Earthquakes, war deaths, and book sales exhibit *long-tailed distributions*: they consist mostly of tiny events but include the occasional whopper. Californians experience over 10,000 earthquakes each year. Unless you are staring at the quivering petals of a jasmine blossom, you would not notice them. Occasionally, though, the earth opens up, highways collapse, and cities tremble.

Knowing whether a system produces a normal or long-tailed distribution matters for any number of reasons. We want to know whether a power grid will suffer massive outages, or whether a market system will produce a handful of billionaires and billions of poor people. With knowledge of distributions, we can predict the likelihood of floodwaters that exceed a levee's walls, the probability that Delta flight 238 arrives in Salt Lake City on time, and the odds that a transportation hub costs double its budgeted amount. Knowledge of distributions is also relevant in design. Normal distributions imply no large deviations, so airplane designers need not create leg space for the eighteen-foot human. An understanding of distributions can also guide actions. As we learn later, preventing riots depends less on reducing average levels of discontent than on appeasing people at the extreme.

In this chapter, we adopt a *structure-logic-function* organization. We define normal distributions, describe how they arise, and then ask why they matter. We apply our knowledge of distributions to explain why good things come in small samples, to test for significance of effects, and to explain Six Sigma process management. We then go back to the logic question and ask what happens if we multiply rather than add random variables. We learn that we obtain a *lognormal distribution*. Lognormal distributions include larger events and are not symmetric about their means. It follows that multiples of effects lead to more inequality, an insight that has implications for how policies for increasing salaries affect income distributions.

The Normal Distribution: Structure

A distribution assigns probabilities to events or values. The distribution of daily rainfall, test scores, or human height assigns a probability to every possible value of the outcomes. Statistical measures condense the information contained in a distribution into single numbers, such as the *mean*, the average value of the distribution. The mean height of a tree in Germany's Black Forest might be eighty feet, and the mean time spent in the hospital following open-heart surgery might be five days. Social scientists rely on means to compare economic and social conditions across countries. In 2017, the United States per capita GDP of $57,000 exceeded that of France, which equaled $42,000, while mean life expectancy in France exceeds that of the United States by three years.

A second statistic, *variance,* measures a distribution's dispersion: the average of the squared distance of the data to the mean.[1] If every point in a distribution has the same value, the variance equals zero. If half of the data have value 4 and half have value 10, then, on average, each point lies distance 3 from the mean, and the variance equals 9. The *standard deviation* of a distribution, another common statistic, equals the square root of the variance.

The set of possible distributions is limitless. We could draw any line on a piece of graph paper and interpret it as a probability distribution. Fortunately, the distributions we encounter tend to belong to a few classes. The most common distribution, the normal distribution, or bell curve, is shown in figure 5.1.

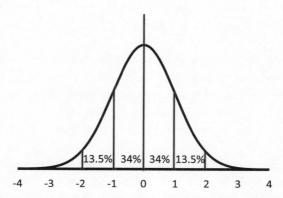

Figure 5.1: Normal Distribution with Standard Deviations

Normal distributions are *symmetric* about their mean. If the mean equals zero, the probability of a draw larger than 3 equals the probability of a draw less than -3. A normal distribution is characterized by its mean and standard deviation (or, equivalently, its variance). In other words, graphs of normal distribution all look identical, with approximately 68% of all outcomes within one standard deviation of the mean, 95% of all outcomes within two standard deviations, and more than 99% lying within three standard deviations. Normal distributions allow for any size outcome or event, though large events are rare. An event five standard deviations from the mean occurs about once in every 2 million draws.

We can exploit the regularity of normal distributions to assign probabilities to ranges of outcomes. If houses in Milwaukee, Wisconsin, have a mean square footage equal to 2,000 with a standard deviation of 500 square feet, then 68% of houses have between 1,500 and 2,500 square feet and 95% have between 1,000 and 3,000 square feet. If the 2019 fleet of Ford Focuses can travel, on average, 40 miles per gallon with a standard deviation of 1 mile per gallon, then more than 99% of Focuses will get between 37 and 43 miles per gallon. As much as a consumer might hope, her new Focus will not run 80 miles on a gallon of gasoline.

The Central Limit Theorem: Logic

No end of phenomena exhibit normal distributions: physical sizes of flora and fauna, student test scores on exams, daily sales at convenience stores, and the life spans of sea urchins. The *central limit theorem*, which states that adding or averaging random variables produces a normal distribution, explains why (see box).

Central Limit Theorem

The sum of $N \geq 20$ random variables will be approximately a normal distribution provided that the random variables are independent, that each has finite variance, and that no small set of the variables contributes most of the variation.[2]

One remarkable aspect of this theorem is that the random variables themselves need not be normally distributed. They could have any distribution so long as each has finite variance and no small subset of them contributes most of the variance. Suppose that data on the purchasing behaviors of the people in a small town of population 500 shows that each person spends on average $100 a week. Some of those people might spend $50 one week and $150 the next. Others might spend $300 every third week, while others might spend random amounts between $20 and $180 each week. So long as each person's spending has finite variation and no small subset of people contribute most of the variation, the sum of the distributions will be normally distributed with a mean of $50,000. Aggregate weekly spending will also be symmetric: as likely to be above $55,000 as it is below $45,000. By the same logic, the number of bananas, quarts of milk, or boxes of taco shells that people buy will also be normally distributed.

We can also apply the central limit theorem to explain the distribution of human heights. A person's height is determined by a combination of genetics, the environment, and interactions between the two. The genetic contribution could be as high as 80%, so we will assume that height depends only on genes. At least 180 genes contribute to human height.[3] One gene may contribute to having a longer neck or head and another to a longer tibia. Though genes interact, to a first approximation, we can assume that each contributes independently. If height equals the sum of the contributions of the 180 genes, then heights will be normally distributed. By the same logic, so too will the weights of wolves and the length of pandas' thumbs.

Applying Our Knowledge of Distributions: Function

Our first application of the normal distribution reveals why exceptional outcomes occur far more often in small populations, why the best schools are small, and why the counties with the highest cancer rates have small populations. Recall that in a normal distribution 95% of outcomes lie within two standard deviations and 99% lie within three standard deviations, and that by the central limit theorem, the mean of a collection of independent random variables will be normally distributed (with the caveats about variance). It follows that we can be pretty confident that population averages on test scores and the like will be normally distributed. The standard deviation of the average of the random variables, however, does not equal the average of the variables' standard deviations, nor does the standard deviation of the sum equal the sum of the standard deviations. Instead, those formulae depend on the square roots of the population sizes (see box).

The Square Root Rules

The standard deviations of the mean σ_μ and of the sum σ_Σ of N independent random variables each with standard deviation σ are given by the following formulae:[4]

$$\sigma_\mu = \frac{\sigma}{\sqrt{N}} \qquad\qquad \sigma_\Sigma = \sigma\sqrt{N}$$

The formula for the standard deviation of the mean implies that large populations have much lower standard deviations than small ones. From this, we can infer that we should see more good things and more bad things in small populations. And in fact we do. The safest places to live are small towns, as are the least safe. The counties with the highest rates of obesity and cancer have small populations. These facts can all be explained by differences in standard deviations.

Failure to take sample size into account and inferring causality from outliers can lead to incorrect policy actions. For this reason, Howard Wainer refers to the formula for the standard deviation of the mean the "most dangerous equation in the world." For example, in the 1990s the Gates Foundation and other nonprofits advocated breaking up schools into smaller schools based on evidence that the best schools were small.[5] To see the flawed reasoning, imagine that schools come in two sizes—small schools with 100 students and large schools with 1,600 students—and that student scores at both types of schools are drawn from the same distribution with a mean score of 100 and a standard deviation of 80. At small schools, the standard deviation of the mean equals 8 (the standard deviation of the student scores, 80, divided by 10, the square root of the number of students). At large schools, the standard deviation of the mean equals 2.

If we assign the label "high-performing" to schools with means above 110 and the label "exceptional" to schools with means above 120, then only small schools will meet either threshold. For the small schools, an average score of 110 is 1.25 standard deviations above the mean; such events occur about 10% of the time. A mean score of 120 is 2.5 standard deviations above the mean; an event of that size should occur about once in 150 schools. When we do these same calculations for large schools, we find that the "high-performing" threshold lies five standard deviations above the mean and the "exceptional" threshold lies ten standard deviations above the mean. Such events would, in practice, never occur. Thus, the fact that the very best schools are small is not evidence that smaller schools perform better. The very best schools will be small even if size has no effect solely because of the square root rules.

Testing Significance

We also use the regularity of the normal distribution to test for significant differences in mean values. If an empirical mean lies more than two standard deviations from a hypothesized mean, social scientists reject the hypothesis that the means are the same.[6] Suppose we advance a hypothesis that commute times in Baltimore equal those in Los Angeles. Suppose that our data show that commute times in Baltimore averaged 33 minutes, compared to 34 minutes in Los Angeles. If both data sets have standard deviations of the mean equal to 1 minute, then we could not reject the hypothesis that the commute times are the same. The means differ, but only by a single standard deviation. If instead commute times in Los Angeles averaged 37 minutes, then we would reject the hypothesis because the means differ by four standard deviations.

Physicists, though, might not reject the hypothesis, at least not if the data came from a physics experiment. Physicists impose stricter standards because they have larger data sets—there are a lot more atoms than people, and cleaner data. The evidence physicists relied on for the existence of the Higgs boson in 2012 would occur randomly less than once in 7 million trials were the Higgs boson not to exist.

The drug approval process used by the United States Food and Drug Administration (FDA) also uses tests of significance. If a pharmaceutical company claims that a new drug reduces the severity of eczema, that company must run two randomized controlled trials. To construct a randomized controlled trial the company would create two identical populations of eczema sufferers. One of the populations receives the drug. The other population receives a placebo. At the end of the trial, the average severity as well as average rates of negative side effects are compared. The company then runs statistical tests. If the drug significantly reduces eczema (measured in standard deviations) and does not significantly increase side effects, the drug can be approved. The FDA does not use a hard-and-fast two-standard-deviation rule. The statistical bar will be lower for a drug that cures a fatal disease and exhibits only minor side effects than for a drug that cures toenail fungus but has a higher-than-expected incidence of bone cancer associated with its usage. The FDA also cares about the *power* of the statistical test—the probability that the test shows that the drug works.

Six Sigma Method

As our final application, we show how normal distributions inform quality control through the *Six Sigma method*. Developed in the mid-1980s by Motorola, the Six Sigma method reduces errors. The method models product attributes as drawn from a normal distribution. Imagine a company that produces bolts for door handles that must fit snugly into knobs made by another manufacturer. Specifications call for the bolts to be 14 millimeters in diameter, though any bolt between 13 and 15 millimeters in diameter will function properly. If the diameters of the bolts are normally distributed with a mean of 14 millimeters and a standard deviation of 0.5 millimeter, then any bolt that differs by more than two standard deviation fails. Two-standard-deviation events occur 5% of the time—far too high a rate for manufacturers.

The Six Sigma method involves working to reduce the size of a standard deviation to lower the probability of a failure. Companies can reduce error rates by tightening quality control. On February 26, 2008, Starbucks closed down over seven thousand shops for over three hours to retrain employees. Similarly, checklists used by airlines and now hospitals reduce variation.[7] Six Sigma reduces the standard deviation so that even a six-standard-deviation error avoids a malfunction. In our bolt example, that would require reducing the standard deviation of a bolt's diameter to one-sixth of a millimeter. Six standard deviations implies an error rate of 2 per billion cases. The actual threshold used assumes an unavoidable rate of one and a half standard deviations. Thus, a six-sigma event actually corresponds to a four-and-a-half sigma event, and an allowable error rate of about 1 per 3 million.

The application of the central limit theorem (and therefore an implicit model of additive error) in the Six Sigma method is so subtle as to almost go unnoticed. The bolt manufacturer likely does not perform a precise measurement of the diameter of every bolt. It may sample a few hundred. From that sample, it estimates a mean and a standard deviation. Then, by assuming that variations in diameter result from the sum of random effects such as machine vibrations, variation in the quality of metals, and fluctuations in the temperature and speed of a press, they can invoke the central limit theorem and infer a normal distribution of diameters. The manufacturer then has a benchmark standard deviation that it can seek to reduce.

Lognormal Distributions: Multiplying Shocks

The central limit theorem requires that we add or average independent random variables in order to get a normal distribution. If the random variables are not added but interact in some way, or if they fail to be independent, then the resulting distribution need not be normal. In fact, generally it will not be. For example, random variables that are the product of independent random variables produce *lognormal* rather than normal distributions.[8] Lognormal distributions lack symmetry because products of numbers larger than 1 grow faster than sums ($4 + 4 + 4 + 4 = 16$, but $4 \times 4 \times 4 \times 4 = 256$) and multiples of numbers less than 1 decrease faster than sums ($\frac{1}{4} + \frac{1}{4} + \frac{1}{4} + \frac{1}{4} = 1$, but $\frac{1}{4} \times \frac{1}{4} \times \frac{1}{4} \times \frac{1}{4} = \frac{1}{256}$). If we multiply sets of twenty random variables with values uniformly distributed between zero and 10, their product will consist of many outcomes near zero and some large outcomes, creating the skewed distribution shown in 5.2.

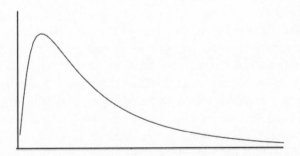

Figure 5.2: A Lognormal Distribution

The length of the tail in a lognormal distribution depends on the variance of the random variables multiplied together. If they have low variance, the tail will be short. If they have high variance, the tail can be quite long because, as noted, multiplying together a sequence of large numbers produces a very large number. Lognormal distributions arise in a wide range of examples, including the sizes of

British farms, the concentration of minerals in the earth, and the time from infection with a disease to the appearance of symptoms.[9] Income distributions within many countries approximate lognormal distributions, though many deviate from lognormal at the upper end by having too many people with high incomes.

A simple model that can explain why income distributions are closer to lognormal than normal links policies about salary increases to their implied distributions. Most organizations assign raises by percentages. People who perform above average receive high-percentage raises. People who perform below average receive low-percentage raises. Instead, organizations could assign raises by absolute amounts. The average employee could receive a $1,000 raise. Those who perform better could receive more, and those who perform worse could receive less. The distinction between percentages and absolute amounts may appear semantic, but it is not.[10] Allocating raises by percentages based on employee performance when performances from year to year are independent and random produces a lognormal distribution. Differences in income become exacerbated in future years even with identical subsequent performance. An employee who has performed well in the past and earns $80,000 will receive $4,000 from a 5% raise. Another employee, who earns only $60,000, receives only $3,000 from the same 5% raise. Inequality begets more inequality even with identical performance. Had the organization allocated raises by absolute amounts, the two employees would receive the same raise and the resulting distribution of incomes would be closer to a normal distribution.

Summary

In this chapter, we covered the structure, logic, and function of normal distributions. We saw that normal distributions can be characterized by a mean and a standard deviation. We described the central limit theorem, which shows how normal distributions arise whenever we add up or average independent random variables with finite variance. And we described formulae for the standard deviations of the mean and sum of random variables. We then showed the consequences of those properties. We learned that small populations will be far more likely to produce exceptional events and how when we lack that insight we make improper inferences and take unwise actions. We learned how assumption of normally distributed random variables allows scientists to make claims about the significance and power of statistical tests, and how process management can predict the likelihood of failure using an assumption of normality.

Not every quantity can be written as the sum, or the average, of independent random variables. Thus, not every distribution will be normal. Some quantities are products of independent random variables and will be lognormally distributed. Lognormal distributions only take on positive values. They also have longer tails, which means more large events and many more very small events. Those tails become long when random variables multiplied together have high variance. Long-tailed distributions imply less predictability, whereas normal distributions imply regularity. As a rule, we prefer regularity to the potential for large events. Therefore, we benefit from knowing the logic that creates the various distributions. In general, we would prefer that we add random shocks rather than multiply them together so as to reduce the likelihood of large events.

6. *Power-Law Distributions: Long Tails*

Every fundamental law has exceptions. But you still need the law or else all you have is observations that don't make sense. And that's not science. That's just taking notes.

—Geoffrey West

In this chapter, we cover power-law distributions. Often described as long- or heavy-tailed distributions, when graphed these distributions produce a long tail running along the horizontal axis corresponding to large events. The distributions of city populations, species extinctions, the number of links on the World Wide Web, and firm sizes all have long tails, as do the distributions of videos downloaded, books sold, academic citations, war casualties, and floods and earthquakes. In other words, all of these distributions include large events: Tokyo has 33 million residents, J. K. Rowling's Harry Potter books have sold in the neighborhood of half a billion copies, and the great Mississippi flood of 1927 covered an area larger than the state of West Virginia under thirty feet of water.[1]

Contemplating a power-law distribution of human heights reveals how much power-law distributions differ from normal distributions. If human heights were distributed by a power law similar to that of city populations, and if we calibrate the mean height at 5 feet 9 inches, then the United States would include one person the height of the Empire State Building, over 10,000 people taller than giraffes, and 180 million people less than 7 inches tall.[2]

To produce a long-tailed distribution requires non-independence, often in the form of positive feedbacks.[3] Book sales, forest fires, and city populations, unlike trips to the grocery store, are not independent. When one person buys a Harry Potter book, she induces others to buy it. When a single tree catches on fire, that fire can spread to neighboring trees. When a city increases in population, it adds amenities and job opportunities, making it more attractive to others. The sociologist Robert Merton referred to the tendency for those who have more to also receive more as the *Matthew effect*: "For unto every one that hath shall be given, and he shall have abundance: but from him that hath not shall be taken even that which he hath" (Matthew 25:29).

Given the variety of domains in which we find power-law distributions, it would be remarkable if a single mechanism could explain them all, and none does. It would be even more remarkable if each instance of a power-law distribution had a unique explanation. That is also not true. Instead, we possess a collection of distinct models that produce power laws, each capable of explaining different phenomena.

In this chapter, we focus on two models: the preferential attachment model, which explains city sizes, book sales, and web links, and the self-organized critical-ity model, which explains traffic jams and war deaths, as well as earthquake, fire, and avalanche sizes. In Chapter 12 when we cover entropy, we learn a third model in which a power-law maximizes uncertainty given a fixed mean. And in Chapter 13, we show that return times in a random walk model also satisfy a power law. Still other models show that power laws result from optimal encodings, random stopping rules, and combining distributions.[4] The remainder of the chapter covers the structure, logic, and functions of power-law distributions, followed by a discussion. The discussion reconsiders the implications of large events and describes the limits of our ability to prevent and plan for them.

Power Laws: Structure

In a *power-law distribution*, the probability of an event is proportional to its size raised to a negative exponent. So for example, the familiar function $\frac{1}{x}$ describes a power law. In a power-law distribution, the probability of an event is inversely related to its size: the larger the event, the less likely it occurs. Power-law distributions, therefore, have many more small events than large ones.

Power-Law Distributions

A **power-law distribution**[5] defined over the interval $[x_{min}, \infty)$ can be written as follows:

$$p(x) = Cx^{-a}$$

where the **exponent** $a > 1$ determines the length of the tail, and the constant term $C = (a-1)x_{min}^{a-1}$ ensures the distribution has a total probability of one.

The size of the power law's exponent determines the likelihood and size of large events. When the exponent equals 2, the probability of an event is proportional to the square of its size. An event of size 100 occurs with probability proportional to $\frac{1}{100^2}$, or 1 in 10,000. When the exponent increases to 3, the probability of that same event is proportional to $\frac{1}{100^3}$. For exponents of 2 or less, a power-law distribution lacks a well-defined mean. The mean of data drawn from a power-law distribution with an exponent of 1.5 never converges. It increases without limit.

Figure 6.1 shows an approximate graph of the distribution of the number of links to webpages on the World Wide Web.

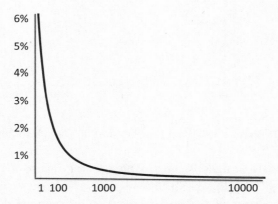

Figure 6.1: Approximate Power-Law Distribution of Webpage Links

The potential for large events distinguishes power-law distributions from normal distributions, from which we practically never see large events. For a long-tailed distribution, though rare, they occur at sufficient frequency to merit attention and preparation. Even one-in-a-million events are worth considering. For example, earthquake sizes approximately satisfy a power law with exponent near two. Suppose that for a region an earthquake larger than size 9.0 on the Richter scale, the size of an earthquake that topples buildings and changes the local topography, occurs each day with a probability of one-in-a-million. Within a century, an earthquake of that size would occur with probability 3.5%.[6]

To see the difference between the probabilities of one-in-a-million events in normal and long-tailed distributions, we can use the distribution of deaths due to terrorist attacks, which follow a power-law distribution with an exponent of 2.[7] A one-in-a-million event consists of nearly 800 deaths. If deaths due to terrorist attacks followed a normal distribution with mean 20 and a standard deviation of 5, a one-in-a-million event would involve fewer than 50 deaths.

71

A power-law distribution has a precise definition. Not all long-tailed distributions are power laws. Plotting a distribution on a log-log scale creates a crude test of whether the distribution is a power law. A log-log plot transforms event sizes and their probabilities to their logged values and transforms a power-law distribution into a straight line.[8]

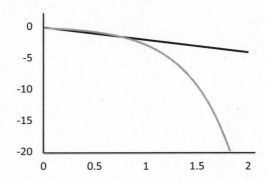

Figure 6.2: Power Law (Black) vs. Lognormal (Gray) on Log-Log Scale

In other words, a straight line on a log-log plot is evidence of a power law, while an initially straight line that gradually falls off is consistent with a lognormal (or an exponential) distribution. The rate at which a lognormal distribution curves downward depends on the variation of the variables that produce the distribution.[9] As we increase the variance in a lognormal distribution, the tail increases, making it closer to linear on a log-log plot.[10]

The special case of power laws with exponents equal to 2 are known as *Zipf distributions*. For power laws with exponents of two, an event's rank times its probability will equal a constant, a regularity known as *Zipf's Law*. Words satisfy Zipf's Law. The most common English word, *the,* occurs 7% of the time. The second most common word, *of,* occurs 3.5% of the time. Notice that its rank, 2, times its frequency of 3.5% equals 7%.[11]

Zipf's Law

For power-law distributions with an exponent of 2 ($a = 2$), the rank of an event times its size equals a constant.

$$\text{Event Rank} \cdot \text{Event Size} = \text{Constant}$$

The populations of cities in many countries, including the United States, are distributed approximately in this way. Using 2016 city population data, each city's rank multiplied by its population produces a value near 8 million.

Rank	City	2016 Population	Rank × Population
1	New York, NY	8,600,000	8,600,000
2	Los Angeles, CA	4,000,000	8,000,000
3	Chicago, IL	2,700,000	8,100,000
4	Houston, TX	2,300,000	9,200,000
5	Phoenix, AZ	1,600,000	8,000,000

Models That Produce Power Laws: Logic

We now turn to models that produce power laws. Lacking models, power-law distributions remain unexplained patterns.

Our first model, the *preferential attachment model,* assumes entities that grow at rates relative to their proportions. It captures Merton's Matthew effect: more begets more. The model considers a population that grows through arrivals. A new arrival either joins an existing entity or creates a new one. If the latter, the probability of joining an existing entity is proportional to the size of that entity.

Preferential Attachment Model

A sequence of objects (people) arrive one after another. The first arrival creates an entity. Each subsequent arrival applies the following rule: With probability p (small), the arrival forms a new entity. With probability $(1-p)$, the arrival joins an existing entity. The probability of joining a particular entity equals its size divided by the number of arrivals to date.

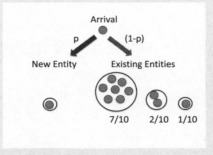

73

Imagine students coming onto a college campus. The first student creates a new club. With some small probability, the second student creates her own club. More likely, she joins the first student's club. The first ten students might create three clubs: one with seven members, one with two members, and one containing a single member. The eleventh arrival will, with small probability, create a fourth club. If not, she will join an existing club. When joining an existing club, she chooses the club with seven students 70% of the time, the club with two students 20% of the time, and the club with one student 10% of the time.

The preferential attachment model helps explain why the distributions of links on the World Wide Web, city sizes, firm sizes, book sales, and academic citations are power laws. In each setting, an action (say, a person buys a book) increases the likelihood others will do the same. If the probability of buying from a firm is proportional to its current market share, and if new firms enter at a low rate, then the model predicts that the distribution of firm sizes will be a power law. The same logic applies to book sales, music downloads, and city growth.

Our second model, the *self-organized criticality model,* produces a power-law distribution through a process that builds interdependencies in a system until the systems reaches a critical state. A variety of self-organized criticality models exist. The *sand pile model* assumes that someone drops grains of sand onto a table from a spot several feet above. As the grains accumulate, a pile forms. Eventually, the pile attains a *critical state* where additional grains can cause avalanches. At this critical state, additional grains often have no effect or cause at most a few grains to fall. These are the many small events in a power-law distribution. Sometimes the addition of a single grain results in a large avalanche. These are the large events.

A second model, the *forest fire model,* assumes a two-dimensional grid on which trees can grow. Trees can also be hit by random lightning strikes. When the density of the trees is low, any fires caused by lightning will be small, affecting at most a few cells. When the density of trees is high, a fire started by a lightning strike will spread across much of the grid.

Self-Organized Criticality: Forest Fire Model

The forest consists initially of an empty N by N grid. Each period a random site on the grid is chosen. If empty, with probability g the site grows a tree. If the site contains a tree, with probability $(1 - g)$ lightning hits the site. If the site contains a tree, the tree catches fire, and the fire spreads to all connected sites with trees.

Notice that in the forest fire model, the probability of a lightning strike equals one minus the probability of the growth rate. This construction allows us to vary the relative rate of growth and lightning. It is a simplification that reduces the number of parameters in our model. Experimenting with the growth rate of trees, we find that for growth rates close to one, the density of trees increases to a critical state: a relatively dense forest of trees, where lightning strikes can wipe out a huge swath of forest. At this critical state, the distribution of the sizes of patches in the forest, and therefore the size of fires, satisfies a power-law distribution. Moreover, the forest naturally tends to this density level. If it is less dense, density increases because fires are small. If density exceeds the threshold, any fire will wipe out the entire forest. Therefore, the tree density *self-organizes* to a critical state.[12]

In both the sand pile model and the forest fire model, a macro-level variable— the height of the pile or the density of the forest—has a critical value. That macro-level variable's value decreases when events occur (avalanches and fires). Variants of this model can explain the distributions of solar flares, earthquakes, and traffic jams. An increasing macro-level variable that decreases when events occur, though necessary, is not sufficient for self-organized criticality. Equilibrium systems also have that property. Water flows into and out of lakes through streams, yet because outflows are smooth, lake levels change gradually. The key assumptions for self-organization to critical states is that pressure increases smoothly, like water flowing into the lake, and that pressure decreases in bursts, including possibly large events.

The Implications of Long Tails

We cover three implications of long-tailed distributions: their effect on equity, catastrophes, and volatility. By definition, a long tail means a few big winners (large collapses, earthquakes, fires, and traffic jams) and many losers as compared to a normal distribution, which is symmetric about a mean. Long-tailed distributions can also contribute to volatility, as random fluctuations in larger entities will have larger effects.

Equity

A person who writes a better book, catchier song, or better academic paper than another should garner more sales and credit. It is not equitable if a person who performs only a little better or who happens to be lucky earns a lot more. As we saw in the preferential attachment model, positive feedbacks create big winners due to the Matthew effect. For positive feedbacks to occur in a market, people must know what others buy, and people must be able to buy the product. For weightless information goods, such as smartphone applications, the latter assumption makes perfect sense. For an iPhone application, no production constraints slow the positive feedbacks as they do for, say, trucks. Ford can only increase production of F-150 trucks by so much. In contrast, Intuit can sell as many copies of TurboTax as people are willing to download.

Empirical studies show that social effects create bigger winners. In the *music lab experiments*, college students could sample and download songs. In the first treatment, subjects did not know what songs others downloaded, and the distributions of downloads had a shorter tail—no song received more than two hundred downloads and only one song received fewer than thirty. In a second treatment, students knew what others downloaded. The tail of the distribution grew: one song received more than three hundred downloads. Perhaps more telling, over half received fewer than thirty. The tail became longer. Social influence increased inequality. This inequality is not a concern if social influence leads people to download better songs. However, correlations between downloads in the two treatments were not strong. If we interpret the number of downloads of a song in the first treatment as a proxy for the song's quality, social influence did not result in people downloading better songs. The big winners were not random, but they were not the best.[13]

We must be careful not to draw too strong an inference from a single study. We can, though, infer that while an author who sells 50 million books or an academic whose work receives 200,000 citations deserves accolades, such extreme success suggests that the central limit theorem is not holding. People are not buying books or making citations independently. Amazing success probably implies positive feedbacks, and perhaps a bit of luck. We return to these ideas when discussing the causes of income inequality in the book's final chapter.[14]

Catastrophes

Long-tailed distributions include catastrophic events: earthquakes, fires, financial collapses, and traffic jams. Even though the models cannot predict earthquakes, they provide insight into why their distribution satisfies a power law. That knowledge tells us the likelihood of earthquakes of various sizes. We know what to expect, if not when.[15]

The forest fire model does guide action. We can prevent large fires by selectively harvesting trees in a forest to lower the density of trees. Or we might build firebreaks. One could argue that we do not need a model to tell us to thin a forest or build firebreaks. That is surely true. The model makes us aware that there exists a critical density. That density may vary by forest. It could depend on the type of tree, the prevailing wind speeds, and the topography. The model explains why forests may self-organize to critical states.

We can also use the model as an analogy. Recall that in Chapter 1 we discussed the failures of financial institutions across networks. We can apply the forest fire model to that setting by representing banks and other financial institutions as trees on a checkerboard and allowing adjacencies to correspond to outstanding loans. In that model, a bank failure would be equivalent to a tree catching on fire. That failure could then spread to neighboring banks.

This naive application of the forest fire model to banks would portend large-scale failures as banks become more connected. As we explore that analogy, we see four shortcomings. First, the financial network is not embedded in physical space. Banks can differ in their number of connections. One bank may have dozens of financial obligations while another may have a mere one or two. Second, trees in a forest cannot take actions to reduce the probability of fire spreading. Banks can. They can increase their level of reserves. Third, the more connected a bank is, the less likely that its failure spreads as its losses will be dispersed across more banks. For example, if a bank defaults on a $100,000 loan borrowed from a single other bank, that second bank may well go under. If the first bank borrowed the money from a consortium of twenty-five other banks, no single bank takes a large hit. The systems may well absorb the default without collapsing.[16] Last, the spread of a failure from one bank to another depends on the banks' portfolios. If two connected banks hold similar portfolios, then if one fails the other probably is likely already weak. The worst-case scenario occurs if all of the banks in the network hold identical portfolios. In this case, when one bank fails, widespread failure would be likely.[17] If, though, each bank holds a distinct portfolio, poor performance by one need not imply poor performance of another. Bank failures may not spread. A useful model must therefore take into account the assets in the various portfolios. Without this information, knowing which banks have obligations to other banks will be insufficient to predict or prevent failures, and the net effect of greater connectedness of banks will not be clear.

Volatility

Last, we consider a more subtle implication of long-tailed distributions. If the entities that make up a power-law distribution fluctuate in size, then the exponent of the power law becomes a proxy for system-level volatility. It follows that the firm size distribution should influence market volatility. For this exercise, think of a country's gross domestic product (GDP) as the aggregate production of thousands of firms. If production levels are independent and have finite variation, then, by the central limit theorem, the distribution of GDP will have a normal distribution. It also follows that the greater the variation in production levels across firms, the greater the aggregate volatility. If a longer-tailed distribution of firm sizes produces greater variation in production levels, then it will also correlate with greater aggregate volatility.

An examination of volatility patterns in the United States shows that volatility rose in the 1970s and 1980s and then fell for the next two decades in what some call the *Great Moderation*.[18] Beginning around 2000, volatility again increased. It is possible to explain these volatility patterns by changes in the distribution of firm sizes.[19] As the distribution of firm sizes becomes longer- (shorter-)tailed, the largest firms have a disproportionally larger (smaller) effect on volatility. In other words, aggregate volatility increases (decreases) as the firm size distribution becomes longer- (shorter-)tailed. In 1995, when volatility was low, Walmart had revenues of $90 billion, which corresponded to 1.2% of GDP. By 2016, Walmart's revenues had increased to $480 billion, or 2.6% of GDP. Walmart's share of GDP more than doubled. In 2016, an increase or decrease in Walmart's revenue would contribute twice as much to aggregate volatility.

No one refutes the logic of this argument. The relevant question becomes whether a calibrated model produces effects with magnitudes that correspond to actual volatility levels. The calibrated fit proves quite close. Firm size distributions correlate nicely with the historical evidence of the Great Moderation. That correlation does not prove that it is changes in firm size distribution (instead of effective government management of the economy or better inventory control) that caused the moderation, but it does prevent us from rejecting the model.[20] The evidence also provides reason to keep this model in our quiver when we evaluate fluctuations in the future.

Contemplating a Long-Tailed World

In long-tailed distributions, large events occur with sufficient probability to be of concern. In the models we covered, long-tailed distributions arise because of feedbacks and interdependencies. We should pay heed to that observation. As our world becomes more interconnected and feedbacks increase, we should see more long tails. And the current long tails that we see may get stretched even further. Inequities may increase, catastrophes grow larger, and volatility become more pronounced. None of these is desirable.

So far, we have discussed these possibilities at macro levels. They also occur at smaller scales. Boston's "Big Dig," a three-and-a-half-mile tunnel through the center of the city, provides an example of a moderate-scale catastrophe. The project cost taxpayers $14 billion, more than three times the original estimate, and it became the most expensive highway project in the history of the United States. Model thinking frames the Big Dig not as a single project but as an aggregate of subprojects: digging a trench, pouring a concrete tunnel, engineering a drainage system, and building walls and a roof. The project's total cost equals the sum of the subprojects' costs.

If the costs of each subproject had been additive, then the distribution of costs for the project would have been normally distributed.[21] However, the subprojects' costs were connected. When the epoxy used to glue the roof into place proved inadequate, it was replaced with a costlier, stronger epoxy and, therefore, raised the cost of the project. The failure of the first epoxy created additional costs associated with removing and replacing the collapsed roof. Those efforts in turn required redoing several other parts of the project. Overall costs more than doubled because each project had to be undone and then redone. Interdependencies led to a large, and costly, event.

The potential for large events makes planning difficult. The distribution of natural disasters such as earthquakes satisfy a power law. Thus, most events will be small, but some will be large. If catastrophic events follow a power-law distribution with an exponent near 2, then governments need to keep a very large amount of money in reserve or at least at the ready. They need to prepare for a very rainy day. If governments do so by maintaining huge surpluses in an emergency fund, they may be able to stop themselves from spending that money or cutting taxes if no large event occurs.

Search and Opportunity

We can apply our knowledge of distributions within a class of search models to explain why the number of opportunities a person receives may correlate strongly with success. We embed one class of models, our distribution models, within a second class, search models. When we search, whether for a new pair of shoes, a career, or a vacation spot, we do not know our choice's value until we try it, though we may know something about the distribution of values, such as the mean, standard deviation, and whether the distribution is normal or has a long tail.

Here, we model choice of profession as a search process. Given a profession, a person tries a career path, which we model as a draw from a distribution. We assume that she can either stick with that career or try again. Trying again corresponds to another draw from the distribution. Consider, for example, the choice of profession for a talented young scientist. She could go to medical school or do research in quantum computing. Medical school offers the safer path. Choosing to work on quantum computing involves becoming an entrepreneur and taking on more risk. To account for these differences, we represent the salary distribution for doctors as a normal distribution with mean $250,000 and a standard deviation of $25,000, and the salary distribution for the entrepreneurial career as a power law with an exponent of 3 with an expected salary of $200,000.[22]

Within each profession, our scientist can try multiple careers. She can search. A doctor can switch from oncology to radiology. A failed entrepreneur can pick up the pieces from her start-up and try anew. Each career switch entails a cost. For a doctor, it means more training. For an entrepreneur in quantum computing, it means more long nights of work with little to no compensation.

We assume that our young scientist finds the two professions equally stimulating and makes her choice based on salary. Our model reveals that the better choice depends on how many times she can afford to try new careers. If she must stick with her first career choice, becoming a doctor offers the higher expected salary. If she has sufficient resources to continue trying to be an entrepreneur, eventually she will get a high-paying draw from the long tail. The figure below shows the average largest salary across twenty trials assuming one, two, five, and ten career searches within each profession. If she has the opportunity to try her hand at quantum computing start-ups ten times, her salary will be nearly double what she would earn had she chosen medical school and experimented with ten careers.

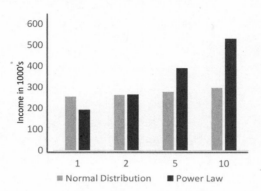

Average Income as a Function of Number of Opportunities

If wealth and family support correlate with the number of opportunities a person has to try new careers, our model predicts that wealthier people will choose riskier professions.[23] Evidence on patents aligns with the model. The probability that someone writes a patent correlates with that person's mathematical abilities. People in the top 1% of math ability are far more likely to hold a patent. Among the top 1%, those from families in the top 10% of the income distribution are even more likely to hold a patent.[24] At least two models could explain the disparity. One model could assume that poorer talented students never attend college. They may be working routine jobs and never have the choice between medical school or quantum computing. Or, perhaps poorer students choose safer careers.

The logic that an increase in opportunities creates an incentive for risk applies widely. Venture capitalists take risks because they make multiple investments. An early investment in a single unicorn, a billion-dollar company, more than compensates for many losers. Pharmaceutical research laboratories also take risks, spending billions on drug research. We can apply the same logic when deciding where to eat lunch. When driving cross-country and stopping in an unfamiliar town, we may want to eat at a chain restaurant. If moving to that town, we should experiment.

7. Linear Models

I'm lying, yes, but why do you force me to give a linear explanation; linear explanations are almost always lies.

—Elena Ferrante

Often models posit specific functional relationships between variables. That relationship could be linear, concave, convex, or S-shaped, or it could include threshold effects. Of these, linear models are the simplest, the most widely used, and the focus of this chapter. The effects of education on income, of gains in life expectancy from exercise, and of income on voter turnout can all be measured using linear models.

   We begin the chapter with a refresher on linear functions with a single variable. We then show how regression fits data to a linear function, revealing the sign, magnitude, and significance of effects. We also discuss why errors, noise, and heterogeneity mean that data do not fall exactly on the regression line. We then expand the linear model to allow for more variables and discuss how to fit multivariable linear models. To build intuition for multiple variable models, we describe a model of success as a linear function of skill and luck. The chapter concludes with an observation of how relying on data and regressions to guide action limits mistakes but can also produce marginal, conservative actions. This big-coefficient thinking can stifle innovation. To identify more innovative options, we might consider constructing other, more speculative models.

Linear Models

In a linear relationship, the amount of change in one variable due to a change in a second variable does not depend on the value of the second variable. If the height of a tree is linear with the tree's age, the tree grows by the same amount each year. If the value of a house increases linearly in its square footage, a 200-square-foot addition increases a house's value by double that of a 100-square-foot addition. A 400-square foot addition increases the house by four times as much.

Linear Models

In a **linear model** changes in the **independent variable**, x, result in linear changes in a **dependent variable**, y, as follows:

$$y = mx + b$$

where m equals the *slope* of the line and b equals the *intercept*, the value of the dependent variable when the independent variable equals zero.

A *linear regression* model finds the line that minimizes the distance to the data points. Linear regression can explain variation in crime, washing machine sales, and even wine prices.[1] Suppose that we have data for adults ranging in age from twenty to sixty and the distances they walk each week and find the following regression equation:

$$\text{Miles Person}_i = -0.1 \cdot \text{age}_i + 12 + \epsilon_i$$

This regression equation tells us the *sign* of the effect (distance decreases with age) and the *magnitude* of that effect (each year of age reduces distance by one-tenth of a mile). In this example, the intercept has no relevance because it lies outside our data range, that is the data includes no one with an age near zero. Based on the equation, we expect a forty-year-old to walk eight miles per week and a fifty-year-old to walk seven. The data used to produce a regression will not fall exactly on the regression line. Figure 7.1 shows hypothetical data used to produce our regression line. The person represented by the gray circle, Bobbi, is age forty and walks eleven miles per day. She exceeds the model's estimate by three miles. To make the data consistent with the model, the equation includes an *error term* for each data point. The error term, denoted by ϵ, equals the difference between what the model estimates and the actual value of the dependent variable. Bobbi's ϵ term equals +3 miles.

In social and biological contexts, we do not expect perfect linear fits. Outcomes depend on many variables, and a single-variable regression, by definition, includes only one variable. Predicted values can deviate from the actual values because of these *omitted variables*. Bobbi may walk more than expected because, as a botany

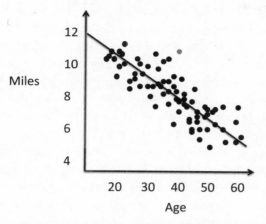

Figure 7.1: A Scatterplot and Regression Line

professor, she takes her students out for walks in the woods. The model does not include profession as a variable, which contributes to why the data in 7.1 do not lie on the line. The ϵ term could also result from *measurement error*. Fitness data collected by smartphones will contain errors if people forget to carry their smartphone or loan their phones to others. Error can also arise from *environmental noise*—people may earn extra distance for bumpy car rides to work.[2]

The closer the regression line lies to the data, the more of the data the model explains and the larger the model's R-squared (the percentage of variation explained). If all data lie exactly on the regression line, the R-squared equals 100%. All else equal, we prefer models with higher R-squared values.

Sign, Significance, and Magnitude

Linear regression tells us the following about *coefficients* of independent variables:

Sign: The correlation, positive or negative, between the independent variable and the dependent variable.

Significance (p-value): The probability that the sign on the coefficient is nonzero.

Magnitude: The best estimate of the coefficient of the independent variable.

In a single-variable regression, the closer fit to the line and the more data, the more confidence we can place in the sign and magnitude of the coefficient. Statisticians characterize the *significance* of a coefficient using its *p-value*, which equals the probability, based on the regression, that the coefficient is not zero. A p-value of 5% means a one-in-twenty chance that the data were generated by a process where the coefficient equals zero. The standard thresholds for significance are 5% (denoted by *) and 1% (denoted by **). Significance is not all we care about. A coefficient can be significant yet of small magnitude. If so, we can be confident of the correlation but the variable has little effect. Or a coefficient can be large though not significant. This often occurs with noisy data or data with many omitted variables.

To see how to use regressions to guide action, imagine a company that ships spices. This company offers over a hundred types of spices. Customers buy packages of six, twelve, or twenty-four spices, which employees pack and ship. A regression estimating the number of orders shipped per eight-hour shift as a function of the number of years an employee has worked produces the following:

$$\# \text{ Orders Filled} = 200 + 20^{**} \cdot \text{Years}$$

The coefficient on years, 20, is significant at the 1% level. We can be confident it is positive. If the relationship is causal (see below), the model can be used to predict the number of orders that each employee can fill per shift as a function of years of work and we can use the model to project how many orders the current employees will fill next year. Here we have an instance of a model both making a prediction and guiding an action.

Correlation vs. Causation

Regression only reveals correlation among variables, not causality.[3] If we first construct a model and then use regression to test if the model's results are supported by data, we do not prove causality either. However, writing models first is far better than running regressions in search of a significant correlate, a technique known as *data mining*. Data mining runs the risk of identifying a variable that correlates with other causal variables. For example, data mining might find a significant positive correlation between vitamin D levels and general health. People absorb vitamin D from sunlight, so the effect could be due to the fact that people with active lifestyles spend more time outdoors and have better health. Or a regression might find that a school's academic performance correlates strongly with the number of students on its equestrian team. Equestrian teams likely have no direct causal effect but they correlate with family income and school funding levels which do.

Data mining can also result in spurious correlations, where just by chance two variables are correlated. We might find that companies with longer names earn higher profit or that people who live near pizza restaurants are more likely to get the flu. With a 5% significance threshold, one in every twenty variables we test will be significant. So, if we try enough variables, we will surely find significant (and spurious) correlations.

We can avoid reporting spurious correlations by creating *training sets* and *testing sets*. A correlation found on the training set that also holds on the testing set is far more likely to be true. We still have no guarantee of a causal relationship, however. To prove causality, we need to run an *experiment* where we manipulate the independent variable and see if the dependent variable changes. Or we look for a natural experiment where this has happened by chance.

Multivariable Linear Models

Most phenomena have multiple causal and correlative variables. A person's happiness can be attributed to health, marital status, offspring, religious affiliation, and wealth. The value of a house depends on square footage, lot size, the number of bathrooms, the number of bedrooms, the type of construction, and the quality of local schools. All of these variables can be included in a regression to explain housing values. We must keep in mind, though, that as we add more variables, we need more data to obtain significant coefficients.

Before discussing multiple-variable regression, we build intuition for multiple-variable equations by introducing Mauboussin's *skill-luck equation*.[4] The equation writes success, be it in work, sports, or games, as a weighted linear function of skill and luck.

The Success Equation

$$\text{Success} = a \cdot \text{Skill} + (1 - a) \cdot \text{Luck}$$

where a in $[0, 1]$ equals the relative weight on skill.

If we can assign relative weights to skill and luck, perhaps by using a regression if we had data, we could use the model to predict outcomes. If the manager of a team of recreational vehicle salespeople finds that success, measured in sales, has a large luck component, he would expect *regression to the mean*: salespeople who did well this month would be likely to be about average the next month. The manager could then use the model to guide action. He might not want to match a higher salary offer from a competitor for a salesperson who had two good months in a row. If instead the regressions showed that luck played almost no role, performance in two months would be a good predictor of performance in future months. In this case, the manager would want to match an outside offer for the best salesperson.

The same insight applies to CEO pay. A board of directors should not pay bonuses to CEOs who work in industries where luck determines success. An oil company's profits depend on the market price of crude oil, a variable that lies outside the company's control. An oil company's board should therefore be reluctant to reward a CEO for a good year. An advertising company would be wise to do the opposite—to award a large bonus to a CEO if the company performs well. In brief, pay for skill; do not pay for luck. Better-run corporations do in fact pay less for luck.[5]

Even the simplest of models, such as this one, produce subtle insights. By thinking about the equation, we see that even in a context that depends almost entirely on skill, such as running, biking, swimming, chess, or tennis, if skill differences are small, luck largely determines who wins. We might expect that in the most competitive environments, like the Olympics, skill differences are small, and thus luck matters. Mauboussin calls this the *paradox of skill*. Michael Phelps, the greatest swimmer in history, has been on both ends of the paradox. In the 2008 Olympic Games, Phelps trailed Milorad Cavic at the end of the 100-meter butterfly. Yet by a stroke of luck, Phelps touched the wall first. In the 2012 Olympic Games, Phelps led Chad le Clos at the finish, but le Clos touched first. Yes, Phelps has incredible skill, but that one win and that one loss were the products of luck.

Multiple-Variable Regression

Multiple-variable linear regression models fit linear equations with many variables and also minimize the total distance to the data. These equations include coefficients for each independent variable. The equation below shows a hypothetical regression output for student performance on a math test as a function of hours studied (HRS), family socioeconomic status (SES), and the number of accelerated classes (AC).

$$\text{Math Score} = 21.1 + 9.2^{**} \cdot \text{HRS} + 0.8 \cdot \text{SES} + 6.9^* \cdot \text{AC}$$

According to the regression, a student's score increases by 9.2 points for every extra hour spent studying. The coefficient has two *'s, so it is significantly different from zero at a 1% level. This implies strong correlation, though not causality.

The equation also shows that a student scores almost seven points higher for each accelerated class. That coefficient is significant as well, but only at the 5% level. Family socioeconomic status (SES), a variable that takes on values from 1 (low) to 5 (high), has a coefficient that is positive but not significantly different from zero, so we can assume it probably has little causal effect.

With this or any regression output, we can predict outcomes. The model predicts that a student who spends seven hours studying and takes one other accelerated class should score in the 90s. The model can also guide actions, though we must be cautious, as we cannot infer causality. The data show that students who study and take accelerated classes perform better. One reason studying more or taking those classes may not help is *selection bias*. It might be that the students who study more and those who take accelerated classes are better at math.

Even though regressions cannot prove what causes patterns in data, they can rule out explanations. Take the large wealth disparity by race in the United States: in 2016, the average wealth of white families (approximately $110,000) was more than ten times that of African American and Latino families. Any number of causes might explain that gap, including institutional factors, differences in income, savings behavior, or marriage rates. Regressions support some explanations and rule out others. For example, regressions reveal no significant relationship between marital status and wealth among African Americans, so marital status cannot be a cause. Income differences, though substantial, also prove insufficient to explain the gap.[6]

The Big Coefficient and New Realities

As already stated, linear regression models play prominent roles in scientific research, policy analysis, and strategic decision-making, in part because they are easy to estimate and interpret. With the increased availability of data, they have become even more widely used. The phrase "In God we trust. Everyone else must bring data" is often heard in business and in the halls of government. A reliance on data—and that often means linear regression models—can steer us toward marginal actions and away from big new ideas. A business, government, or foundation that gathers data, fits a linear regression model, and finds the variable with the largest statistically significant coefficient almost cannot stop itself from adjusting that variable and taking the marginal gain.

When taking an action, it is better to choose the variable with the *big coefficient* than a variable with a small coefficient. At the same time, big-coefficient thinking builds in conservatism. It focuses attention on certain modest improvements and pulls attention away from novel policies. A second problem with big-coefficient thinking is that the magnitude of the big coefficient corresponds to the marginal effect given existing data. Often, as we see in the next chapter, effect sizes diminish as we increase the value of a variable. If so, the big coefficient becomes smaller as we try to exploit it.

The Big Coefficient vs. the New Reality

Linear regressions reveal the magnitude of correlations of independent variables with the variable of interest. If that correlation is causal, changes to the variable with a **big coefficient** will have large effects. Policies based on big coefficients guarantee improvements but rule out *new realities* that involve more fundamental changes.

The alternative to big-coefficient thinking is *new-reality thinking*. Big-coefficient thinking widens roads and builds high-occupancy vehicle lanes to reduce traffic. New-reality thinking builds train and bus systems. Big-coefficient thinking subsidizes computers for low-income students. New-reality thinking gives everyone a computer and reduces mail delivery to three days a week. Big-coefficient thinking changes the width of airline seats. New-reality thinking creates an airplane interior that can be filled with interchangeable pods. Big coefficients are good. Evidence-based action is wise, but we must also keep our eyes open to big new ideas as well. When we encounter them, we can use models to explore whether they might work. A regression on teenage traffic accidents may find that age has the largest coefficient, implying that states might want to raise the driving age. That may work, but so too might more novel policies such as curfews that prohibit nighttime driving, automated monitoring of teenage drivers through smartphones, or limits on the number of passengers in teenagers' cars. These new-reality policies might produce larger effect sizes than riding the big coefficient.

Summary

To summarize, linear models posit constant effect sizes. Linear regression offers a powerful tool for taking a first cut at data, enabling us to identify the sign, magnitude, and significance of variables. If we want to know the health effects of coffee, alcohol, or soda consumption, we can run regressions. We may find that coffee consumption reduces the risks of cardiovascular disease and that so do modest levels of alcohol consumption. That said, we should be skeptical of extrapolating linear effects too far outside of the existing data range. We should not infer that thirty cups of coffee, much less six glasses of wine, would be a good idea. Nor should we make linear projections too far ahead in time. California's population grew at a rate of 45% from 1880 to 1960. Had we made a linear projection, we would have pegged California's population in 2018 at 100 million people, more than double its actual level.

Keep in mind we are just getting started. Most phenomena of interest are not linear. For that reason, regression models often include nonlinear terms such as age squared, the square root of age, or even the log of age. To account for nonlinearities, we can also arrange linear models end to end. These concatenated linear models can approximate a curve in much the same way as we can use straight-edged bricks to construct a curved path. Though linearity may be a strong, and unrealistic, assumption, it offers a good place to start. If given data, we can use linear models to test our intuitions. We can then construct more elaborate models in which the effect of a variable dampens as it increases (diminishing returns) or becomes more powerful (positive returns). These nonlinear models are the focus of the next chapter.

Binary Classifications of Data

In an era of Big Data, organizations use algorithms informed by models to classify their data. A political party might want to learn who votes, an airline might want to learn the attributes of their frequent flyers, and an event organizer might want to learn about the event's attendees. In each case, the organization classifies people into two sets: those who buy, contribute, or enroll are labeled as *positives* (+'s) and those who are not are labeled as *negatives* (-'s).

Classification models apply *algorithms* to partition the people into categories based on attributes such as a person's age, income, education level, or hours spent on the internet. Different algorithms imply different underlying models of the relationship between attributes and outcomes. Applying multiple algorithms—using many models—will produce an even better classification.

Linear classifications: In figure M1, positives (+) represent voters and negatives (-) represent nonvoters. A linear function of a person's age and education level can be used classify whether or not a person votes. The data show that more educated people are more likely to vote and that older people are more likely to vote. In this example, a straight line classifies nearly perfectly.[7]

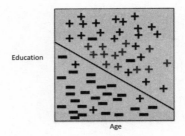

Figure M1: Using a Linear Model to Classify Voting Behavior

Nonlinear classifications: In figure M2, positives (+) represent frequent flyers, consumers who fly more than 10,000 miles per year, and negatives (-) represent all other customers of an airline. People of middle age and higher income are more likely to fly. To classify these data requires a nonlinear model, which could be estimated using *deep-learning* algorithms, such as neural networks. Neural networks include more variables so that they can fit almost any curve.

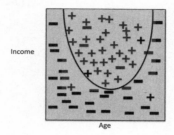

Figure M2: Using a Nonlinear Model to Classify Frequent Flyers

Forests of decision trees: In figure M3, positives (+) represent people who attended a science fiction convention based on their age and the hours per week they spend on the internet. Here we classify the data using three *decision trees*. Decision trees make classifications based on sets of conditions on the attributes. The figure shows three trees:

Tree 1: *If (age < 30) and (internet hours per week in [15, 25])*

Tree 2: *If (age in [20, 45]) and (internet hours per week > 30)*

Tree 3: *If (age > 40) and (internet hours per week < 20)*

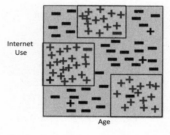

Figure M3: A Forest of Decision Trees Classifying Conference Attendees

The collection of trees are called a *forest*. Machine learning algorithms create trees randomly on a training set and then keep those that classify accurately on the testing set and on a training set.

8. *Concavity and Convexity*

To say nonlinear science is akin to saying non-elephant zoology.

—John von Neumann

We now introduce nonlinear models and nonlinear functions. Nonlinear functions can curve downward or upward, they can form S-shapes, they can kink, jump, and squiggle. In time, we cover all of these possibilities. We start here with models that rely on convexity and concavity. We show how growth and positive feedbacks produce convexity and how diminishing returns and negative feedbacks produce concavity. Most disciplines contain models of both types. Economic production models assume that delivery and inventory costs decrease with a firm's size, making profits per unit sold a convex function of a firm's size, which explains why Walmart earns such large profits.[1] Economic models of consumption assume that the utility (or value) is concave, that we enjoy the fifth piece of pizza less than the first. In ecosystems, when a new species invades and confronts no predators, its population grows at a constant rate, producing a convex function. As that population grows, it has less food. Fitness, as a function of population size, is therefore concave.

The chapter consists of three parts. The first part covers models of population growth and decay. The second part covers concavity. In it, we see how concavity implies risk aversion and a preference for variety. In the third part, we study a series of growth models from economics that combines concave functions and linear functions.

Convexity

Convex functions have an increasing slope: the function's value increases by a larger amount as we increase a variable's value. The number of possible pairs of people is a convex function of the group size. A group of three people includes three unique pairs. A group of four people includes six unique pairs, and a group of five includes ten unique pairs. Each increase in group size increases the number of pairs by a larger amount. Similarly, each time a chef adds a new spice to his repertoire, he increases the number of spice combinations by a larger amount.

Our first model of convexity, the *exponential growth model*, describes the amount of a variable, often a population or a resource, as a function of its initial value, a growth rate, and the number of periods.

Exponential Growth Model

A value of a resource at time t, V_t, that has an initial value of V_0 and grows at a rate R can be written as follows:

$$V_t = V_0(1 + R)^t$$

This single-equation model plays central roles in finance, economics, demography, ecology, and technology. When applied to finance, the variable is money. Using the equation, we can calculate that a $1,000 bond paying 5% annual interest increases in value by $50 in year one and by more than $100 in year twenty. To draw clean inferences, we assume a constant growth rate. Given that assumption, we can manipulate the exponential growth equation to derive the *rule of 72*.

Rule of 72

If a variable grows by a percentage R (less than 15%) each period, then the following provides a good approximation:

$$\text{Periods to Double} \approx \frac{72}{R}$$

The rule of 72 quantifies the cumulative effect of higher growth rates. In 1966, Zimbabwe had a per capita GDP of $2,000, twice that of Botswana. Over the next thirty-six years, Zimbabwe experienced little growth. Botswana, meanwhile, averaged 6% growth, meaning that Botswana's GDP doubled every twelve years. In thirty-six years, it doubled three times, an 8-fold increase. Thus, in 2004, Botswana's per capita GDP of $8,000 was four times that of Zimbabwe.

This same formula reveals why housing bubbles must end and technological progress need not. In 2002, home prices in the United States rose by 10%. That would imply a doubling every seven years. Had that trend continued for thirty-five more years, prices would have doubled five times—a 32-fold increase. A house costing $200,000 in 2002 would cost $6.4 million in 2037. Prices cannot rise at that rate. The bubble had to burst. In contrast, *Moore's law* states that the number of transistors that can fit on an integrated circuit doubles every two years. Moore's law has persisted because spending on research and development has generated a near constant rate of improvement.

Demographers apply the exponential growth model to human populations. A population that grows at 6% a year doubles in size in twelve years. In thirty-six years, it doubles three times, and in one hundred years, it doubles eight times (increasing 256-fold). In 1798 British economist Thomas Malthus noticed that the population was growing exponentially and wrote a model showing that if the economy's ability to produce food only increased linearly, then a crisis loomed. The short version goes as follows: Population was growing like 1, 2, 4, 8, 16, 32, ... Food production was growing like 1, 2, 3, 4, 5, ... Malthus foresaw disaster. Fortunately, birth rates fell, and the arrival of the Industrial Revolution increased productivity. Had nothing changed, Malthus would have been correct. But he ignored the potential for innovation—the focus of models later in this chapter. Innovation subverted the trend.

The exponential growth model can be applied to the growth of species as well, and not just to rabbits. When you acquire a bacterial infection, tiny bacteria reproduce at incredible rates. Bacteria in human sinuses grow at around 4% a minute. By applying the rule of 72, we can calculate they double every twenty minutes. In a single day, each initial bacterial cell spawns over a billion offspring.[2] Their growth stops when the physical constraint of your sinuses leaves them no room. Food constraints, predators, and lack of space all reduce growth. Some species, such as deer in suburban America or the hippos brought to Colombia by drug lord Pablo Escobar, encounter few constraints on growth and their population grows rapidly, though not at bacterial rates.[3]

A convex function with a positive slope increases at an increasing value. A convex function with a negative slope becomes less steep. A convex function with an initially large negative slope will flatten. That is true for the equation in the *half-life model,* which captures decomposition, depreciation, and forgetting.

In the model, every H periods half of the quantity decays. Hence, H is known as the *half-life* for that process. For some physical processes, the half-life is constant. All organic matter contains two forms of carbon: an unstable isotope, carbon-14, and a stable isotope, carbon-12. In living organic matter, these isotopes are present in a constant ratio. When an organism dies, the carbon-14 in its body starts to decompose with a half-life of 5,734 years. The amount of carbon-12, on the other hand, does not change. Willard Libby, a physical chemist, realized that by measuring the ratio of carbon-14 to carbon-12, one can estimate the age of a fossil or artifact, a technique known as radiocarbon dating. Paleontologists apply radiocarbon dating to the remains of dinosaurs, woolly mammoths, and prehistoric fish. Archeologists use it to adjudicate claims of authenticity. The remains of Ötzi the Iceman, discovered in the Italian Alps, were estimated to be five thousand years old. The Shroud of Turin, first displayed in 1357 and claimed to be Christ's burial shroud, was found to date from the fourteenth century and not the time of Christ.

Half-Life Model

If every H periods half of the remaining quantity decays, then after t periods the following holds:

$$\text{Proportion Remaining} \approx \left(\frac{1}{2}\right)^{\frac{t}{H}}$$

A novel application of the half-life model comes from psychology. Early psychological studies showed that people forget information at a near-constant rate. Our half-life of remembering depends on the salience of the event.[4] In 2016, the film *Spotlight* won the Academy Award for Best Picture. If people's memory of Oscar winners has a half-life of two years, in 2018, $\frac{1}{4}$ of people will have remembered that fact, but by 2026, only $\frac{1}{1024}$ will recall it. The recollection of any particular event varies across people. Tom McCarthy, who directed and cowrote *Spotlight,* will likely never forget the year he won the Academy Award.

Concave Functions

Concave functions are the opposite of convex functions. Concave functions have slopes that decrease. Concave functions with positive slopes exhibit *diminishing returns*: the added value of each extra thing diminishes as we have more of that thing. Our utility or value from almost all goods exhibits diminishing returns. The more leisure, money, ice cream, or even time spent with loved ones, the less we value having more of it. Evidence for this can be found in the fact that the more we consume of just about anything, including chocolate, the less we enjoy it and the less we are willing to pay for it.[5]

Diminishing returns can explain a variety of phenomena, including why long-distance relationships are often so happy. If you see your partner just a few hours each month, every additional minute is wonderful. After a month of uninterrupted togetherness, the slope of the happiness curve flattens, and those few extra moments matter less.[6] It explains why developers invite people for free weekend visits to their beachfront condominiums. During a short weekend, you cannot get enough time on the beach. You are inclined to buy. After ten days on the beach, though, you may become bored.

When we assume concavity, we imply a *preference for diversity* and *risk aversion*. To show the former requires a concave function with multiple arguments. If our happiness is concave and increasing in both leisure and money, we prefer some leisure and some money to all leisure and no money or all money and no leisure. Risk aversion means a preference for a sure thing over a lottery. A risk-averse person prefers a certain payoff of $100 to a lottery that pays $200 half of the time and nothing the other half of the time. A risk-averse person prefers a double-dip ice cream cone to having either no ice cream or an unwieldy four-scooper.

Figure 8.1 shows why concavity implies risk aversion. The figure plots happiness for values for three outcomes: a high outcome (H), a low outcome (L), and the mean of those outcomes (M). Given the downward-shaped curve, happiness at the mean outcome exceeds the average happiness of the low outcome and the high outcome. The opposite holds for convex functions. Convexity implies *risk-loving*: we prefer the extremes to the average. The amount of a stock you can buy is a convex function of its price. Therefore, buyers of stocks prefer price volatility. If prices go up and down, buyers end up with more stocks than if prices stay constant.[7]

Economic Growth Models

We next construct a series of economic growth models. These models reveal the causes of growth and can explain and predict growth patterns across countries. They can also guide actions such as increasing the savings rate. To lay the foundation for our study of growth models, we introduce a standard economic production model in which output depends on labor and physical capital. Empirical evidence and logic support concavity of output in both labor and capital. Holding the amount of capital fixed, labor should be worth less as more is added. Similarly, adding more machines or computers adds less value given a fixed number of workers. Logic also suggests that output should be linear in scale. Doubling both the number of workers and the amount of capital should double output. A broom-making company with sixty workers and one factory that builds a second factory and hires sixty additional workers should double its output. The *Cobb-Douglas model,* one of the most widely used models in economics, includes both properties. Output is concave in labor

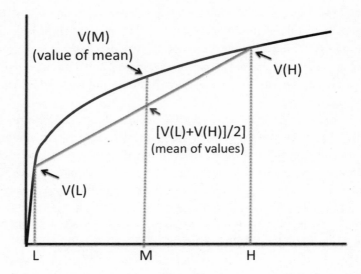

Figure 8.1: Risk Aversion: Value(Mean) > Mean of the Values

and capital and linear in scale. This model can be applied to capture production by single firm or by an entire economy.[8]

Cobb-Douglas Model

Given L workers and K units capital, the total output equals:

$$\text{Output} = \text{Constant} \cdot L^a \, K^{(1-a)}$$

where a is a real number between 0 and 1 capturing the relative importance of labor.

We use the Cobb-Douglas model to construct models of economic growth. To simplify, we assume 10,000 workers in the economy and ignore wages and prices, allowing us to focus on how the number of machines affects total output. We can then connect investment in capital to growth. To make the model as simple as possible, we assume that output takes the form of a single commodity, coconuts. The coconuts provide flesh and rich milk for food. However, the coconuts grow high in trees, so the workers require machines to pick them. We then make the very unrealistic assumption that the machines are constructed from coconuts. This simplifies the model but maintains the key trade-off between consumption today and investment in the future As a special case of the Cobb-Douglas model, we write output as the square root of the number of workers times the square root of the number of machines.

$$\text{Output} = \sqrt{\# \text{ Workers}}\sqrt{\# \text{ Machines}} = 100\sqrt{\# \text{ Machines}}$$

If the economy has one machine, output equals 100 tons. If people consume all 100 tons of coconuts, they invest in no new machines. Output will be unchanged in the next year. The economy exhibits no growth. If they invest 1 ton of coconuts to build a second machine, output increases to 141 tons, a 41% growth rate. If they build a third machine, output grows to 173 tons.[9] Through a constant investment, the economy grows at a decreasing rate. Output is a concave function.

Simple Growth Model

Production Function: $O(t) = 100\sqrt{M(t)}$

Investment Rule: $I(t) = s \cdot O(t)$

Consumption-Investment Equation: $O(t) = C(t) + I(t)$

Investment-Depreciation Equation: $M(t+1) = M(t) + I(t) - d \cdot M(t)$

$O(t) =$ output, $M(t) =$ machines, $I(t) =$ investment, $C(t) =$ consumption, $s =$ savings rate, and $d =$ depreciation rate

---

Now that we have the basic idea of how investment drives growth, we can construct a more elaborate model that includes an investment rule. We can write investment as *savings rate* times output and assume a fixed *depreciation rate* on the machines, such as that the number of machines that are no longer useful at the end of the year equals a fixed proportion of the number of machines. We can then write the total number or machines in the next year as last year's machines plus the investment in new machines minus the machines lost to depreciation. The complete *simple growth model* consists of four equations.

If we assume the economy has 100 machines, a savings rate of 20%, and a depreciation rate of 10%, output equals 1,000 tons of coconuts, consumption equals 800 tons, and new investment equals 200 machines. A total of 10 machines will be lost to depreciation, leaving 290 machines at the start of the new year. Similar calculations show that in the second year, outcome will equal 1,702 tons and in the third year it will equal almost 2,500 tons.[10] In the first three years, output increases

101

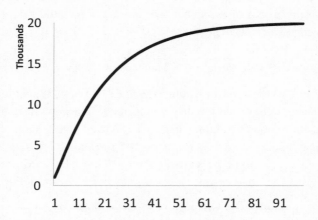

Figure 8.2: Output in the Basic Growth Model for One Hundred Years

at an increasing rate. This initial convexity is a result of the small number of machines in the first few years implies almost no effect of depreciation. Over time the
number of machines grows and depreciation starts to matter making output concave. In the long run it ceases altogether, as shown in figure 8.2. By analyzing the
model we can see why. Investment is linear in output: the number of new machines
added grows linearly with output. Output is concave in the number of machines, so
as the economy grows, investment will also be concave in the number of machines.
Depreciation, though, is linear in the number of machines, and eventually the linear
depreciation catches up with the concave increases in production.

In the *long-run equilibrium* of the economy the number of new machines created
by investment equals the number lost to depreciation. In our model, the equilibrium
occurs when economy has 40,000 machines and produces 20,000 tons of coconuts.
At that point, the economy invests 20% or 4,000 coconuts, in new machines and
loses exactly that many machines to depreciation (10% of the 40,000). Thus, the
number of new machines lost to depreciation equals the number of new machines
created through investment and growth stops.[11]

The Solow* Growth Model

We now construct a more general model that is a simplification of the *Solow growth
model* (thus the asterisk). We replace machines with physical capital and include labor as a variable. We also add a technology parameter that increases output linearly.
Innovations increase this parameter. As in the previous model, the long-run equilib-

rium occurs when investment equals depreciation. Here, though, the equilibrium-level output depends on the amount of labor and on the technology parameter, as well as the savings and depreciation rates.[12]

***

Solow* Growth Model

Total output in the economy is given by the following equation:

$$\text{Output} = A\sqrt{L}\sqrt{K}$$

where L denotes the amount of labor, K denotes the amount of physical capital, and A represents the level of technology. The long-run equilibrium output, O^*, is given by the equation[13]

$$O^* = A(t)^2 L \frac{s}{d}$$

***

Long-run equilibrium output increases in the amount of labor, the growth in technology, and the growth in the savings rate. It decreases with a rise in the depreciation rate. None of these results is surprising. More workers, better technology, and more savings increases output, and faster depreciation reduces output. The fact that output increases linearly with labor and savings is less intuitive. Labor produces diminishing returns, so without working through the model, we might expect long-run output to be concave in the amount of labor. However, as the amount of labor increases, so too does output, which in turn increases investment, leading to more output. The positive feedback from investment exactly offsets the decreasing returns. Last, equilibrium output is convex with the depreciation rate. Lowering the depreciation by 20% increases output by 25%.

Finally, long-run equilibrium output increases as the square of the technological improvements. Innovation therefore increases output more than linearly. We can use the model to show why. If we start with an economy in a long-run equilibrium and increase the technology parameter by 50%, output increases by 50%, and so too does investment. Investment then exceeds depreciation, so the economy continues to grow. Investment continues to outpace depreciation until the economy has grown another 50%, at which point the capital lost to depreciation offsets investment. These calculations reveal that innovation has two effects, creating an *innovation multiplier*. First, innovations directly increase outputs. Second, they indirectly lead to more capital investments creating an additional increase in output. Innovation, therefore, is the key to sustaining growth.[14]

These increases in output do not occur instantaneously. When a breakthrough occurs, the technology parameter changes slowly. The direct effects unfold over time. Old physical capital must be replaced by new physical capital with the better technology. A company's computers do not get faster when technology changes; they get faster when technology changes and the company buys new computers. The second-order increase that results from the increased investment in physical capital takes place over an even longer time frame. Lags between technology and its effects on growth can imply that an innovation produces growth over a period of decades. Trains were invented in the early 1800s. The Gilded Age did not begin until the latter part of that century, a gap of over fifty years. The internet boom took place three decades after the creation of the ARPANET.[15]

Why Nations Succeed and Fail

We can apply our growth models to big policy questions such as whether backward countries can catch up, why some countries succeed and some fail, and the role of government in promoting growth. Those investigations show the value and limits of our models. We can begin with the ability of low-GDP countries to achieve fast growth. The models show that building up capital can produce fast growth, as will investing in technology. A backward country with less physical capital that could jump to the technological frontier with new capital outlays could experience incredible growth.[16]

The necessity of innovation for long-term growth, as shown in the second model, implies the limits of one-time imports of new technology. Continued growth requires innovation. Thus, when the Soviets dismantled German factories and rebuilt them in the Soviet Union following World War II, they could produce short-term growth, so much so that on November 18, 1956, Soviet premier Nikita Khrushchev, speaking at the Polish embassy in Moscow to ambassadors from Western nations, proclaimed, "Mi vas pokhoronim!," or "We will be present at your funeral!" They did not. They failed to do so because the Soviet Union did not innovate.[17] They limited freedom and stifled entrepreneurs.

The models also show how extraction and corruption, the taking of output from the economy for government use, will reduce growth through reduced savings. Cross-country comparisons of growth rates support both findings: reducing extraction and corruption and promoting innovation enhance growth. Achieving those aims requires a strong but limited central government that promotes pluralism. The strong center establishes property rights and rule of law. Pluralism prevents capture by the elite, who often prefer the status quo and may not embrace innovation, which can be destructive.

As an example of destructive innovation, consider the website Craigslist, which posts for-sale and help-wanted ads. In the early 2000s, Craigslist contributed to the loss of hundreds of thousands of newspapers jobs in the United States. At that time, Craigslist itself employed only a few dozen workers. Though people lost jobs, Craigslist made the economy more efficient by increasing the technology parameter. In a less pluralistic society, the newspaper industry might have lobbied the government to stop Craigslist. Doing so would have slowed growth.

~~Japanese~~ Chinese Economic Dominance

Linear model + rule of 72: From 1960 to 1970 Japan's GDP grew at a 10% annual rate. A linear projection of continued 10% increases would result in a doubling of the Japanese economy every seven years (using the rule of 72). In 1970, Japanese per capita GDP was approximately $2,000 in current US dollars. Had that trend continued, by 2012 per capita GDP would have doubled six times, resulting in a per capita GDP of $128,000.

Growth model: This model explains Japanese growth as due to investments in physical capital. The model predicts *concave* growth rates over time. The growth model predicts that as Japan's GDP approached that of the United States and Europe, its growth rate should decrease to the historical cross-country average of 1–2%.[18] The evidence supports this. From 1970 to 1990 Japan's GDP grew at around 4% annually. From 1990 to 2017, it grew at 1% or less.

Chinese growth: China's GDP grew at nearly a 10% rate from 1990 to 2010. In 2016, the per capita GDP in China reached approximately $8,000, and as predicted by the growth model, growth has slowed, with GDP growing at closer to 6% from 2013 to 2017. In China as well, sustained 10% growth rates run afoul of the rule of 72. If Chinese economic growth averaged 10% for the next century, per capita GDP would exceed $100 million.

It's a Nonlinear World After All

We construct nonlinear models because few phenomena of interest are linear. In this chapter we saw how diminishing and increasing returns are common features of economic, physical, biological, and social phenomena. We also saw some of the implications of including curvature in our models. Most important, perhaps, we saw how functional forms structure our thinking and then how fitting functional forms to data allows us to make precise statements. Scientists can compute the age of artifacts using carbon-14 data. Economists can estimate the long-term effects of small increases in growth.

A central takeaway from this chapter is that intuition becomes insufficient once we include nonlinearities. Intuition tells us the direction of effects: growth is increased by a rise in savings, an increase in labor, and technological innovation. Models reveal the shape and form of those effects. Savings, as we would expect, have a linear effect. Increases in labor do as well in the long run, even though the model assumes short-run diminishing returns. Increases in innovation produce a multiplier effect: we get the square of those effects. The first increase is the direct effect of the innovation. The second increase in output arises from the increase in capital.

Insights such as these become clear with the help of models. Without models, we can usually infer what goes up and what goes down, but we lack understanding of the shape of functional relationships. As a result, we often make linear extrapolations—China's economy will soon take over the world. With models, we can better think through the logic that produces nonlinear effects. The set of nonlinear functions is enormous. The concave and convex models we covered in this chapter represent but a small dip in that vast sea. If we hope to improve our capacity to reason, explain, and act in a complex world, we need an even deeper dive into nonlinear phenomena.

9.   *Models of Value and Power*

Your value will be not what you know; it will be what you share.

—Ginni Rometty

In this chapter, we cover models that quantify the value and power of individual actors. Some cases are easy. When a group produces output equal to the sum of individual contributions, each individual's value equals her contribution. When the collective output cannot be separated into individual components, such as when a team of computer programmers writes a software program or a group of entrepreneurs proposes creative uses for a new technology, assigning credit becomes difficult. Assigning power to political parties creates similar problems; the number of seats a party controls correlates with power, but not perfectly.

In this chapter, we define two measures of value and power: last-on-the-bus value, which equals an actor's marginal contribution given that the group has already formed, and Shapley value, which equals an actor's average marginal contribution across all possible sequences of adding people to a group. In a group of three people, we average a person's added value when she joins the group first, second, and third. We define these measures within the structure of cooperative game models, which consist of a set of players along with a value function that assigns a collective payoff to every possible subset of the players.

The chapter consists of four parts. In the first part, we define cooperative game models, last-on-the-bus value, and Shapley value, and work through some examples. In the second part, we describe axiomatic foundations for the Shapley value. We show it to be the unique measure satisfying four conditions. One condition is that a player who never adds value must be assigned value zero. A second condition is that the sum of the player's values must equal the total value of the game. In the third part, we apply Shapley value to a group performing a creative task. Each person thinks up ideas. We show how in this context, the measure produces an intuitive measure of value. In the fourth part, we consider the special case of applying the Shapley value to voting games. We use it to distinguish between voting power and vote percentage. We find that they need not always agree. A party might hold 20% of the seats and have no power in one case and a third of the total power in another.

Cooperative Games

A *cooperative game* consists of a set of players and a *value function* that assigns a value to every possible subset, often called a coalition, of players. Cooperative games are meant to capture collective work and joint projects. In the model, we assume that people participate so that we can focus attention on how to assign value to their participation.

Cooperative Games

A **cooperative game** consists of a set of N **players** and a **value function** that assigns a value to any subset $S \subseteq N$, $V(S)$. These subsets are called **coalitions**. The value of the coalition consisting of no players equals zero, $V(\emptyset) = 0$; the value of all N players, $V(N)$, equals the **total value** of the game.

In a cooperative game, a player's *last-on-the-bus (LOTB) value* equals the value she adds if she is the last to join the group. LOTB values capture players' values at the margin. If four people are hired to move a table, and moving the table produces a value of 10, and all four are needed, then each has a LOTB value of 10. If only three are required, then each has a LOTB value of zero. Notice that LOTB values need not sum to the total value of the game. In particular, if the value function exhibits diminishing returns to scale, then LOTB values sum to less than the total value, and if added values exhibit increasing returns to scale, then the sum of LOTB values exceeds the total value.

A player's *Shapley value* equals her marginal contribution when she is added to a coalition averaged across all possible orderings in which the coalition of everyone forms. In other words, we imagine adding the players to the coalition in sequence and calculating a player's added value for each sequence. Consider a small firm that operates in Spain and France and requires one French speaker and one Spanish speaker to conduct daily business. The firm has three employees: a Spanish speaker, a French speaker, and a bilingual person capable of speaking both French and Spanish.

Suppose that our cooperative game assigns a value of $1,200 to any set of workers capable of speaking French and Spanish. This amount equals the daily revenue of the firm if it is able to operate. If any two employees show up at the office, the third is not needed. Therefore, each player has an LOTB value of zero.

To compute the Shapley value for the French speaker, we consider all six orderings in which people could arrive to work. In only one of these orderings, the one in which the Spanish speaker arrives first and the French speaker arrives second, does the French speaker add value. Her Shapley value equals $\frac{1}{6}$ times $1,200, or $200. The Spanish speaker adds value only if he arrives second and the French speaker arrives first, so his Shapley value also equals $200. In the other four orderings, the bilingual person arrives either first or second and adds value. Her Shapley value therefore equals $800. The sum of the Shapley values equals $1,200, the total value of the game.

Shapley Value

Given a cooperative game $\{N, V\}$, the **Shapley value** is defined as follows: let O represent all $N!$ orderings in which the N players could arrive and be added to a group. For each ordering in O, define the **added value** of player i to be the change in the value function that occurs when player i is added. Player i's **Shapley value** equals the average of her added values over all orderings in O.

Now that we have the idea, we construct a more complicated example. Imagine a crew team that requires four rowers and a coxswain—a smaller person who manages the stroke rate and steers. Our crew team (the players in the cooperative game) consists of six individuals: five tall, powerful rowers and a smaller person who has been trained as a coxswain. To enter a race, the team needs four rowers and a coxswain. A team of five that includes the smaller, trained coxswain will be competitive and has value 10. A team of five rowers without the coxswain could enter a race, but would perform poorly because of the extra weight. We assign that team of five a value of 2.

To compute Shapley values, we imagine the players arriving in every possible order. If the smaller coxswain arrives first, second, third, or fourth, she adds no value. If she arrives fifth, which occurs one-sixth of the time, she adds value 10. If she arrives sixth, she replaces one of the rowers as coxswain and her added value equals 8. Averaging across all of these possibilities, we find the coxswain's Shapley value equals 3.

Each rower adds value if and only if she arrives fifth, which occurs one-sixth of the time. If the coxswain has not arrived, the rower who arrives fifth adds value 2. If the coxswain has arrived, the rower adds value 10. Given the one-in-five chance the coxswain is last among the five other players, and the four-in-five chance the coxswain arrives among the first four, we arrive at a Shapley value of $\frac{7}{5}$ for each rower.[1] Intuitively, the coxswain's value should be more than the value of a single rower and, given that the rowers can compete, albeit poorly, without the coxswain, less than the combined value of all of the rowers. There are an infinite number of ways to assign values that satisfy those constraints. Shapley values assign specific values: 3 for the coxswain and 7 total for the five rowers.

Axiomatic Basis for the Shapley Value

We now describe a set of axioms that Shapley values uniquely satisfy. That result explains why we might privilege Shapley values over other possible measures. First, note that we calculate Shapley values by averaging a player's marginal contribution across all possible orderings, so any player who never adds value has a Shapley value of zero. Moreover, any two identical players—that is, two players who, for each coalition, contribute the same amount—must be assigned the same Shapley value. And given that the sum of the added values equals the total value of the game for any ordering, Shapley values must also sum to the value of the game. These will be three of the four axioms. Notice that LOTB values satisfy the first two axioms but not the third.

To these three properties, we add a fourth, *additivity*, which requires that if the value function of a cooperative game can be decomposed into two value functions, each assigned to a different cooperative game, a person's value in the combined game should equal the sum of her values in the two constituent games. A moment's reflection reveals that Shapley value satisfies this property as well. That those four properties uniquely characterize the Shapley value is less obvious.

Showing that a measure uniquely satisfies a set of axioms places the measure on logical foundations. Without the axioms, a measure may be intuitive but could be seen as arbitrary, as one of several plausible measures. The theorem also tells us that if we choose any other measure, we must abandon one of the axioms. This does not mean that Shapley value is the only reasonable measure. Lloyd Shapley, an economist and mathematician, may have first written down the measure and only after the fact constructed axioms that it uniquely satisfies. Which came first is of little relevance. Even if the axioms had been backward-engineered, if we accept the axioms, we should embrace the measure. The appropriateness of the measure hinges on the reasonableness of the axioms. In this case, the first three are difficult to dispute. The fourth, additivity, though more complicated than the others, can be supported on the grounds that if it did not hold, players would have incentives to split up or form coalitions.

Shapley Value: Axiomatic Basis

The **Shapley value** uniquely satisfies the following axioms:

Zero property: If a player's added value equals zero for any coalition, the player's value equals zero.

Fairness/Symmetry: If two players have the same added value for any coalition, then those players have the same value.

Full allocation: The sum of the values of the players equals the total value of the game, $V(N)$.

Additivity: Given two games defined over the same set of players with the value functions V and $\hat{V}$, the value of a player in the game $(V + \hat{V})$ equals the sum of that player's values in V and $\hat{V}$.

Shapley Values and the Alternative Uses Test

We now apply Shapley values to a cooperative game based on the *alternative uses test*. In the test, each person must come up with novel uses of a common object, such as a brick. The test measures a person's creativity based on the number of uses or categories of uses that she generates. When we calculate Shapley values, we find that they produce an intuitive scoring rule.

Imagine three players, Arun, Betty, and Carlos, who each think up alternative uses for blockchain, a distributed ledger technology, shown in figure 9.1. Arun and Carlos each think of six ideas, giving each a creativity score of 6, and Betty thinks of seven, making her score 7. The group's total creativity equals 9, as there are nine unique ideas. To compute the Shapley values, we could write down all six possible orders in which the group could form, give individuals credit only for unique ideas added to the group, and then average over all six cases. Or we can notice that when we are computing Shapley values, the probability of someone getting credit for an idea equals 1 divided by the number of people who propose the idea. Anyone who proposes a unique idea always receives full credit. In the

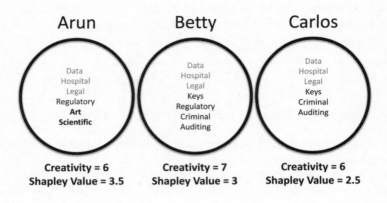

Arun

Data
Hospital
Legal
Regulatory
Art
Scientific

Betty

Data
Hospital
Legal
Keys
Regulatory
Criminal
Auditing

Carlos

Data
Hospital
Legal
Keys
Criminal
Auditing

Creativity = 6
Shapley Value = 3.5

Creativity = 7
Shapley Value = 3

Creativity = 6
Shapley Value = 2.5

Group Creativity = 9

Figure 9.1: Shapley Values and the Alternative Uses Test

figure, we denote those ideas, such as Arun's idea of art transactions, in **bold** font. If two people think of an idea, each has a one-in-two chance of joining the group first. Similarly, if all three people think of an idea, each has a one-in-three chance of joining first. It follows that allocating credit equally among people who thought of an idea produces Shapley values. Thus, it is the unique way to assign values that satisfies the four axioms. These values show that Arun, though he did not have the most ideas, adds the most value.[2]

The Shapley-Shubik Index

We next apply Shapley values to a class of voting games. In a voting game, each player (representing a political party or official) controls a fixed number of seats or votes, and a majority of those seats or votes are necessary for taking an action. In voting games, the Shapley value is referred to as the *Shapley-Shubik index of power*.[3] By calculating the index, we find there does not exist a direct translation between the percentage of seats (votes) a party controls and its power.

To compute power indices, we consider all possible orderings of parties being added to a coalition. If a party joins the coalition and creates a strict majority, the party's added value equals 1. In those cases, the party is said to be *pivotal*. Otherwise, the party adds no value. Consider a parliament with 101 seats allocated across four political parties as follows: party A controls 40 seats, party B controls 39 seats, and parties C and D each control 11 seats. In this example, party A cannot be pivotal if it arrives first or last. If party A arrives second or third, it is always pivotal. Therefore, it has a power index of $\frac{1}{2}$. If party B arrives first or last, it also adds no value. If B arrives second, it is pivotal only if party A arrived first. If party B arrives third, the only way that it can be pivotal is if party A arrives last. Each of those combinations of events also occurs with probability $\frac{1}{12}$. Therefore, party B's power index equals $\frac{1}{6}$. Parties C and D are pivotal in a similar set of cases as party B. Neither can be pivotal if it arrives first. Each is pivotal if it arrives second only if party A arrived first. Each is also pivotal if it arrives third when party A arrives last. Thus, each of those parties also has a power index of $\frac{1}{6}$.

Party	Seats	Power
A	40	$\frac{1}{2}$
B	39	$\frac{1}{6}$
C	11	$\frac{1}{6}$
D	11	$\frac{1}{6}$

Figure 9.2: The Disconnect Between Seats and Power

The example reveals a possible disconnect between the percentage of seats a party controls and its power. Parties A and B control almost identical numbers of seats, but A has three times the power of party B, which has no more power than party C or party D. Similar allocations of seats occur often in real-world parliamentary systems. As a result, parties with few seats can often control substantial power. Israel's parliament, the Knesset, has 120 seats. In 2014, a coalition led by the Likud party had 43 seats. The opposition coalition had 59 seats (just shy of a majority), and an Orthodox coalition held eighteen seats. All three parties have the same Shapley-Shubik index. This does not mean that the small Orthodox parties have the same power in practice—all models are wrong. It does suggest that the Orthodox parties had more influence than would be anticipated from their seat count.

An even more stunning disconnect between seats and power occurred with the Nassau County Board of Supervisors in New York in the mid-1960s. At that time, the board consisted of six members, and each controlled votes proportional to the population of the districts she represented, as shown in figure 9.3. A majority vote required 58 or more of the 115 votes. Notice that any two of the three largest districts constituted a majority. It follows that the votes of the other three districts could never be pivotal. The other district representatives therefore had no power.

The North Hempstead representative controlled 21 votes, more than 18% of the total yet could not influence a voting outcome.

District	Votes	Power
Hempstead 1	31	$\frac{1}{3}$
Hempstead 2	31	$\frac{1}{3}$
Oyster Bay	28	$\frac{1}{3}$
North Hempstead	21	0
Long Beach	2	0
Glen Cove	2	0

Figure 9.3: Votes but No Power

The Shapley-Shubik index of power can be applied to any situation with unequal allocations of seats or votes, such as the European Union or the Electoral College. That does not mean that it is necessarily an appropriate measure in all cases. The fifty states can be arranged in 50! (3×10^{64}) different orders. Given regional correlations in voter preferences, not all coalitions are possible. Mississippi may not be likely to form a coalition with New York. To make a more useful measure of power we would need to privilege some coalitions over others or rule out some coalitions altogether. Later in the book, we describe Myerson values, which allow us to do the latter, to rule out some coalitions.

Summary

An individual's Shapley value corresponds to her average added contribution to coalitions as they form. It is a measure of added value. In voting games, Shapley value can also be interpreted as a measure of power. It may not always be the best measure of power. An individual's LOTB value may be the better measure of power in situations where a group has already formed, as that measures how much each individual could extract through a threat to leave, assuming that threat is credible.

In those cases, the coalition wants to reduce LOTB values. Creating a coalition with a high value but low LOTB values can be accomplished by increasing the coalition size. Adding extras makes existing members expendable and drives their LOTB values to zero. We see this in practice. Employers hire excess workers to reduce worker power. Manufacturing firms rely on multiple competing suppliers of intermediate goods. Governments award contracts to keep multiple contractors in business.

The same intuition applies to the creation of coalitions in legislatures. Congressional lobbyists and party leaders want to pass legislation (an outcome of value) but restrict the power of individual representatives and senators.[4] If a lobbyist makes contributions to the minimal number of representatives and senators necessary to win a vote, then each representative and senator has an enormous LOTB value. Any could switch his or her vote and flip the outcome of the bill. The lobbyist can reduce their LOTB values by buying a supermajority of representatives and senators. The same logic implies that a party that holds a slim majority may be difficult to lead. Every member has a large LOTB value. Within a strong majority, no representative or senator has much power.

If we broaden our perspective and contemplate power in the modern connected world, we find it useful to apply both LOTB values and Shapley values. The power of an individual, organization, corporation, government, or terrorist group depends partly on how much damage it could do by deviating from a cooperative regime (LOTB value). A sophisticated computer hacker, a person capable of destroying a substantial amount of wealth, has enormous power. This holds even though the hacker lacks the ability to add value.

In thinking about the value of corporations or other multinational organizations, Shapley value may be a better measure. In these cases, exit may not be a viable option. An energy company participates in an energy generation game, an energy distribution game, a real estate game, an environmental game, an employment game, and so on. The company's total added value equals the sum of its added values across the various domains.

Thinking of power and value through the lens of cooperative game theory provides powerful, basic insights. It also points to where we should look next. In politics and business, not all coalitions are plausible. The model assumes that they are. A richer model would take into account the connectedness of the world. Consulting companies and financial firms buy software from tech companies. Tech companies and consulting companies invest and borrow through financial firms. And financial firms and tech companies hire consultants. Within those webs, each actor adds value and wields power. To calculate power in these settings requires models of networks, where we turn next.

10. Network Models

Network theory is a whole branch of science, but it's relatively new in terms of the last 20 or 30 years. We haven't had a chance to take all that theory out of the universities and apply it to ask: "What kinds of networks should we build, and for what purposes?"

—Anne-Marie Slaughter

In this chapter, we cover models of networks. A comprehensive study of networks would require multiple books. We have more modest goals. We want to understand the basics of networks, to be able to name their parts, and to ask why they matter for modeling. The answer we arrive at will be that networks almost always matter. Any model we construct, be it of a market, the spread of a disease, or the transmission of information, can be enriched by embedding the actors in a network.[1]

Networks are ubiquitous. People talk of trade networks, terrorist networks, and networks of volunteers. Species organize into food webs, a form of network. Firms build supply chain networks. As already noted, the financial system is usefully thought of as a network of promises to pay. Networks have always been important to understanding social relations. During much of human history, social networks were constrained by geography and difficult to map. Due to technological advances, many social interactions and economic transactions now take place over virtual networks which can be analyzed using models.

The organization of this chapter follows the same structure-logic-function format we applied to distributions. We first characterize the structure of networks using statistical measures of degree, path length, clustering coefficient, and community structure. Then we discuss common classes of networks: random networks, hub-and-spoke networks, geographic networks, small-world networks, and power-law networks. After that we turn to the logic of how networks form. We construct micro-level processes that produce the network structures we see. Last, we take up function, the question of why network structure matters. Here we focus on five implications. We begin with the friendship paradox, and then describe the six degrees

117

of separation phenomenon and the strength of weak ties property. Last, we take up the robustness of networks to node or edge failure and the aggregation of information over networks. The chapter concludes with a discussion of how networks influence model outcomes.

Network Structure

A network consists of *nodes* and *edges* that connect them. We refer to nodes connected by an edge as *neighbors* and to a network as *connected* if it is possible to get from any one node to any other along edges. Networks can be represented as graphs, as lists of edges, or as matrices of zeros and ones, where a one in row *A* and column *B* denotes an edge between node *A* and node *B*. Though people prefer graphical representations of networks, lists and matrices are better for representations for calculating network statistics.

The edges in a network can be *directed*—that is, pointing from one node to another. In an information network, a directed edge denotes that one person gets information from another. In an ecosystem network, a directed edge from a red-tailed hawk to a gray squirrel represents that the hawk eats the squirrel. Edges can also be *undirected*. Edges that connect friends are drawn in this way. In an undirected network, the *degree of a node* equals the number of edges that connect to it. Networks are characterized by a set of network statistics. For each statistic, we can compute the network average and the distribution across all nodes. The *average degree* of a friendship network tells us, on average, how many friends each person has. The *degree distribution* tells us if some nodes are more connected than others. Social networks have more equal distributions than the World Wide Web, the internet, and citation networks, all of which have long tails.

Network Statistics

Degree: The number of neighbors (also the number of edges) of a node.

Path length: The minimum number of edges that must be traversed to get from one node to another.

Betweenness: The number of paths of minimal length connecting two other nodes that pass through a node.

Clustering coefficient: The percentage of a node's pairs of neighbors that are also connected by a edge.

Path length, the minimal distance between two nodes, varies inversely with degree. As we add edges, we shorten the average length between nodes. In an airline flight network, path length corresponds to the number of flights, on average, a person needs to take to get from one city in the network to another. Given a choice between two airlines, all else equal (namely, prices), a traveler would prefer the one with lower average path length. Average path length also correlates with information loss. Information that passes through several people is more likely to suffer distortion than information passed between only two people. The nodes on minimal paths play critical roles in networks. If information takes the shortest route, then it goes through the nodes on a minimal path. A node's *betweenness* score equals the percentage of minimal paths that go through a node. In a social network, people with high betweenness scores know more information and wield more power.

The final statistic, the *clustering coefficient,* equals the proportion of a node's pairs of neighbors who are also neighbors of one another. For example, a person with 10 friends has 45 pairs of friends. If 15 of those 45 pairs are themselves friends, then the person's clustering coefficient equals $\frac{1}{3}$. If all 45 friendships existed, then the person's clustering coefficient would equal 1, the maximal possible value. The clustering coefficient for the entire network equals the average of the clustering coefficients of the individual nodes.

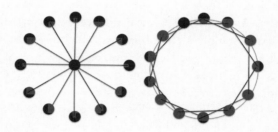

Network	Hub and Spoke		Geographic
Measure	Hub	Outer	Network
Degree	12	1	4
Average Path Length	1	$1\frac{11}{12}$	2
Betweenness	1	0	$\frac{1}{12}$
Clustering Coefficient	0	0	$\frac{1}{2}$

Figure 10.1: A Hub-and-Spoke Network and a Geographic Network

Figure 10.1 shows two networks with thirteen nodes: a hub-and-spoke network and a geographic network. In the *hub-and-spoke network,* the hub has degree 12 and all other nodes have degree 1, for an average degree of less than 2. The degree distribution is unequal. The hub has a distance of 1 to every node. All other nodes have a distance of 1 to the hub and a distance of 2 to the other nodes. It follows that average path length is also less than 2. The hub, which lies on every minimal path between any two other nodes, has a betweenness score of 1. The spoke nodes do not lie on any minimal paths connecting other nodes, so they have a betweenness of 0. Finally, in the hub-and-spoke network, no nodes connected to a node are connected to one another. Therefore, the network has a clustering coefficient of zero.

In the *geographic network,* each node is connected to the two nodes to its right and its left, so the average degree equals 4. Each node is distance 1 from four nodes, distance 2 from four nodes, and distance 3 from four nodes. So the average distance equals exactly 2. The degree and distance distributions for this graph are *degenerate*—every node has the same degree and the same average distance. It can be shown that the betweenness of each node equals $\frac{1}{12}$.[2] Each node has four neighbors, creating six pairs. Of those six pairs, exactly three are connected: the two nodes to the immediate left and right are each connected to an outer node and to each other. Therefore, the clustering coefficient equals $\frac{1}{2}$.

An alternative method for capturing clustering is to partition the nodes into *communities*. In a junior high friendship network, the communities might correspond to teenagers interested in the arts, athletics, or science. Or they could be defined by race and gender. A network of political alliances might partition into regional or ideological allies. Multiple methods exist for determining communities. One approach sequentially removes edges with the highest betweenness, as edges with high betweenness are more likely to connect distinct clusters. Other approaches take the number of communities as given and seek an optimal partitioning given an objective function such as minimizing the number of edges between the communities or maximizing the proportion of edges within communities.[3]

We can use community detection algorithms to ask questions of network data. Studies show that people may reside in *online bubbles*. That is, we may belong to communities of people who get their news from similar sources. If so, that has implications for social cohesion. Prior to the creation of the internet, that may have been true as well, but demonstrating it with data would have been hard. Now data scientists can scrape the web to identify the news sources that people frequent and tell us that, yes, in fact we do live in bubbles, to an extent. Models provide the formal definitions of communities. Data tells us the strength of those communities. Using judgment we can make wise inferences based on what the data say.

Common Network Structures

In analyzing networks, we encounter a problem of variety. A handful of network statistics are incapable of pinning down the specific network structure: one can construct billions of distinct networks with ten nodes and an average degree of 2. An alternative approach to characterizing a network is to test whether its statistical measures differ significantly from those of a common network structure. For example, a scholar might gather data on judicial citations and put it in network form by drawing an edge when one judge cites another judge's opinions. The graph of that network may appear to possess interesting structures and clusters. We can test whether a network is random by comparing the network's statistics to those of a random network that has the same number of nodes and edges. A *random network* clustering coefficient equals the probability of a random edge because two neighbors of a node are no more likely to contain an edge than any other randomly chosen node.

Monte Carlo Method for Random Networks

To test whether a network with N nodes and E edges is random, we create a large number of random networks with N nodes and E edges and calculate distributions for degree, path length, clustering coefficient, and betweenness. We then perform standard statistical tests to accept or reject the hypothesis that the network's statistics could have been drawn from the simulated distributions.[4]

Theoretical models often assume a particular network structure. Many assume random networks, while others assume regular geographic networks such as when the nodes are arranged in a circle and each is connected to the nearest nodes in each direction. Other geographic networks arrange the nodes on a checkerboard and connect each node to its neighbors to the north, south, east, and west. Most of the common geographic networks have low degree—they connect to only the local neighbors—and relatively high average path length. On geographic networks, betweenness and clustering coefficient have no variation.

A third common type of network, a *power-law network,* has a power-law degree distribution. A handful of nodes has many connections, but most nodes have very few networks. A fourth type of network, a *small-world network*, combines features of geographic and random networks.[5] To construct a small-world network, we begin with a geographic network and then "rewire" it by randomly selecting an edge and replacing one of the nodes it connects with a random node. If the rewiring probability equals zero, we have a geographic network. If it equals 1, we have a random network. In between, we have a small-world network, distinguished by small clusters from the geographic network connected by random links to other clusters. Social networks look similar to small worlds. Each person has a cluster of friends as well as random friends.

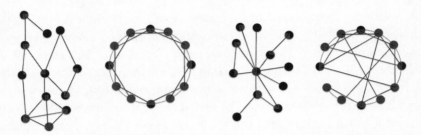

Figure 10.2: Random, Geographic, Power-Law, and Small-World Networks

Network Formation: Logic

We now briefly describe models of network formation. These models provide logic to explain network structures. Most of the network structures that we encounter *emerge* from choices of individual actors to make connections. That is true of friendship networks, the World Wide Web, and power grids. These networks are not planned. Other networks, such as supply chain networks, do result from planning. We would expect planned networks to be robust to the failure of nodes. The fact that emergent network structures are robust is more of a puzzle.

We have already discussed how to create random networks and small-world networks. We create the former by randomly creating a set of nodes and then drawing edges connecting random pairs of nodes. We create a small-world network by first constructing a regular geographic network (often by arranging nodes in a circle and connecting k neighbors in each direction) and then randomly "rewiring" a proportion of the edges.

Models of the formation of the power grid rely on economic and engineering principles. The network must deliver power to homes, businesses, and the government. Whether the producers are for-profit companies or public utilities, they have little incentive to create high clustering, as it would be inefficient. This lack of clusters reduces the robustness of the network. Economic and engineering considerations also rule out long leaps: connections that reach far across the network.

Power companies do not build direct connections from Chicago to Dallas. People and businesses, however, do. A Chicagoan might strike up a friendship with someone from Dallas. A firm in Singapore might trade with a firm in Detroit. As we see in the next section, these long leaps contribute to network robustness.

To create a network with a long-tailed distribution, we can apply a version of the preferential attachment model. We create nodes randomly and then draw edges from new nodes to existing nodes. If we let the probability of connecting to a node be proportional to its degree, we produce a power-law degree distribution. In that model, early-arriving nodes will be far more likely to be of high degree. A shortcoming of the model is that it does not allow for any difference in node quality. Higher quality nodes should have higher degree. The *quality and degree network formation model* corrects that omission while also producing a long-tailed distribution.

Quality and Degree Network Formation Model

Create d disconnected nodes. In each period t create a new node with quality Q_t drawn from a distribution F. Connect that node to d other nodes based on the degree of those nodes. If D_{it} denotes the degree of node i at time t, the probability of choosing node i given N nodes equals:

$$\frac{D_{it} + Q_{it}}{\sum_{j=1}^{N}(D_{jt} + Q_{jt})}$$

If the quality of new nodes has a low mean and low variance, the model resembles the standard preferential attachment model. If the quality distribution has a long tail, then new nodes of very high quality can grow to have large degree.[6]

Why Networks Matter: Function

In Chapter 1, we mentioned the friendship paradox, the fact that on any network, on average, people cannot have more friends than their friends do. The logic for why this holds can be shown using the hub-and-spoke network. In that network, twelve people have one friend and one person has twelve friends. The twelve people with one friend are all connected to the hub, and the hub has twelve friends. That feature—the fact that high-degree people are connected to more people—drives the result. On the hub network, people, on average, have fewer than two friends. Yet, on average, each person's friends have more than eleven friends.

The friendship paradox holds for any network: academic citation networks, email networks, sexual contact networks, banking networks, and international trade networks. On average, the references cited by an academic article receive more citations than the article itself; a country's trading partners, on average, trade with more countries than the country itself; and the multiple species connected to a single species in a food network have, on average more connections than the single species itself. The disparity between the number of friends and the number of friends of friends becomes more pronounced on networks with degree distributions that are more dispersed. One analysis of friendships on Facebook found that the average person has around two hundred friends and their friends, on average, have more than six hundred friends.[7]

The Friendship Paradox

If any two nodes in a network differ in their degree, on average a node has lower degree than its neighbors. In other words, on average, people's friends are more popular than they are.[8]

The logic of the friendship paradox extends to any attribute that correlates with the number of friends. If active, happy, intelligent, wealthy, and kind people have, on average, more friends, then a person's friends will be, on average, more active, happier, more intelligent, wealthier, and more beautiful.[9] Imagine a network in which 90% of unhappy people have four friends and 10% have ten friends. Reverse the proportions for happy people: 10% have four friends and 90% have ten friends. People's friends will disproportionately consist of people with ten friends. A large majority of those people will be happy, so most people's friends will be happier than they are.

We now show the *Six Degrees of Separation* phenomenon, the claim that any two people on the earth can be connected through six friends or fewer. While the friendship paradox holds for any network, six degrees of separation only holds for some types of networks. The phenomenon's name derives from an experiment carried out by Stanley Milgram in the 1960s. Milgram sent packets to 296 individuals in Omaha, Nebraska, and Wichita, Kansas, that were to be forwarded to an individual in Boston, Massachusetts. The recipients had to follow the same rules. These participants were only allowed to send the packets via mail to people that they knew personally and whom they believed might have a greater chance of knowing the target person in Boston, with instructions to do the same. Individuals signed a roster to record the path and mailed postcards to the researchers so the researchers could track breaks in the chain. Sixty-four of the letters arrived in Boston. Of those that did, the average path length was slightly less than six, hence the phrase "six degrees of separation."

A second experiment run fifty years later on a much grander scale using email created eighteen global targets and sent them to more than 20,000 people. The median path length of email chains was between five and seven, depending on the geographic distance between the source and the target. The length of paths found does not equal the minimal path length between participants. The evidence therefore suggests that most people are linked by fewer than six degrees.[10]

We construct a simplified version of the small-world network to give intuition about the six degrees of separation phenomenon. Our version assumes that individuals have a small cluster of *clique friends,* who all know one another, and that they also have friends outside of those cliques, whom we call *random friends.*[11] Figure 10.3 shows an individual (denoted by the black circle) with five clique friends and two random friends. It also shows a selection of friends of the node's friends (light gray circles).

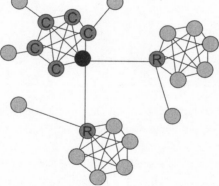

Figure 10.3: A Node's Clique Friends (*C*) and Random Friends (*R*)

These random friends might also be thought of as *weak ties*—people who connect you to other communities of people. Our weak ties, the random friends in our network, play an important informational role by connecting communities with diverse interests and information. Hence, sociologists speak of the *strength of weak ties.*[12]

This construction allows us to calculate the number of neighbors of degree two (the friends of friends), by adding up all of the friends of random friends but only adding the random friends of the clique friends. We do not count the clique friends of the clique friends, as they are members of the node's clique. We calculate the number of *friends of friends of friends* similarly. We add in all of the clique friends' random friends' friends, but we do not add in a random friend's clique friends' clique friends, as they have already been counted as neighbors of degree two. To produce the six degrees of separation phenomenon, we apply the same logic to a network with 100 clique friends and 20 random friends.

Six Degrees of Separation

Assume each node has 100 **clique friends** (C), all of whom are friends with one another, and 20 **random friends** (R), who have no friends in common with the node.

Degree one: $C + R = 120$

Degree two: $CR + RC + RR = 2000 + 2000 + 400 = 4400$

Degree three: $CRC + CRR + RCR + RRC + RRR = 328,000$

Degree four: $17,360,000$ [13]

Degree five: > 1 billion

Degree six: > 20 billion

By assuming no overlap in the friends of the random friends, the model implicitly assumes an infinite population. An actual social network will have overlap in friends as the degree increases. In a network that includes overlap and other realistic features such as heterogeneity in the number of friends, the values will differ from those calculated above. The relative magnitudes of the number of neighbors of each degree will remain similar. A person will have many more neighbors of degree three (friends of friends of friends) than of degree two (friends of friends).

The large number of friends of degree three, over a quarter million in our example, can be consequential. Unlike a person's clique friends, a person's friends of degree three tend to live in different cities, attend different schools, and have different information. They are more diverse. They are also near enough for trust to be established: a friend of a friend of a friend could be your roommate's mother's coworker, or your sister's boyfriend's aunt. The number of friends of degree three, their diversity, and their relative proximity make them an important asset. They can provide new information and job opportunities. These are the people most likely to help a person find a job, facilitate a move to a new city, or become a life or business partner.

Network Robustness

Our last implication of network structure evaluates the *robustness* of network properties, or how close the network is to node (or edge) failure. The most essential property of a network is whether it remains connected. We can use models to calculate the probability that a network remains connected as a function of the number of nodes removed. We could also ask what happens to average path length as nodes are removed. Applied to an airline network, an analysis of path length robustness would tell us how many extra flights would be needed if an airport were to shut down due to weather or a power failure.

Here, we consider the question of how the size of the largest connected component of the network, the *giant component,* changes as nodes randomly fail. Figure 10.4 shows the size of the giant component for a large random network and a large small-world network. In the random network, the size of the giant component falls linearly at first. At a critical value where the probability of an edge equals 1 divided by the number of nodes, the size of the largest component falls to an arbitrarily small proportion of the original network size. The small-world network shows no such abrupt change. A majority of connections exist within the geographic clusters. Each cluster can withstand the failure of multiple nodes. This feature combined with the random links prevents the entire network from collapsing.

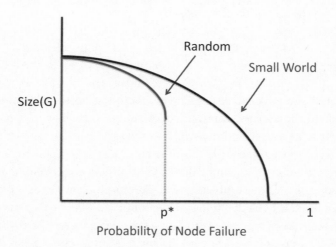

Figure 10.4: Size of the Giant Component (G) as a Function of Node Failure

From figure 10.4 we can infer that sparse networks that lack local clustering are susceptible to failure. We can apply that insight to the power grid. It lacks the long leaps and the tight clusters that make the small-world network robust. In a power grid, the failure of a node or a link cannot be overcome by other links in a cluster or by a long connection to a working node far away. Local failures can cascade through the network.[14] In contrast, the internet, which has a long-tailed degree distribution, is robust to random node failure. The degree distribution implies that the vast majority of nodes have few connections. Even if they fail, the network remains connected.

Up to now, we have assumed random node failure. We can also consider strategic node removal. Networks with long tails, like the internet, now become non-robust. Strategic removal of the nodes of highest degree destroys the network. The logic can be seen by considering the hub-and-spoke network. When nodes are removed randomly, the network remains connected unless the hub node is removed, a low-probability event. Strategic removal, wiping out the hub, disconnects the network in one step.

With some networks, such as terrorist networks and drug supply networks, we might want to disconnect a network. If those networks are sparse, like the power grid, or have a long-tailed degree distribution, they can be disconnected through strategic node removal. For the terrorist network, this would entail arresting the most connected members. If those networks resemble small worlds, they will be robust, even to the strategic removal of nodes. Attempts to cut off any geographic segment of the network will fail because of the random reconnections that connect the segment to the rest of the network.

Summary

When we build network models of people, we often do so to capture social influences, where the success, behavior, information, or beliefs of a person in the network influence the success, behavior, information, or beliefs of their friends. Behavior can be contextual or intrinsic; so too can a person's value or contribution to a collective enterprise. A person's value or contribution could be due to properties of that person, such as her brilliance, her effort level, or her good fortune. A person's success could also be due to his network of friends and colleagues. This is an age-old question: Does success depend on what you know or whom you know?

Imagine a group of scientists working together in a research lab. They share advice, ideas, and knowledge. The number of academic papers, patents, or scientific breakthroughs produced by a scientist depends on what she knows, but it can also be influenced by whom she knows, on her interactions with other scientists. By thinking in terms of contextual features (friendship networks) as well as intrinsic attributes (individual abilities), we can determine how much of a scientist's success to attribute to each.

Investment firms that hire away superstar fund managers based on the belief that investment success depends mostly on talent have not had very promising results. Empirical evidence shows that top investors also depend on networks of colleagues who provide them with specific types of information.[15] That specific finding can be viewed within the lens of a much larger literature (some of which is model-based) showing how a person's position in an organization influences success.

Success still correlates with ability. A business idea that makes its investors millions was probably a good one. A scientist who publishes hundreds of papers and receives numerous awards has high ability. At the same time, those best positioned in the network make the largest contributions. We can measure a person's position using betweenness and other measures of centrality. The people who occupy high-betweenness positions in a network fill what Ron Burt calls *structural holes* between communities, which we can identify using algorithms.[16] Access to information and ideas from multiple communities gives people who fill structural holes power and influence. Filling a structural hole requires certain talents and abilities. A person cannot just jump in and fill any hole. She must build trust and understanding within each community. And she must be conversant in the knowledge base of each community.

We can apply nearly identical logic to assess the value of firms and assign power to countries. We can see a firm's value as intrinsic and take a balance sheet perspective by looking at assets and liabilities. We can also look at the context in which the firm operates, such as its position in the supply chain. Similarly, the power of a country depends on its resources and its alliances. For both firms and countries, intrinsic attributes and connectedness correlate. Those who occupy powerful positions in the network also possess important attributes.

Our analysis as well as most of the literature considers the node as the unit of analysis. The edges can be critical as well. Taking an even broader perspective, the network itself may be an appropriate unit of analysis. For example, teacher networks that allow ideas and information to flow between classrooms can improve educational outcomes, and so a well-connected administrator can effectively coordinate curriculum reform. Similarly, a second-grade teacher knows a lot of information about the students from his class who are going on to third grade; that information could help the third-grade teacher. A mathematics teacher knows what concepts students have yet to grasp; that information could help the science teacher structure her lessons. Good schools, therefore, have strong faculty networks. This is just one example of how network models can improve our thinking.[17]

Myerson Value and Burt's Structural Holes

People who fill structural holes connect communities in a network and have more influence. A variety of statistical measures of a network, such as betweenness, should correlate with occupying a structural hole. An alternative measure of influence in a network, *Myerson value*, relies on the logic of Shapley values. To compute Myerson values, we construct a cooperative game on a network but only allow coalitions that include connected components.

Consider three individuals arranged in a line. Assume that their locations represent political ideologies with person B in the center, as shown below. If we restrict coalitions to immediate neighbors, then A, the left-most person, cannot connect to C, the right-most person, unless person B also belongs to the coalition. To compute each player's Myerson value, we first assign added values to all feasible coalitions. We then compute Shapley values for each possible coalition, treating each as a distinct game. Last, we add up the Shapley values for each coalition game to obtain Myerson values.

Possible Orderings: ABC, BAC, BCA, CBA
Ruled Out: ACB, CAB

As an example, suppose that any coalition of two players produces an output of value 10 and all three players together produce output of value 14, we obtain the following: Players 1 and 3 have Myerson values of 3, and Player 2 has a Myerson value of 8.[18]

Centrality measures such as betweenness are based only on the network. Myerson values depend on a value function. By having both measures we can disentangle the dependence of power on a person's position in the network and on the functions she performs. In our example, the Myerson values for the three players, $(3, 8, 3)$, correlate perfectly with their betweenness scores $(0, 1, 0)$. That will not always be the case, particularly for more complicated networks and value functions.

11. *Broadcast, Diffusion, and Contagion*

As contagion of sickness makes sickness, contagion of trust can make trust.

—Marianne Moore

In this chapter, we model the spread of information, technologies, behaviors, beliefs, and diseases throughout a population using models of broadcast, diffusion, and contagion. These models play central roles in communication, marketing, and epidemiology. All three models partition the population into people who know or have some thing and those who do not. Over time, people move between those two groups. Someone moves from being susceptible to a disease to being infected, or from being uninformed about a new product or idea to being informed.

Empirical plots of the number of people who over time catch a disease, buy a product, or know a piece of information (the *adoption curve*) tend to be either concave or S-shaped. How people learn the information or catch the disease—that is, whether it spreads by broadcast or diffusion—determines the shape of that graph. One contribution of this chapter will be to link the micro-level processes of how ideas and diseases spread to the shape of these adoption curves. The chapter begins with an analysis of the broadcast model, which applies when people hear of an idea or catch a disease from a single source. This model produces plots with an r-shape. We then cover the diffusion model, in which spread occurs from contact, as when a disease spreads from person to person. This model produces an S-shaped curve.

Many products, programs, ideas, and pieces of information spread by both broadcast and word of mouth. We can model these environments by allowing for both broadcast and diffusion. The resulting model, known as the Bass model, plays a central role in marketing. Whether it produces more of an r-shape or S-shape depends on the strengths of the two processes. The last model we cover, the SIR model of contagion from epidemiology, includes a rate of recovery. This assumption could capture an immune system fighting off a disease, behaviors or styles dropping out of fashion, or information becoming less worthy of passing on to others. The SIR model produces a tipping point, where small changes in the attributes of the product or a disease spell the difference between failure and success. A slight reduction in virulence can transform a mass infection into a minor outbreak. A small increase in the probability of spreading word of a hot new band can be the difference between the Beatles and a band that played pubs in Liverpool for a few months in the 1960s.

The Broadcast Model

All of the models we cover in this chapter assume a *relevant population*, denoted by N_{POP}. This consists of those people who could potentially catch the disease, learn the piece of information, or adopt the product. The relevant population is not the entire population of, say, a city or country. If we are modeling the spread of a continuous aortic suture method, the relevant population is heart surgeons, not everyone in the city of Philadelphia.

At any moment in time, some people have the disease, know the information, or adopt the behavior. We refer to these people as either the *infected* or the *informed* (denoted by I_t). The remaining members of the relevant population are *susceptible* (denoted by S_t). These people could catch the disease or learn the information or behavior.[1] The relevant population equals the sum of the number of people infected (or informed) plus the number of susceptible people: $N_{POP} = I_t + S_t$.

Broadcast Model

$$I_{t+1} = I_t + P_{broad} \cdot S_t$$

where P_{broad} denotes the **broadcast probability**, and I_t and S_t equal the number informed and susceptible at time t.

Initially, $I_0 = 0$ and $S_0 = N_{POP}$.

The *broadcast model* captures the spread of ideas, rumors, information, or technologies through media like television, radio, or the internet. Knowledge of most current events spreads through broadcast. The model captures processes in which a source, which could be the government, a corporation, or a newspaper, spreads information. It could also capture contaminations that spread through a water supply.

The model does not apply to diseases or ideas that spread from person to person. As the broadcast model better fits the spread of ideas and information than disease, we refer to the number of people informed, as opposed to the number infected.

The number of informed people in a given time period equals the number informed in the previous period plus the probability that a susceptible person hears of the information multiplied by the number of susceptible people (see box). By convention, the initial population contains only susceptible people. Calculating the number of informed people in all future periods involves plugging the number of informed and susceptible people into the difference equation. The result will be an r-shaped adoption curve.

Imagine that the mayor of a city with 1 million residents announces a new tax policy. Prior to the announcement, no one could have known about the policy. If we assume the probability that someone hears the news on any given day equals 30% ($P_{broad} = 0.3$), then 300,000 people hear about it the first day. On the second day, 30% of the remaining 700,000 people, or 210,000, hear about it. In each period, the number of informed people increases and does so at a decreasing rate, as shown in figure 11.1.

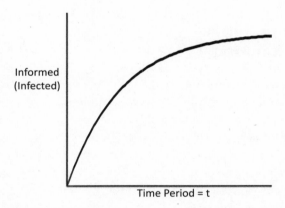

Figure 11.1: The r-Shaped Adoption Curve Produced by the Broadcast Model

In the broadcast model, everyone in the relevant population learns the information or buys the product. Using initial sales data, we can therefore estimate the relevant population size. Suppose that a company introduces a new line of shoes for people who practice tai chi, and in the first week, it receives orders for 20,000 pairs of shoes. If in the second week it receives orders for 16,000 pairs, we can make a crude estimate of eventual total sales, the size of the relevant population, to be 100,000.

> # Fitting the Broadcast Model to Data
>
> **Period 1:** $I_1 = 20,000 = P_{broad} \cdot N_{POP}$
>
> **Period 2:** $I_2 = 36,000 = 20,000 + P_{broad} \cdot (N_{POP} - 20,000)$
>
> **Solution:**[2] $P_{broad} = 0.2$ and $N_{POP} = 100,000$

We should not have a great deal of confidence in any estimate based on two data points. The model leaves out any number of real-world features. People might be hearing by word of mouth as well as through media, some people may have bought multiple pairs, or advertising may have targeted likely buyers. Including these features would change the estimates. Caveats aside, the model provides a rough estimate. The firm should not expect to sell exactly 2 million pairs, but it should be confident that they will sell more than 100,000 pairs. As more data arrives, the estimate can be improved. If week three's sales equals 13,000 pairs (the amount the model predicts), then the firm can place more confidence in the initial prediction.

The Diffusion Model

Most diseases as well as information about many products, ideas, and breakthroughs spread by word of mouth. The *diffusion model* captures such processes. It assumes that when one person adopts a technology or catches a disease, that person has some probability of passing it on to those with whom she comes in contact. In the case of a disease, choice pays no role. The probability a person catches the disease depends on factors such as genetics, the virulence of the disease, and even the temperature. Malaria will spread much faster during a hot, wet season than during a cold, dry one.

The spread of a technology involves a choice on the part of the adopters, so technologies that are more useful will be adopted with a higher probability. We do not explicitly consider choice in the model, however. Therefore, the hipness of the Apple Watch plays the same role as the virulence of a flu strain.

We again emphasize the spread of information, so we will refer to people as informed or uninformed. New people become informed if they meet an informed person and the information spreads between them. These are two distinct events that vary by context. People in cities may have higher contact probabilities than rural people, and information with high salience—say, news that aliens have landed—

has a higher sharing probability than news of the reintroduction of pretzel M&M's. Thus, we write the *diffusion probability* as the product of a *contact probability* and a *sharing probability*. We can write the model in terms of the diffusion probability, but when we estimate or apply the model, we must keep track of the two probabilities independently.

The diffusion model assumes *random mixing,* that is, that any two people in the relevant population are equally likely to make contact. This assumption should raise a red flag. It may be an accurate assumption for a model of disease spread in a preschool where children interact with high frequency. It is problematic to apply it to a city-level population. People do not randomly mix. People live and work in neighborhoods; they belong to work teams, families, and social groups. Their interactions are primarily within those groups. Remember, though, an assumption need not be accurate to be part of a useful model. We proceed with the assumption but keep an open mind toward changing it.

Diffusion Model

$$I_{t+1} = I_t + P_{\text{diffuse}} \cdot \frac{I_t}{N_{\text{POP}}} \cdot S_t$$

where $P_{\text{diffuse}} = P_{\text{spread}} \cdot P_{\text{contact}}$.

In this model as well, in the long run everyone in the relevant population learns the information. However, in this model, the adoption curve for the diffusion model has an S-shape. Initially, few people are informed; I_0 is small. It follows that the number of susceptible people who meet an informed person must also be small. As the number of informed people grows larger, the number of contacts between informed and uninformed people increases, producing larger increases in the number of informed people. When nearly everyone in the relevant population is informed, the number of newly informed people decreases, forming the top of the S-shape. Technological adoption curves often have this shape. For example, adoption curves for hybrid seed in the last century vary by state (Iowans adopted hybrid seeds faster than Alabamans), but all of the curves have an S-shape.[3]

In the broadcast model, estimating the relevant population size from data is straightforward. The initial numbers of adopters correlates strongly with the relevant population. In contrast, estimating the size of the relevant population using data from a diffusion model can be difficult. The same increases in product sales could result from a large diffusion probability among a small relevant population

Day	Phone App #1	Phone App #2
1	100	100
2	136	130
3	183	169
4	242	220
5	316	286
	...	
365	1,000	1,000,000

Figure 11.2: Two Adoption Curves for Sales of an Application

or a small diffusion probability among a large population. Figure 11.2 shows data for two hypothetical smartphone applications. On the first day, one hundred people bought each application. For each of the next five days, the first application realizes both higher total sales and larger increases in sales. Absent a model, we would probably predict the first application to have the larger market. Fitting the model to the two data streams shows the opposite to be true.

The first application fits a diffusion probability of 40% and a relevant population of 1,000 people, while the second application corresponds to a diffusion probability of 30% and a relevant population of 1 million people.[4] Within a few days, we would come to realize that there is a larger relevant population for the second application. Nevertheless, absent the model, we would make the incorrect inference about total sales if we based it on just five days of data.

When using the diffusion model to guide action, we unpack the probability of diffusion as the product of the probability of sharing and the probability of making contact. To increase the speed of an application's sales, its developer could increase the rate at which people meet or increase the probability that they share information about the application. Changing the first probability would be difficult. To increase the second probability, the developer could provide incentives for signing up new buyers, which many developers do. A game developer might give points within the game to users who sign up new buyers. Though this would increase the speed of diffusion, it would not affect total sales, at least not according to the model. As mentioned, total sales equals the relevant population size, regardless of the probability of sharing. Increasing the rate of sales produces no long-term effect.

Most consumer goods and information spread through both diffusion and broadcast. Our next model, the *Bass model*, combines the two processes in a single model.[5] The difference equation in the Bass model equals the sum of the difference equations from the broadcast model and the diffusion model. The adoption curve of the Bass model will be more S-shaped the larger the probability of diffusion. The adoption curves for televisions, radios, cars, computers, telephones, and cell phones all are combinations of r-shapes and S-shapes.

Bass Model

$$I_{t+1} = P_{\text{broad}} \cdot S_t + P_{\text{diffuse}} \cdot \frac{I_t}{N_{\text{POP}}} \cdot S_t$$

where P_{broad} = probability of broadcast and P_{diffuse} = probability of diffusion.

The SIR Model

In the models that we have covered so far, a person who adopts a technology never abandons it. That makes sense for the adoption of technologies like electricity, the dishwasher, and television; we never reverse our adoptions. That assumption does not hold for all things that spread by diffusion. After we catch a disease, we recover. When we adopt a fashion or fad, such as a particular style of dress or a dance step, we can abandon it. Following convention, we refer to people who drop an adoption as *recovered*. The resulting model, the *SIR model* (susceptible, infected, recovered), occupies a central position in epidemiology.

Given the model's origins and given that recovery occurs more naturally in diseases, we describe the model using the spread of a disease as an example. To avoid overcomplicating the mathematics, we assume that people who recover reenter the susceptible pool, that being cured of the disease does not create future immunity.

SIR Model

$$I_{t+1} = I_t + P_{\text{contact}} \cdot P_{\text{spread}} \cdot \frac{I_t}{N_{\text{POP}}} \cdot S_t - P_{\text{recover}} I_t$$

where P_{spread}, P_{contact}, and P_{recover} equal the probability of spreading the disease, the probability of contact, and the probability of recovery.

Epidemiologists keep separate track of the probability of contact and the probability of spreading, so we will as well. Contact rates depend on how the disease passes from one person to another. HIV spreads through sexual contact. Diphtheria spreads through saliva. Flu viruses spread through the air. Thus flu has a higher contact probability than diphtheria, which has a higher contact probability than HIV. Once contact occurs, the probability of spread also varies. Pertussis (whooping cough) transfers to another person more readily than SARS.

The SIR model produces a *tipping point* at what is known as the *basic reproduction number* (R_0), the ratio of the probability of contact times the probability of spread to the probability of recovery. A disease with an R_0 greater than 1 can spread through the population. Diseases with R_0's less than 1 dissipate. In this model, the information, or in this case the disease, need not spread to the entire relevant population. Whether or not it does depends on the value of R_0. Hence, government agencies like the Centers for Disease Control rely on estimates of R_0 to guide policy.[6]

R_0: The Basic Reproduction Number

$$R_0 = \frac{P_{\text{spread}} \cdot P_{\text{contact}}}{P_{\text{recover}}}$$

As shown in the table below, measles, which can spread through the air, has a higher reproduction number than HIV, which spreads through sexual contact and needle sharing. Estimates of R_0 do not assume that people change their behavior in response to a disease. Parents might respond to a lice infestation in a school by keeping their children at home, lowering the probability of contact. Or, they might shave their children's heads, reducing the probability of spreading when contact occurs. Both of these behavioral changes would lower lice's R_0.

	Measles	Polio	HIV	Influenza
R_0	15	6	4	3

In the absence of a vaccine, quarantine is an option, but it is costly.[7] If a vaccine exists, then vaccination can prevent disease spread. Disease spread can be prevented even without vaccinating everyone. The proportion of people who must be vaccinated, the *vaccination threshold,* is given by the formula $V \geq \frac{R_0-1}{R_0}$ which we can derive from the model.[8]

The vaccination threshold increases with R_0. To prevent the spread of polio, which has an R_0 of 6, the vaccine must cover $\frac{5}{6}$ of the population. To stop the spread of measles, which has an R_0 of 15, the vaccine must cover $\frac{14}{15}$ of the population. The mathematical derivation of the vaccination threshold provides guidance to policy makers. If too few people are vaccinated, then the disease will spread, so governments vaccinate more than the threshold amount estimated by the model. For diseases with high basic reproduction numbers, such as measles and polio, governments try to vaccinate everyone.

Some people worry about side effects of vaccines and choose not to participate in vaccination programs. If these people constitute a small percentage of the population, the vaccination of others prevents them from catching the disease. Epidemiologists call this phenomenon *herd immunity*. The people who choose not to get the vaccine *free ride* off the vaccinations of others. We study free riding in greater detail later in the book.[9]

R_0, Superspreaders, and Degree Squaring

The derivation R_0, the basic reproduction number, assumes random mixing: in each time step, individuals in the population randomly meet one another. As noted above, the random mixing assumption may approximate airborne diseases or diseases spread by touch, but it makes less sense for diseases that spread through sexual contact.

If we embed the SIR model on a network, we see the importance of the degree distribution to disease spread. Here, we compare a *rectangular grid network* (a checkerboard)—where each node is connected to the nodes to the north, south, east, and west—to a *hub-and-spoke network* where a hub node connects to all other nodes.

Assume that a disease randomly occurs at a node. We set $P_{contact} = 1$ within the network so that each person makes contact with everyone to whom he is connected. In the next period, the disease potentially spreads to each neighbor independently with a given probability corresponding to the virulence of the disease.

First consider the rectangular grid network. In each period, the disease can spread to any of the four nodes to the north, south, east, and west. If the probability of the disease spreading exceeds $\frac{1}{4}$, we would expect the disease to spread. If we look ahead one period, we see that if one new node caught the disease, then that node has three possible neighbors who could catch the disease. If two neighbors, those to the north and east of the original node, caught the disease, then there exist six nodes to which the disease could spread. This network, therefore, does not seem to have much of an effect on the likelihood of disease spread.

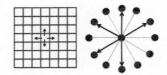

Next, consider the hub-and-spoke network. The first node to get the disease could be the hub or a spoke. If the hub catches the disease, then it could spread the disease to any one of the spokes. We would expect the disease to spread, even for a low probability of spreading. If a spoke caught the disease, then the only possible node that could catch the disease is the hub. And as we just learned, if the hub catches the disease, the disease will spread even for low probabilities of spreading.

For the hub-and-spoke network, R_0 is less informative because if the hub catches the disease, the disease will spread. Epidemiologists refer to high-degree hub people as *superspreaders*. Superspreaders contributed to the early spread of both HIV and SARS.[10] A superspreader need not be extremely social or well connected. A superspreader may have an occupation—tollbooth operator, bank teller, dental hygienist—that puts him in contact with people who belong to distinct social networks. Mary Mallon (Typhoid Mary) worked as a cook in New York at the turn of the twentieth century. She moved from family to family infecting each with typhoid fever. Once discovered as the source, Mary was quarantined against her will.

To derive the effect of high-degree nodes, we note first that a high-degree node is both better able to spread the disease and more likely to catch it. A person with three times as many friends as another will be three times as likely to catch the disease and able to spread it three times as widely. His total contribution to the spread of the disease will therefore be nine times that of the other. Thus, a node's contribution to the spread of a disease (or an idea) correlates with the square of the node's degree. If node A has a degree K times larger than node B, then node A will be K times as likely to spread the disease and spread it to K times as many others as B. Its total effect will be K^2 times larger than B's, a phenomenon known as *degree squaring*.

One-to-Many

Though the SIR model was designed to examine the spread of diseases, we can apply it to social phenomena that spread by diffusion and then fade: books, songs, dance steps, phrases, websites, diets, and exercise regimens. We can estimate probabilities of contact, spread, and recovery and basic reproduction numbers in these contexts as well. The model implies that small changes in these probabilities could spell the difference between success and failure by moving the basic reproduction number above zero. Success can hinge on what John Updike, in describing Ted Williams's last at-bat, called the "tissue-thin difference between a thing done well and a thing done ill."[11] Suppose that you think up a new joke. Making the joke a little bit funnier might push the basic reproduction number above 1 and cause the joke to spread. The same logic applies to the stickiness of ideas. If an idea sticks in people's minds a little longer, the recovery rate will be lower, increasing the basic reproduction number.

Not all cases lie on the threshold. The Beatles had enormous talent. Their reproduction number surely exceeded 1 by a large amount. That is of course conjecture. For current pop stars, we can use internet downloads to estimate basic reproduction numbers. Pop star Justin Bieber had an estimated R_0 of 24, making him more virulent than the measles.[12]

In the SIR model, we derived two critical thresholds, R_0 and the vaccination threshold. These thresholds are *contextual tipping points,* at which small changes in the environment (the context) have large effects on the outcomes. These differ from *direct tipping points,* where small actions at a particular moment in time forever alter the path of a system. Direct tips occur at unstable points, such as when a ball is perched atop a hill. A small push in either direction sends the ball down one side of the hill or the other. That small push is a direct tip.[13]

At a contextual tipping point, a change in a parameter changes how the system behaves. At a direct tipping point, the trajectory of future outcomes takes a sharp turn. A kink, such as the first bend in the S-shaped adoption curve produced by the diffusion model, satisfies neither definition of tipping point. The kink in the adoption curve corresponds to the point where the slope has maximal increase. At that point, the diffusion is well under way. No tip occurs.

Figure 11.3 shows the number of users of Google+ in its first two weeks.[14] A kink in the graph occurs six days after the release. By that time, the process of diffusion was well under way. It is not the case that Google+ struggled early and that a direct tip occurred on day six, with the result that within two weeks Google+ had over 16 million users. This conflation of tips with sharp upturns leads to an

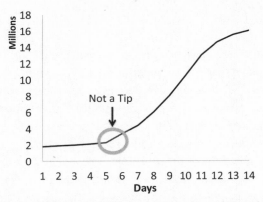

Figure 11.3: A Kink (Not a Tip) in the Number of Google+ Users

overuse of the term *tipping point*. Moments identified in the news media and on the internet as tipping points rarely satisfy the formal definition.

We can also think of obesity as an epidemic. Though people cannot catch obesity the way they might catch a cold, they can be socially influenced to adopt behaviors that contribute to obesity.[15] To reverse the obesity epidemic we must lower its basic reproduction number, which can be accomplished by decreasing the probabilities of contact or sharing or increasing the probability of recovery. The SIR model applied to obesity, school dropout rates, or crime is not better than economic or sociological models. It is a different model, so it produces different explanations and predictions. It also possibly points to different actions or policies. It contributes to an ensemble of models that help us make sense of the world. It is not a golden arrow that will solve the problem.

In applying models of broadcast, diffusion, and contagion to social phenomena, we may find that some assumptions hold and others do not. In the spread of a disease, each contact has an independent probability of spreading the disease. In social domains, contagion may become more likely with more exposure because adoption is a choice. We do not choose the flu. We catch it. We do choose to buy tight-fitting jeans. As more people wear tight jeans, we may become more likely to as well. Similar logic applies to the choice to become involved in a social movement, to adopt a new technology, or to get a tattoo. In these cases, as well as in the contagion of beliefs or of trusting behavior, we may have to emend the model to allow for the possibility that the probability of adoption per exposure increases with more exposures.[16] Such modifications are often necessary when broadening the set of applications of a model.

12. *Entropy: Modeling Uncertainty*

Information is the resolution of uncertainty.

—Claude Shannon

In this chapter, we introduce *entropy*, a formal measure of uncertainty. With it, we can show an equivalence between uncertainty, information content, and surprise. Low entropy corresponds to low uncertainty and little information being revealed. When an outcome occurs in a low-entropy system, such as the sun rising in the east, we experience little surprise. In high-entropy systems, say the drawing of numbers in a lottery, the outcomes are uncertain and when realized, they reveal information. We experience surprise.

Using entropy, we can compare disparate phenomena. We can say whether election outcomes in New Zealand are more uncertain than outcomes of United Nations votes on censure. We can compare the uncertainty of stock prices to the uncertainty of outcomes of sporting events. We can also use entropy to distinguish between the four classes of outcomes: equilibrium, periodicity, complexity, and randomness. We can distinguish complex patterns that appear random from true randomness and discern whether what appears to be a pattern is, in fact, random.

We can also use entropy to characterize distributions. In the absence of a controlling or regulating force, some populations may drift toward maximal entropy. Given constraints, such as a fixed mean or variance, we can solve for maximum entropy distributions. Maximal entropy results can also guide our modeling choices by justifying some distributions over others.

The chapter has five parts. In the first part, we provide intuition for and define information entropy. In the second part, we describe Shannon's axiomatic foundations for a general class of entropy measures. In the third part, we discuss how to use entropy to distinguish equilibrium, order, randomness, and complexity. In the fourth part, we investigate systems that produce maximal entropy given constraints. We conclude by discussing why, sometimes, we prefer complexity to equilibrium.

143

Information Entropy

Entropy measures the uncertainty associated with a probability distribution over outcomes. It therefore also measures surprise. Entropy differs from variance, which measures the dispersion of a set or distribution of numerical values. Uncertainty correlates with dispersion, but the two differ. Distributions with high uncertainty have nontrivial probabilities over many outcomes. Those outcomes need not have numerical values. Distributions with high dispersion take on extreme numerical values.

The distinction can be seen in stark relief by comparing a distribution that has maximal entropy with one that has maximal variance. Given outcomes that take values 1 through 8, the distribution that maximizes entropy places equal weight on each outcome.[1] The distribution that maximizes variance takes value 1 with probability $\frac{1}{2}$ and value 8 with probability $\frac{1}{2}$, as shown in figure 12.1.

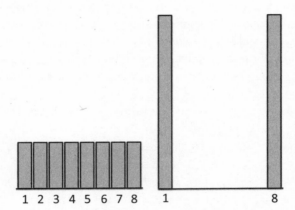

Figure 12.1: Maximal Entropy and Maximal Variance

Entropy is defined over probability distributions. It can therefore be applied to distributions over nonnumerical data such as the species of birds in a forest or the market shares of flavors of jam. The formal expression for entropy is written as minus the sum of products of probabilities and their logarithms. That sounds complicated, but it will become intuitive.

We begin with the special case of *information entropy,* which measures uncertainty in terms of number of random flips of a fair coin. Suppose that every family has exactly two children and that boys and girls are equally likely. The sexes of a family's children (listed by birth order) are equivalent to two coin flips. The distribution over outcomes therefore has an information entropy of 2 because it cor-

144

responds to 2 random events. The information content also equals 2 because we could learn the outcomes by asking 2 yes-or-no questions.

Similarly, the sexes of the children in families of size three are equivalent to 3 coin flips. To learn about a family's children, we would need to ask 3 questions. The same logic applies for any number of children. In the general case, to learn the sexes of N children, we would need to ask N questions.

Notice that those N questions distinguish among 2^N possible birth orders. That mathematical relationship is the key to understanding the entropy measure: N binary random events produce 2^N possible outcome sequences, and, equivalently, we could learn the outcome sequence by asking N questions. For this reason, information entropy assigns an uncertainty level (and an information content) of N to an equal distribution over 2^N outcomes.

To capture that relationship in formal mathematics, we first note that each of the outcome sequences has a probability of $\frac{1}{2^N}$. To convert this to N requires the rather complicated expression $N = -\log_2(\frac{1}{2^N})$.[2] We can generalize this construction to arbitrary probabilities. If an outcome sequence arises with probability p, then we assign an uncertainty $\log_2(p)$ which approximates the number of yes-or-no questions required to identify the sequence To compute the information entropy of a distribution, we average the expected number of questions across all outcomes, or, as in the example, sequences of outcomes.

Information Entropy

Given a probability distribution $(p_1, p_2, ...p_N)$, the **information entropy**, H_2, equals:

$$H_2(p_1, p_2, ...p_N) = -\sum_{i=1}^{N} p_i \log_2(p_i)$$

Note: the subscript 2 denotes the use of the base 2 logarithm.

At first, the mathematical representation complicates more than it clarifies. Working through an example makes the formula more intuitive. Imagine that families who first have a girl stop having children, and that families who first have a boy have two more children. Half of all families will have a single girl. The half will be split evenly among four outcomes: three boys, two boys followed by a girl, a boy followed by two girls, and a boy followed by a girl followed by another boy. Each of those four outcomes occurs with probability $\frac{1}{8}$.

Information entropy equals the expected number of questions we must ask to learn the family's children. We would first ask if the first child is a girl. With probability $\frac{1}{2}$ the answer is yes, and we need not ask more questions. Thus, half of the time, we ask one question. We can write this as $-\frac{1}{2}\log_2(\frac{1}{2})$. If the answer is no, we must ask two more questions for a total of three questions. Each of those four cases occurs with probability $\frac{1}{8}$, so each contributes $\frac{1}{8} \times 3$ to information entropy. We write each as $-\frac{1}{8}\log_2(\frac{1}{8})$. Information entropy equals 2, the sum of the five terms.[3] Notation and logarithms aside, the intuition should be clear: information entropy corresponds to the expected number of yes-or-no questions. If we have to ask a lot of questions, the distribution is uncertain. Knowing the outcome reveals information.

Axiomatic Foundations of Entropy

Axiomatic Foundations: Entropy

$$H_a(p_1, p_2, \ldots p_N) = -\sum_{i=1}^{N} p_i \log_a(p_i) \text{ where } a > 0.$$

The above class of **entropy measures** uniquely satisfies the following four axioms:

Symmetric, continuous function: $H(\sigma(\vec{p})) = H(\vec{p})$ for any σ that permutes the probabilities.

Maximization: $H(\vec{p})$ is maximized at $p_i = \frac{1}{N}$ for all N.

Zero Property: $H(1, 0, 0, \ldots, 0) = 0$.

Decomposability: If $\vec{P} = (p_{11}, p_{12} \ldots, p_{nm})$:

$$H(\vec{P}) = H(P_1, P_2, \ldots, P_N) + \sum_{i=1}^{N} H(Q_{P_i})$$

where $P_i = \sum_{j=1}^{m} p_{ij}$ and $Q_{P_i} = (\frac{p_{i1}}{P_i}, \frac{p_{i2}}{P_i}, \ldots, \frac{p_{im}}{P_i})$.

To arrive at a general expression for entropy, we take an axiomatic approach. Claude Shannon imposed four conditions on his measure. The first three are easy to understand. It needed to be continuous and symmetric, maximized when outcomes occur with equal probability, and equal zero for certain outcomes. The fourth condition (decomposability) requires that the entropy of a probability distribution defined over n categories each with m subcategories equals the entropy of the distribution over the categories plus the sum of the entropies of each of the subcategories. This is a natural assumption for products of distributions. For example, in the case where outcomes are the product of two independent events, the assumption implies that the information content of the joint event equals the sum of the information contents of each event separately. Shannon then proved that a general class of *entropy measures* uniquely satisfies those axioms.

As was the case for the axioms that characterize Shapley values, the contribution of these axioms resides less in their existence than in their reasonableness. A clever mathematician can always construct axioms that uniquely define a function. The first two axioms are difficult to question. We might quibble with the arbitrariness of setting the uncertainty of a known distribution at zero, but it is an appropriate benchmark. Another possibility would be to assign 1 as the uncertainty of a known distribution.[4] The decomposability axiom, though complicated to explain, is also difficult to challenge. The uncertainty of two combined random events should equal the sum of the uncertainties of each event. Overall, the axioms are more than defensible. They are, in fact, hard to dispute.

Using Entropy to Distinguish Classes of Outcomes

We now show how the entropy measure can help us to categorize empirical data and model output within Wolfram's four classes: *equilibrium, cyclic (periodic), random,* and *complex.*[5] In Wolfram's classification, a pencil resting on a desk is in equilibrium. The planets orbiting the sun are in a cycle. A sequence of coin flips is random, so are (approximately) stock prices on the New York Stock Exchange, as we shall learn in the next chapter. Finally, the neuronal firings in a person's brain are complex; they do not fire randomly, nor do they fire in a fixed pattern. Figure 12.2 represents these four categories graphically.

Equilibrium outcomes have no uncertainty, and therefore, have an entropy equal to zero. Cyclic (or periodic) processes have low entropy that does not change with time, and perfectly random processes have maximal entropy. Complexity has intermediate entropy—it lies between ordered and random. While entropy gives us a definitive answer in the two extreme cases, equilibrium and random, it does not for cyclic and complex outcomes. We will have to use other measures to distinguish those cases.

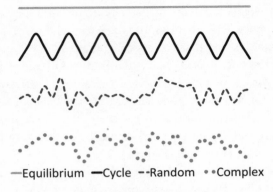

—Equilibrium —Cycle --Random • •Complex

Figure 12.2: Wolfram's Four Classes

To classify a time series of data, we calculate the information entropy across subsequences of different lengths. Suppose that a man keeps track of the type of hat he wears each day—either a beret (B) or a fedora (F). His choices over a year create a binary time series of 365 events. We can first calculate the entropy of sequences of length 1, that is, we calculate the entropy over the probability of wearing each type of hat. If we find that he is equally likely to wear each type of hat, the entropy over sequences of length 1 equals 1. We can therefore rule out equilibrium, as he changes his choices, but any of the other three categories are possible.

To determine the category, we next compute the entropy of sequences of length 2 through 6. If all have maximal entropy, then we can rule out a simple cycle. Suppose that as we consider longer sequences the entropy increases slowly until it reaches a maximum of 8. In other words, no matter how long the subsequence, the entropy never exceeds 8. An entropy of 8 is equivalent to an equal distribution across 256 outcomes. That cannot be a simple cycle. It is more representative of a complex sequence containing structure and patterns. We cannot say for sure that the time series is complex. It might be that the person is trying to be random, yet fails.

Maximal Entropy and Distributional Assumptions

Many of the situations that we model include uncertainty, and, as modelers, we must make assumptions about those distributions. As a rule, we want to avoid making ad hoc assumptions. It may be that we have some understanding of the process that produces the distribution. If so, we can often derive the statistical structure produced by that process using our logic-structure-function approach.

For example, suppose that we want to make an assumption about the distribution of the total value of the items up for auction at an estate sale. The total value equals the sum of the values of the individual items. We can therefore invoke the central limit theorem and assume a normal distribution. We might also assume a normal distribution for the possible values of a house, as the house's value depends on its attributes: the number of bedrooms, bathrooms, and the size of the lot.

A normal distribution may not make sense for the possible values for a piece of art or a rare manuscript. In those cases, we may have little understanding of the process that determines value. One approach is to assume a distribution with maximal uncertainty, that is, the maximal entropy distribution.

The shape of the maximal entropy distribution depends on the constraints. As we have already seen, if we assume a minimal and maximal value, the *uniform distribution* maximizes entropy. Many social science models in textbooks and journals assume uniform distributions. We might question that assumption on the grounds that few distributions in the real world are uniform. However, a *principle of indifference*—if we know nothing other than the range or set of possibilities—can justify the uniform distribution.

In some cases, we may know the mean of the distribution and also know that all values must be positive. Given those constraints, the maximal entropy distribution must have a long tail, and as we spread the distribution across more values, we must balance high values with many low-value outcomes. It can be shown that the entropy-maximizing distribution will be an *exponential distribution*. Thus, if we are writing a model that assumes distribution of website hits or market shares, in the absence of data an exponential distribution is a natural assumption.

Finally, if we fix the mean and the variance (and allow negative values), then the maximal entropy distribution is the normal distribution. The logic here is similar to the previous case. To create more uncertainty, we create extreme values. Here we can balance positive and negative values and not change the mean. However, doing so increases the variance, so we must add more values near the mean, resulting in a bell curve.

We can interpret these maximal entropy distributions within the logic-structure-function framework. If we thought that in a given social, biological, or physical context a micro-level process was maximizing entropy, then we should expect one of these distributions. Alternatively, we might assume a micro-level process and be able to show that entropy increases. If so, one of these distributions would emerge.

> # Maximal Entropy Distributions
>
> **Uniform distribution:** Maximizes entropy given a range, $[a, b]$.
>
> **Exponential distribution:** Maximizes entropy given a mean, μ.
>
> **Normal distribution:** Maximizes entropy given a mean, μ, and a variance, σ^2.

We can also interpret these results as exploratory. We may encounter data that is exponentially or normally distributed. Though we are not obliged to ask if some underlying behavior is increasing entropy subject to a constraint, we might gain a novel insight by doing so. Previously, we explained the normal distribution of heights, weights, and lengths of species by an appeal to the central limit theorem. Here we present a different, model-based explanation. If mutation maximizes entropy (to best explore niches), and if average size and total dispersion are fixed, then the distribution of sizes will be normal. The point is not that the maximal entropy approach offers a better explanation, but that maximizing entropy given constraints results in a normal distribution. So, when we see a normal distribution, it could be the result of entropy maximization.

Positive and Normative Implications of Entropy

We have seen how entropy measures uncertainty, information, and surprise, how it differs from variance, which measures dispersion, and how it can help us classify and compare classes of outcomes. Later, in Chapters 13 and 14, when we study random walks and path dependence, we use entropy to identify randomness and to measure the extent of path dependence. We can put the entropy measure to use in any number of real-world applications. We can measure whether an intervention in financial markets increases or decreases uncertainty. We can test whether or not outcomes in elections, sporting events, or games of chance are random.

In each of these applications, entropy functions as a positive measure. It tells us what the world is, not what it should be. Entropy in a system is not intrinsically bad or good. How much entropy we desire depends on the situation. In constructing a tax code, we might want an equilibrium pattern of behaviors. We would not want randomness. In designing a city, we may seek complexity. Equilibrium or even cycles would be dull. We would prefer a city to be teeming with life, to offer opportunities for fortuitous meetings and interactions. More entropy would be better, but only to a point. We would not want randomness. Randomness would make planning difficult and possibly overwhelm our cognitive abilities. Ideally, the world produces some complexity and we live in interesting times.

The architect Christopher Alexander shows how geometric properties such as strong centers, thick boundaries, and non-separateness can produce complex, living buildings, neighborhoods, and cities.[6] Alexander argues for complexity in cities and in living space. Central bankers may be less fond of complexity. They may prefer predictable equilibrium outcomes and stable growth paths. A central takeaway from this chapter is that we often care whether a system goes to equilibrium, produces a pattern or randomness, or whether it results in complex, novel sequences of patterns. By using models, we can perhaps see which will arise and, in some cases, design systems that produced the class of outcome we desire, whether that be complexity or equilibrium.

13. *Random Walks*

A drunk man will find his way home, but a drunk bird may get lost forever.

—Shizuo Kakutani

In this chapter, we learn two classic models from probability and statistics: the Bernoulli urn model and the random walk model.[1] Both models describe random processes even if it may appear that they are producing complex structures. Randomness can be hard to discern without gathering data. We often think we see patterns in election outcomes, stock prices, and scoring in sporting events, but instead, to borrow Nassim Taleb's lovely phrase, we are being fooled by randomness.[2]

The Bernoulli urn model describes random processes that produce discrete outcomes, like the flip of a coin or the roll of a die. Developed centuries ago to explain the odds of winning at gambling, it now occupies a central position in probability theory. The random walk model builds on that model by keeping running totals of the number of heads and tails. The model can capture the movement of particles in liquids and gases, the movement of animals in physical space, and growth in human height from birth to childhood.[3]

The chapter begins with brief coverage of the Bernoulli urn model along with an analysis of the length of streaks. We then describe the random walk model. We learn that one-dimensional and two-dimensional random walks return to their starting point infinitely often, while a three-dimensional random walk need not return home at all. We also learn that the time between returns to zero for a one-dimensional random walk will follow a power-law distribution. This finding, which we might be tempted to dismiss as a mathematical curiosity, can explain the life spans of species and firms. In the final section, we use the random walk model to evaluate the efficient market hypothesis and to determine the size of a network.

The Bernoulli Urn Model

The *Bernoulli urn model* consists of an urn containing gray and white balls. Draws from the urn represent the outcomes of random events. Each draw is independent of previous and future draws, so we can apply the *law of large numbers*: in the long run, the proportion of balls drawn of each color will converge to its proportion in the urn. That does not mean that a thousand draws from an urn containing seven white balls and three gray balls will produce exactly seven hundred white outcomes, only that the proportion of white balls will converge to 70%.[4]

Bernoulli Urn Model

Each period, a ball is randomly drawn from an urn containing G gray and W white balls. The outcome equals the ball's color. The ball is returned to the urn prior to the next period's draw. Let $P = \frac{G}{(G+W)}$ denote the proportion of gray balls. Given N draws, we can calculate the expected number of gray balls chosen, N_G, and its standard deviation, σ_{N_G}:

$$N_G = N \cdot P \quad \text{and} \quad \sigma_{N_G} = \sqrt{N \cdot P \cdot (1-P)}$$

Outcomes in the Bernoulli urn model produce streaks of predictable lengths. In an urn with equal numbers of gray and white balls, the probability of drawing a white ball equals $\frac{1}{2}$. The probability of drawing two consecutive white balls equals $\frac{1}{2}$ times $\frac{1}{2}$. In general, if a proportion P of the balls in the urn are gray, the probability of drawing N consecutive white balls equals P^N. By calculating probabilities, we can assess whether a streak was likely, amazing, or so improbable that we should expect fraud. When a basketball player makes a three-point shot nine times in a row, does he have a hot hand, or should we expect a random sequence of that length? The math shows that in a ten-year career, a good three-point shooter would be as likely as not to make nine in a row.[5]

We can make similar calculations to decide whether an investor has been lucky, good, or fraudulent. Berkshire Hathaway, the conglomerate run by Warren Buffett, outperformed the market forty-two out of fifty years from 1965 to 2014. A dollar invested in Berkshire Hathaway in 1964 was worth over $10,000 in 2016, while a dollar invested in the S&P 500 was worth about $23. If Berkshire had a 50% chance of beating the market, it should have outperformed the market twenty-five times during that fifty-year period, with a standard deviation of 3.5 years ($3.5 \approx \sqrt{50 \cdot \frac{1}{2} \cdot \frac{1}{2}}$).

The actual number of years Berkshire beat the market lies about four standard deviations above the mean, a one-in-a-million event. We can rule out luck. Given that Berkshire reveals its investments, we can also rule out fraud. Bernie Madoff did not reveal his investments. His proclaimed streak of successes—decades of consecutive positive returns—was so unlikely that his clients should have demanded transparency.[6]

Random Walk Models

Our next model, the *simple random walk model,* builds on the Bernoulli urn model by keeping running totals of past outcomes. We set the initial value, the state of the model, to be zero. If we draw a white ball, we add 1 to the total. If we choose a gray ball, we subtract 1. The state of the model at any time equals the sum of the previous outcomes (i.e., the total number of white balls drawn minus the number of gray balls drawn).

A Simple Random Walk

$$V_{t+1} = V_t + R(\text{-}1, 1)$$

where V_t denotes the value of the random walk at time t, $V_0 = 0$ and $R(\text{-}1, 1)$ is a random variable that is equally likely to equal -1 or 1. The expected value of a random walk in any period equals zero and has a standard deviation of $\sqrt{t}$, where t equals the number of periods.[7]

Figure 13.1 shows a simple random walk. The graph appears to have a pattern: a long downward trend followed by an upward trend and then a modest crash when the process crosses the zero line. That pattern happened by chance.

A simple random walk is both *recurrent* (it returns to zero infinitely often) and *unbounded* (it exceeds any positive or negative threshold). If we wait long enough, a random walk exceeds 10,000 and falls below negative 1 million. It also crosses zero infinitely often. In addition, the distribution of the number of steps required to return to zero satisfies a power law.[8] Most returns to zero occur in a few steps. Half of all walks return in two steps. Other walks, though, take a long time to return. That must be true given the unboundedness of random walks. A walk that crosses a threshold of 1 million would requires more than 2 million steps to reach that value and then return back to zero.

Figure 13.1: Plot of a Simple Random Walk for 300 Periods

The power law distribution result has unexpected applications. If we model firms' sales (or employees) as a random walk, firm life spans will be a power law. To be more precise, if we assume that when sales are strong, a firm adds an employee, that when sales are poor, a firm fires an employee, and that the firm closes when it no longer has employees, then the distribution of return times will equal the distribution of firm life spans, which will be a power-law distribution. And, to a first approximation, firm life spans are a power law.[9] We can apply the same logic to predict the life spans of biological taxa (kingdom, phylum, class, order, family, genus, and species). If the number of members of a taxon follow a random walk— for example, if the number of species in a genus goes up and down randomly—the taxon sizes should satisfy a power law. Here again, data are supportive.[10]

We can apply the model as an analogy by thinking of the random walk as a glacier moving along the ground. That model would predict that the distribution of sizes of glacial lakes would satisfy a power-law distribution. Each time the glacier falls below the land mass's surface and returns to the top, it creates a lake with a diameter that corresponds to the return time. Once again, data roughly align.[11]

The basic random walk model can be modified in several ways. We can create a *normal random walk* whose value in each period changes by an amount drawn from a normal distribution. A normal random walk will not return to zero exactly, though it will cross zero infinitely many times.

We can also make one outcome more likely than the other, producing a *biased random walk*. We can use biased random walks to predict the odds of winning in games of chance. In roulette, the probability of winning a bet on a red outcome equals $\frac{9}{19}$.[12] We can model the aggregate winnings (or losses) of a sequence of bets as a random walk that increases by 1 with probability $\frac{9}{19}$ (about 47.4%) and decreases by 1 with probability $\frac{10}{19}$. After 100 bets, the expected losses are $5 with a standard deviation of $10. We can be 95% confident of losing no more than $25 and winning no more than $15. After 10,000 bets, expected losses equal $526 with a standard deviation of 100. Therefore, 95% of the time we lose between $325 and $725.[13] Being ahead after10,000 equal bets is an event more than five standard deviations above the mean, a less than one-in-a-million possibility. It follows that to win at roulette, a person should make one big bet rather than many small bets.

Some sporting contests, such as basketball, can be modeled as two biased random walks. Each team has a probability of scoring on each trip down the court. That probability is estimated based on a profile of the team's offensive abilities and the opposing team's defensive abilities. We model a team's trip down the court as a random event. Each team's score corresponds to the value of its random walk. The team with the higher probability of scoring will be more likely to win. Analysis of data from the NBA reveals a close match to the model. Scoring deviates from random only when one team gains a huge lead, at which point the lead becomes more likely to decrease than increase. This phenomena could be explained by the winning team having less incentive to run up the score than the losing team has to make the score respectable.[14]

If we watch basketball, the outcomes seem far from random. Intelligent, athletic players run sophisticated offenses and make clutch plays. That is true, but the effects of effort may wash out. Extra effort to score on offense may be offset by extra effort on defense. A great steal may be wiped out by a player sprinting the length of the court to block a layup from behind. The model also suggests a strategy: stronger teams should speed up the game to create more possessions. Favored teams would rather spin the roulette wheel more often, as drift works to their advantage.

The simple random walk model takes place in a single dimension. We can also model higher-dimensional random walks. A two-dimensional random walk would begin at the origin in the plane, $(0, 0)$, and then walk randomly to the north, south, east, or west in each period. A two-dimensional random walk resembles a squiggly line drawn on a sheet of paper. Two-dimensional random walks also satisfy recurrence and unboundedness. Random search will locate a lost earring in your living room. The mathematical fact of recurrence enables random foraging as a strategy for ants.[15] If the two-dimensional random walk did not recur, ants would need more sophisticated internal maps or stronger pheromone trails to find their nests. In three dimensions, random walks do not satisfy recurrence. A fly skittering around a room and a molecule bouncing in the air return to their starting points a finite number of times—hence Kakutani's quote about drunken men and drunken birds at the start of this chapter.[16]

The lack of recurrence of random walks provides yet another example how models can clarify our thinking. Intuition tells us that recurrence should occur less often as we add dimensions. Logic reveals an abrupt change. In one and two dimensions, a random walk returns to its origin infinitely many times. In three dimensions, it wanders off forever. To arrive at that type of result requires mathematics. Intuition alone will be insufficient.

Using Random Walks to Estimate Network Size

We can exploit the recurrence of low-dimensional random walks to estimate a network's size. The method is straightforward. We select a node at random, start a random walk along the edges of the network, and keep track of how frequently it returns to the original node. The average time between returns correlates with the network's size. To estimate the size of a social network, we could ask someone to name a friend, and then ask the friend to name a friend. We could continue that process and keep track of how often we return to the same person.

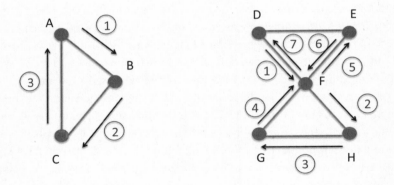

Figure 13.2: Random Walks on Networks

Figure 13.2 shows two networks. The network on the left has three nodes forming a triangle. The network on the right has six nodes forming two triangles. We can start a random walk on the left network at A. Suppose that it moves to B, then C, and back to A. The random walk returns to its starting point in three steps. On the network on the right, a random walk starting at D might follow a seven-step path $F-G-H-F-E-F-D$. If we repeat these experiments many times, the average return times on the network on the left will be shorter. Though unnecessary for small networks such as these, this method becomes useful on larger networks, like the World Wide Web or large email networks.

Random Walks and Efficient Markets

Stock prices prove to be nearly normal random walks with a positive drift to capture gains in the market. Many individual stock prices also are approximately random. Figure 13.3 shows the daily stock price data for Facebook for the year following its initial public offering on May 18, 2012. Facebook was offered at $42 per share. By June 1, 2012, the price had fallen to $28.89. One year later the price had fallen to $24.63. The figure also shows a random walk calibrated to have similar variation.

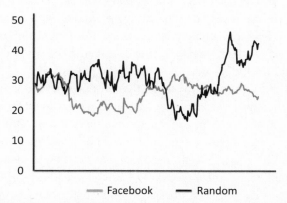

Figure 13.3: Facebook Daily Stock Price June 2012–June 2013 vs. a Random Walk

We can apply statistical tests to the sequence of Facebook share prices to see if it satisfies the assumptions of a normal random walk. First, the price should go up and down with equal probability. In the 249 trading days covered, Facebook's stock price went down on 127 days, or 51% of the time. Second, in a random walk, the probability of an increase should be independent of an increase that occurred in the previous period. Facebook's stock price moved in the same direction on consecutive days 54% of the time. Finally, the expected longest streak of moves in the same direction should be eight days. Facebook's stock price went up on ten consecutive days once during this period. Overall, we cannot reject the claim that Facebook's stock price is consistent with a normal random walk.

The same analysis can be done for daily prices in all stocks. To do so we must first subtract the mean upward trend in stock prices. Studies show that from the 1950s through the 1980s, daily stock prices had a slight positive correlation. After detrending, the probability of an increase following an increase exceeded $\frac{1}{2}$. From 1980 onward, as investors became more sophisticated, the probability of an increase following an increase fell to 50%, consistent with a random walk. The reason stock prices might follow a random walk pattern is that smart investors identify and therefore eradicate patterns. For example, in the 1990s, analysts noticed that stock prices rose at the beginning of each year, a phenomenon called the January effect. Smart investors could buy stocks in December at low prices and sell them in January for a profit. If that strategy seems too good to be true, it is. If investors buy stocks in December, they raise prices, wiping out the January effect. We should not be surprised that the January effect no longer exists.

Economists draw an analogy between recognizable persistent patterns in market prices and hundred-dollar bills on the sidewalk. If someone sees a hundred-dollar bill, she picks it up. When she does, it goes away. The same logic applies to patterns in stock prices: if they exist, they go away. A market with smart investors will therefore contain few predictable price patterns. If prices exhibit no pattern, what remains must be a random walk (with the caveat that the general upward market trend must be subtracted away).

Paul Samuelson wrote an early model that produced a random walk. His model did not require that investors know the value of the stock in all future periods, only that they know the distribution. As Samuelson himself notes, "One should not read too much into the established theorem. It does not prove that actual competitive markets work well."[17] Samuelson's reticence was not shared by everyone. Others extended this thinking to create the *efficient market hypothesis,* which states that at any moment in time the price of a stock captures all relevant information, and future prices must follow a random walk. The efficient market hypothesis rests on paradoxical logic.[18] Determining an accurate price requires time and effort. A financial analyst must gather data and construct models. If prices followed a random walk, those activities would have no expected return. However, if no one expends effort to estimate prices, then prices will become inaccurate and the sidewalk will be covered in hundred-dollar bills. In brief, the *Grossman and Stiglitz paradox* states that if investors believe in the efficient market hypothesis, they stop analyzing, making markets inefficient. If investors believe the market is inefficient, then they perform analyses by applying models, making markets efficient.

In point of fact, price movements are rather close to random walks, although sophisticated statistical techniques do reveal short-run patterns.[19] While there may be no hundred-dollar bills on the sidewalk, there are some four-leaf clovers in grassy fields that one can find by looking hard enough.

Critics of the hypothesis argue that some investors consistently win over longer periods than would be predicted by chance.[20] Furthermore, prices could move randomly for some other reason, such as the aggregation of sophisticated trading rules. Day-to-day price volatility exceeds the amount of information that flows into markets, and the market takes huge jumps and dives when little of relevance appears to be happening in the wider world, suggesting the presence of bubbles. One person's inconvenient facts can be another person's "these caveats notwithstanding." Yes, volatility is high, but small amounts of information can have large effects. And even though the market does takes big jumps and dives, the market could still follow a *longer-tail random walk,* where day-to-day movements come from a longer-tailed distribution.

Though it seems implausible to think that stock prices are accurate at all times, prices cannot diverge wildly from true values in the long run. We can see this by applying the rule of 72. If the economy grew by 3% per year, in half a century, the economy would increase 4-fold. If we go back to 1967, the United States GDP equaled about $4.2 trillion (in 2009 dollars). By 2017, GDP had increased to almost $17 trillion (in 2009 dollars), a 4-fold increase, exactly what we would expect given 3% growth. During that same period, the real value of stocks in the S&P 500 also increased about 4-fold. Had the stock market risen at 12% per year (in real dollars), then stock prices would have increased 256-fold, an impossibility.[21]

In the long run, assuming the efficient market hypothesis or something close to it is a reasonable assumption. In the short run, betting on prices correcting can be risky. The case of Long Term Capital Management (LTCM), a hedge fund whose board of directors included two Nobel Prize winners in economics, proves instructive. In 1996 and 1997, LTCM posted returns in excess of 40% in part by identifying inefficiencies and predicting the market would correct. In 1998, they noticed (correctly) that the price of Russian bonds was out of alignment with prices of US Treasury bonds. They bet big. However, a Russian default, the first since 1917, increased the misalignment in the short term. LTCM lost $4.6 billion and nearly caused a collapse of financial markets. Soon after LTCM was bailed out bond prices did align, though not soon enough. The lesson should be obvious: do not put too much faith in one model.

Summary

In this chapter, we learned the Bernoulli urn model and random walk model. We applied these models widely. We saw how to distinguish randomness from hot streaks, to develop strategies for gambling, to evaluate time series of stock prices, and to make sense of outcomes in basketball games. We also saw how to apply the power-law distribution of return times for a random walk to inform our understanding of the duration of firms and biological taxa.

From these applications, we see how the random walk model provides a useful frame for evaluating time series. We should not be fooled by a few years of success. It need not imply sustained excellence. In *Good to Great: Why Some Companies Make the Leap and Others Don't,* one of the best-selling business books of all time, Jim Collins identified characteristics of consistently successful companies, such as having humble leaders, getting the right people on the team, and maintaining discipline (what Collins called "rinsing your cottage cheese" in homage to six-time Ironman triathlon champion Dave Scott's habit of rinsing his cottage cheese to reduce the fat content). Collins singled out eleven great companies that kept to his principles. In the decade following the publication of his book, only one of the eleven produced strong growth. One was bought out. One was taken over by the government, and the other eight generated zero returns.

The fact that the great firms shared attributes does not imply that those attributes contribute to success. Perhaps the lowest-performing firms also share those attributes. Selecting the best firms and looking at their attributes is not model thinking. Model thinking would derive attributes that cause success, such as talented workers. It would then test those conclusions against data, and if possible look for natural experiments—instances where the relevant attributes change randomly. Other models, such as the dancing and rugged landscape models we cover in Chapter 28, call into question Collins' core assumptions. If the economy is complex, traits that prove successful today need not work in the future. What creates great success now—big rocks first—may not be a good strategy in ten years. As a rule, we should apply many models before making broad pronouncements, lest we risk correspondingly large errors. We should also avoid being fooled by patterns. What appears to be a trend might well be random.

14. *Path Dependence*

No man ever steps in the same river twice, for it's not the same river and he's not the same man.

—Heraclitus

In this chapter, we cover models of path dependence. In any domain in which people base their behavior on the actions of others, be it international affairs, art, music, sports, business, religion, technology, or politics, we should expect some degree of path dependence. A college student's choice of courses point her toward some career paths over others. An endorsement of a candidate may launch a political career. A friendship may lead to other social connections. The clothes we wear, the books we read, the movies we watch, and the activities that consume our time all exhibit some degree of path dependence.

Path dependence also exists on grander scales. Common-law rulings establish and reinforce precedents, influencing future rulings.[1] Early institutional forms impact later institutional choices. The decision in the United States to provide health insurance through private firms resulted in a large private health insurance industry, health maintenance organizations, and a mix of public and private hospitals.[2] Institutions also induce behavioral patterns, such as selfish or cooperative tendencies, that can in turn influence the efficacy of future institutions.[3]

In this chapter, we build dynamic urn models that produce sequences of outcomes that exhibit path dependence. These models extend the Bernoulli urn model by allowing the distribution of balls within the urn to change as a function of past outcomes. With these models to structure our thinking, we then provide a formal definition of path dependence and distinguish path-dependent outcomes from path-dependent equilibria. These formal definitions differentiate path dependence from tipping points, which are more abrupt changes in outcomes.

The chapter consists of four parts. The first two cover the Polya process and the balancing process. The Polya process assumes positive feedbacks and produces

163

both path-dependent outcomes and equilibria. Many of the canonical examples of path dependence, including the growth of the QWERTY typewriter, are based on positive feedbacks, also known as increasing returns. The balancing process assumes negative feedbacks and produces path-dependent outcomes but not path-dependent equilibria. The third part defines a measure of path dependence based on entropy. The final section discusses further applications of the models.

Polya Process

The *Polya process* captures positive feedbacks using an extension of the Bernoulli urn model in which we add a ball to the urn that matches the ball chosen. This process generates *outcome path dependence*, where outcomes in each period depend on previous outcomes. It will also be true that the long-run distribution over outcomes—*equilibrium path dependence*—depends on outcomes.[4] The distinction between these two types of path dependence will be central to what follows. A process that is equilibrium path dependent must be outcome path dependent. If outcomes in the long run depend on the path, then so must outcomes along the way. A process can be outcome path dependent but not equilibrium path dependent. What happens now could depend on the past, but the long-run equilibrium might be determined at the outset.

The Polya Process

An urn contains one white ball and one gray ball. Each period a ball is drawn randomly and returned to the urn along with an additional ball of the same color as the one drawn. The color of the ball drawn denotes the **outcome**.

The Polya process captures a variety of social and economic phenomena. A person's choice of whether to learn to play tennis or racquetball could depend on the choices of others. A person might be more likely to choose tennis if more of her friends also choose tennis, as it increases her chances of finding a game. Similarly, a person's decision about what type of software to buy, language to learn, or smartphone to buy could also depend on earlier choices by friends. Similar logic also applies to choices by firms over which technological standards to adopt. They may base their choices on the actions of others.

The model captures these social influences by changing the distribution of balls. If gray balls represent people who choose tennis and white balls represent people who choose racquetball, then as more people choose tennis, the urn contains more gray balls, causing subsequent people to be more likely to choose tennis as well. This increasing pull toward the outcome that more people choose creates path dependence.

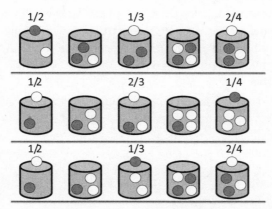

Figure 14.1: Outcomes Consisting of Two White Balls and One Gray Ball

We can derive two unexpected properties of the Polya process. First, any sequence with the same number of white outcomes occurs with equal probability. Second, every distribution of white and gray balls occurs with equal probability. The second property implies extreme path dependence. Anything can happen. Everything is equally likely. After 1,000 periods, the probability that the urn contains 40% white balls equals the probability that it contains 2% white balls.

To see why, consider all possible sequences of outcomes in the first three periods. The first period outcome is gray with probability $\frac{1}{2}$. If so, we add a gray ball, increasing the probability that the second outcome will be gray to $\frac{2}{3}$. If that outcome is also gray, we add a third gray ball, increasing the probability that the third outcome is gray to $\frac{3}{4}$. It follows that the total probability of three gray balls (or three white balls) equals $\frac{1}{2}$ times $\frac{2}{3}$ times $\frac{3}{4}$, which equals $\frac{1}{4}$.

The three sequences in which the first three outcomes consist of two white balls and one gray ball are shown in figure 14.1. In the top row, the order of the outcomes is gray, white, and then white. The probability of this sequence is $\frac{1}{12}$, as is the probability of other sequences. It follows that the probability of getting one of the three sequences equals $\frac{1}{4}$. By symmetry, the probability of choosing two gray balls and one white ball also equals $\frac{1}{4}$. Therefore, each set of outcomes—three white, three gray, two white and one gray, and two gray and one white—occurs with the same probability of $\frac{1}{4}$. Moreover, sequences of two white and one gray also occurs with equal probability. Similar results can be shown for any number of periods.[5]

If we extend the *Polya process* to add balls of additional colors, extensions of both regularity properties still hold. Any proportion of the colors can arise and is equally probable. These results create a conundrum for producers of consumer products. Long-run consumer preferences for some product attributes may be random. Knowledge that an outcome cannot be predicted can still inform action. Ford would not want to build 40,000 yellow pickup trucks and later find that red emerged as the favorite color from a path-dependent process. The potential for unsold inventory in unwanted colors points to two potential actions. A company could construct its supply chain so that color choices come last; for example, a clothing company might wait to dye sweaters until popular colors become clearer. Or a company could choose to not give people a choice. Henry Ford offered his customers any color Model T they desired, so long as it was black. Apple did the same when it rolled out the first iPhone: you could get black, or, for the same price, you could get black.

The Balancing Process

Our second model, the *balancing process,* makes the opposite assumption of the Polya process. After drawing a ball of one color, we add a ball of the opposite color. If we draw white balls in the first two periods, the urn will contain three gray balls and only one white ball, resulting in a $\frac{3}{4}$ probability of drawing a gray ball. This process produces path-dependent outcomes, in that the likelihood of an outcome in any period depends on the history of past outcomes. However, it does not produce path-dependent equilibria. In the long run, the urn converges to equal proportions of each color ball.[6]

> ## The Balancing Process
>
> An urn contains one white ball and one gray ball. Each period a ball is drawn randomly and returned to the urn along with an additional ball of the color opposite to the color drawn. The color of the ball denotes the outcome.

The balancing process captures sequences of decisions or actions that include pressures toward equal allocation. Parents with two children may try to give equal time to each. Spending an afternoon with one child creates a desire to spend more time with the other child. The balancing process could even model organizational efforts to achieve equity. The International Olympic Committee (IOC) would like every region of the world to host games. In 2013, the IOC announced that Tokyo had been selected as the host city for the 2020 Summer Olympic and Paralympic Games. Two European cities, Istanbul and Madrid, lost. Four years later, the IOC awarded Paris the 2024 games and a North American city, Los Angeles, the 2028 games. Tokyo won the 2020 games in part on the strength of its proposal and in part because the Summer Games had not been held in Japan since 1964. Geographic fairness appears to exert sway. Europe, Asia and Oceania, and the Americas have hosted the games approximately equal numbers of times in the period following World War II. Europe has been awarded the games eight times, the Americas six times, and Asia and Oceania seven times.

Path Dependence or Tipping Point

Path dependence, a gradual effect on outcomes, differs from a tipping point, an abrupt change in outcomes. The growth of Microsoft provides a good example of path dependence. Founded in 1975, Microsoft developed interpreters for the BASIC computer language. In 1979, Microsoft inked a deal with International Business Machines (IBM) to provide the operating system for IBM's personal computer. This deal set Microsoft on a path that transformed a company with forty employees into one of the most valuable companies in the world.

The IBM contract contributed to Microsoft's upward path but did not guarantee long-term success. At the time, the personal computer market was small. The internet did not exist, nor did sophisticated word processing, business software, or video games. Moreover, the success of the personal computer depended in part on the DOS operating system that Microsoft developed. As the personal computer market grew, other companies developed software compatible with DOS, providing more positive feedbacks. These events—the success of DOS, the growth of the personal computer market, and the development of software running on the DOS platform—can be thought of as one color of ball being consistently drawn from the urn. Each outcome made the next more likely. The computer age may have been inevitable, but Microsoft's central role and the growth of the personal computer represent one of many potential paths.

We can contrast the path dependence of Microsoft's growth with the assassination of Archduke Franz Ferdinand on June 28, 1914, which many see as a tipping point that led to World War I. Six years prior to the assassination, Austria-Hungary had annexed Bosnia and Herzegovina. Among the Serbians unhappy with that development was Gavrilo Princip, who shot and killed Franz Ferdinand and his wife, Sophie. Austria-Hungary blamed Serbia, a near-inevitable reaction, and then turned to Germany's Kaiser Wilhelm for assurance as they prepared for war against

Serbia. Tensions escalated. Serbia had an alliance with Russia, which in turn had alliances with France and the United Kingdom. By August 2, Germany had declared war on France. On August 3, Belgium refused to grant Germany free passage into France, and full-scale war began. This vastly simplified version of events suggests that given the alliances, the killing of the archduke tipped the world toward war.

We can measure path dependence and tipping points through changes in the probabilities of the possible outcomes.[7] For the Polya process, the initial probability distribution is uniform over all distributions in the urn. This is the maximum entropy distribution. As events unfold, the distribution slowly narrows, indicative of path dependence: what might happen changes as outcomes occur. The reduction in entropy is gradual. At a tipping point, the probability distribution changes abruptly. Entropy may fall quickly. Figure 14.2 demonstrates the difference in two processes, each with two possible outcomes. After an event occurs—the contract for Microsoft or the killing of the archduke—the probabilities of each change. Subsequent events also change the probabilities. The process with the tipping point has a sharp kink. The path-dependent process changes slowly.

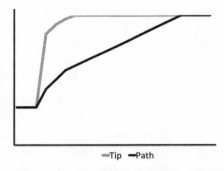

Figure 14.2: A Tipping Point vs. Path Dependence

Further Applications

In real situations, path dependence may not be as extreme as in the Polya process. Nevertheless, we can infer from the model that when behavior has a large social component, almost anything can happen. On one college campus, most students may wear black winter parkas; on another, they may wear blue peacoats. Model thinking suggests that differences could be the result of social influence as much as distinct underlying preferences. That holds in any context where people choose among a fixed set of options and their choices depend on the choices made earlier by others. Examples include democratic elections, which movie to see, and which technology to purchase.

The model can be extended to allow social influence to vary by the alternative chosen. Vanilla ice cream may have a constant level of feedback. The more exotic green tea ice cream may generate more variation in feedback: a friend may not like it and discourage you from trying it, or a friend may love it and encourage you to order it. It can be shown that less variation in feedbacks increases the likelihood of choosing an outcome.[8] The model can also be changed so that people differ in their susceptibility to social influence; some people give more or less weight to the balls added to the run.

In any variation of the model, we can measure (or estimate) the extent of path dependence and compare it to other versions. If the assumptions we make in constructing a model of new product introductions shows that outcomes depend on the early part of the path, then entering, intervening, or subsidizing early may be a good strategy. The model provides a logic for companies to rush their products to market or offer steep discounts to generate early adopters. Other assumptions may show that having the better product may matter more than entering early and that the better strategy is to focus on quality. By using models, we can identify the features relevant to a particular situation—the relative importance of individual preferences and social effects, the variation in feedbacks, and the relative differences in quality—and deploy that knowledge to inform strategy and guide data collection.

On a final note, the Polya process shows how positive feedbacks can produce path-dependent outcomes and equilibria. Path dependence arises in a far broader set of contexts. Some degree of path dependence (in outcomes if not equilibria) occurs whenever one action bumps into or interacts with future actions. That is true when making decisions on large public projects.[9] The decision to build a park or a highway constrains future planning decisions. The extent of that path dependence generally will depend on the size of the project. Central Park has had a profound impact on how New York City has developed. While the Polya process reveals the core idea that interactions produce path dependence, we need more realistic models for that insight to guide action.

Value at Risk and Volatility

We can interpret the standard deviation in a time series of data as *volatility*. Investments in stocks, real estate, and privately held businesses all exhibit volatility. *Value at risk (VaR)* measures the probability of a loss of a given amount during a specific time period. An investment with a one-year 5% VaR of $10,000 has a 5% probability of losing more than $10,000 at the end of one year.[10] Banks use VaR calculations to determine the amount of assets that must be kept on hand to avoid bankruptcy. For example, to secure an investment with a two-week 40% VaR of $100,000, an investor may be asked to hold $100,000 in cash.

If an investment follows a simple random walk with an increase or decrease of size M each period, then it has an N period 2.5% VaR of $2M\sqrt{N}$.[11] Thus, an investment that randomly goes up or down $1,000 each day has a nine-day 2.5% VaR of $6,000, and a one-year 2.5% VaR of $38,000. Notice that VaR increases linearly in the size of the steps but that it increases like the square root of the number of periods. We can use the formula for VAR to explain why the FDIC only requires that banks hold around 2% of their assets in cash overnight, but banks require that consumers put down 20% deposits on houses. The duration on the overnight loans is one day. Home loans can last for over a decade. The square root of three thousand and sixty-five days is approximately sixty.

Here, we have assumed a normal random walk. Analysts calculating VaR often consider the past empirical distribution of returns. If the empirical distribution has a longer tail, that is, if it includes more large events, then VAR would increase as large events are more likely.

Though VaR originated in finance, the idea can be applied broadly. A nonprofit that operates a volunteer-led Saturday morning soup kitchen that requires twenty-five volunteers might want to know the likelihood of lacking sufficient volunteers. If the number of volunteers follows a simple random walk that increases or decreases by 1 each week, then using the formula for VaR above, and setting $M = 1$ and $N = 52$, we find a one-year 2.5% VaR of 15, implying the nonprofit has a 2.5% chance of a volunteer shortage.

15. *Local Interaction Models*

Every generation laughs at the old fashions, but follows religiously the new.

—Henry David Thoreau

In this chapter, we study two models of local interactions, the *local majority model* and the *Game of Life*. These models both take place on a checkerboard consisting of cells that can be in one of two states. Otherwise, the models could not be more different. In the local majority model, cells update by matching the state of the majority of their neighbors. In the Game of Life, cells rely on a more complicated rule with multiple thresholds. The outcomes of the models also differ. The local majority model always converges to an equilibrium, while the Game of Life, depending on its initial configuration, can produce any class of outcome: equilibria, cycles, complexity, or randomness.

The local majority model can be used to explain and predict real-world outcomes in social and physical systems. It can represent discrete choices by conforming individuals or capture physical systems such as spin glasses, where magnetic entities align with neighbors. In contrast, the Game of Life is purely exploratory. It was developed to explore how simple rules can aggregate to produce complex phenomena. In the Game of Life, the periodic patterns, complex sequences, and randomness emerge from the interactions. The model shows how the whole can be different in kind from the parts. As a crude analogy, the human brain also produces emergent phenomena such as emotion, cognition, and consciousness from much simpler parts.

We begin by analyzing the local majority model. We show how a standard coordination game provides microfoundations for the behavioral rule assumed in the model. We can thus interpret the actors in the model as either rule-following agents or rational actors applying a best-response strategy. We then describe the Game of Life and show how it produces complexity from simple rules. The discussion at the end of the chapter highlights the value of exploring with local interaction models.

The Local Majority Model

The *local majority model* assumes cells arrayed on a checkerboard.[1] Each cell is in one of two states, which we refer to as *on* or *off*. Initially we assign states randomly; thereafter, a cell's state depends on the states of its neighbors. The neighbors can be defined in several ways. We take the neighbors of cell C to be the four cells to the north, south, east, and west as well as the four diagonally adjacent cells, creating a neighborhood of size eight.

Local Majority Model

Each cell on a two-dimensional square grid is in one of two states: on or off. Each cell has eight neighbors (shown in the diagram below).[2] In each period, a cell is chosen randomly.[3] The cell changes its state if and only if five or more of its neighbors are in the other state.

1	2	3
4	C	5
6	7	8

The local interactions in the local majority model includes positive feedbacks: cells match the state of other cells. Figure 15.1 shows a typical equilibrium configuration of the local majority model.

In equilibrium, every cell's state matches the state of a majority of its neighbors. Equilibrium configurations resemble the black-and-white patchiness of a Holstein cow. While the equilibrium configuration depends on the initial configuration of the cells, the model does not exhibit extreme sensitivity to initial conditions. Switching the state of one cell results in at most small changes in the final configuration. The pattern also depends on the order in which cells are activated. Thus, the model exhibits path dependence. The number of equilibria is enormous. Two equilibria produced by the model look no more alike than two Holsteins in a field.

The model was developed to capture physical systems where each cell's state represents an atomic spin—think of each cell as a magnet with either a negative or positive charge. Each magnet resides in a local magnetic field that physically drives

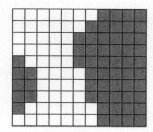

Figure 15.1: Equilibrium Pattern in the Local Majority Model

it to match the spins of its neighbors. The same model can also represent glasses and crystals.

Here, we use the model to capture local coordination or conformity among people. We think of each cell as representing an individual's action. The action could be any convention such as shaking hands or bowing, interrupting or raising one's hand. A person wants to choose an action that matches those of her neighbors. The checkerboard represents the social network. The checkerboard would be an appropriate social network for a homeowner's decision to maintain a clean yard, plant trees, or practice ecological landscaping or for people in an auditorium deciding whether to give performers a standing ovation.[4] While the checkerboard is at best a crude approximation, with it we gain some core intuitions.

If we run the model on a computer, we find it always goes to a patchy equilibrium configuration. In Chapter 16, we learn why. In the physical interpretation of the local majority model, the patchy equilibrium pattern corresponds to a *frustrated state*. Many cells have some neighbors in the on state and some in the off state. If we interpret the model through a social lens, the frustrated state can be seen as a *suboptimal equilibrium*. If being on corresponds to greeting people by shaking hands and being off corresponds to greeting people by bowing, then people on the boundaries of the patches experience awkward interactions with some of their neighbors: they bow when others shake, or they shake when others bow. People would be happier overall if everyone chose the same action—that is, if they solved the coordination game. The suboptimal equilibria, the frustrated state, arises because the interaction effects apply locally. If, instead, cells matched the global majority, then very quickly all of the cells would be in the same state. That insight implies that creating common behaviors may require broad influence networks. If people coordinate with their local neighbors, they create pockets of diverse behaviors. Paradoxically, coordination results in diversity.

Pure Coordination Games

In a *pure coordination game*, each player chooses one of two actions, *A* or *B*. If both players choose the same action, each receives a payoff of 1. If they choose different actions, each receives a payoff of zero.

Actions	A	B
A	**1**, 1	**0**, 0
B	**0**, 0	**1**, 1

A pure coordination game has two efficient equilibria: both players choose *A* or both players choose *B*. It also has an inefficient equilibrium, in which each player randomizes between *A* and *B*. We can reinterpret the local majority model with each cell being a player who must choose a common action to play against her eight neighbors. If players can change their action only when randomly activated, a player could increase her payoff by choosing the action that matches a majority of her neighbors' actions. Such a strategy is called a *myopic best response* because it does not take into account the likely future actions of the neighbors. A player with five neighbors who have chosen *B* could increase her payoff in the short term by switching from *A* to *B*, but if the player and her neighbors are surrounded by a sea of other players choosing *A*, then she might have a higher expected payoff by staying with *A*. The key takeaway is that the behavioral rule in the local majority model, though an assumed rule, can be rooted in a game theoretic model.

The *paradox of coordination* explains differences across groups as idiosyncratic. For some actions—whether your soy sauce or ketchup is stored in the cupboard or in the refrigerator, or whether people wear their shoes in your house or leave them at the door—it is sensible to coordinate with others. The resulting regional variety adds richness to our lives. The tiny ristretto in Italy, the midsized espresso in France, and the enormous *kawa ze smietanka* (coffee with cream) in Warsaw add to the pleasure of traveling around Europe.

Other differences, though, can be inefficient. Variations of electrical plugs—two prongs here, three prongs there—can be maddening. As the world becomes more integrated, technological coordination failures can be costly. The Swedes decided to switch from driving on the left to driving on the right to match the rest of continental Europe. The switchover, known as Dagen H, occurred at 4:45 a.m. on September

3, 1967. Every car in Sweden—and many Swedes were on the road in the early morning hours to participate in the event—came to an abrupt stop, and then, over the next fifteen minutes, all of the cars maneuvered from the left to the right side of the road. At 5:00 a.m., the cars began moving again on the opposite side of the road. Despite the incentives to coordinate, sometimes people fail to do so. The people of England, though connected by tunnel to the continent, continue to drive on the "wrong" side of the road, as do the island inhabitants of some, though not all, of their former colonies.

The Paradox of Coordination

If people coordinate locally, then global configurations will be patchy and diverse.

When applying this model, we must keep in mind that many coordinated cultural practices, such as how people mourn their dead or celebrate the birth of a child, are not idiosyncratic curiosities but components of culture, a coherent constellation of behaviors, practices, and artifacts that define who a people are and give them a sense of meaning and belonging.[5]

As we can with any model, we can experiment with parameters and see how doing so affects the results. For the local interaction model, the size of the patches that form in equilibrium increase faster than the neighborhood size. If we make the neighborhoods, that is, the number of squares that influence a square on the grid, twice as large, the patches become more than twice as large. The model therefore suggests that as technology and urbanization bring us closer together, the force of coordination could result in larger homogeneous patches of behaviors and beliefs.

Experiments also show that if we make the configuration a long, narrow rectangle, the model tends to produces horizontal and vertical stripes, as shown in figure 15.2.[6] The zebra-like stripes are an equilibrium because each on (off) cell has five on (off) neighbors. This type of pattern would also be an equilibrium on the square, though it rarely occurs. Perplexing findings like this can result in deep dives into rabbit holes of little empirical or theoretical value. They can also provide insights that lead to deeper, unexpected discoveries.

In this instance, the "squares produce Holstein-style patterns and skinny rectangles give zebra-style patterns" result all but begs us to ask if models like this could explain patterns on animal hides. A foray into the literature shows that they can.[7]

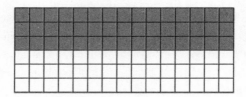

Figure 15.2: Stable Lines in the Local Majority Model

The Game of Life

Our next model, the *Game of Life,* also assumes cells on a checkerboard that are in one of two states. The key differences will be that the cell's rule for updating has two thresholds and that all cells update their states synchronously. Thus we can speak of the initial configuration, the configuration at time 1, the configuration at time 2, and so on. Synchronous updating can be thought of as "marching band dynamics" (update! update! update!).[8]

The Game of Life

Each cell on a dimensional square grid is either alive (on) or dead (off). Each cell's neighbors consist of the eight adjacent cells on the grid. Cells update their states synchronously using two rules:

Life rule: A dead cell with exactly three live neighbors becomes alive.

Death rule: A live cell with fewer than two or more than three live neighbors dies.

Start with three live cells in a horizontal row, as shown in figure 15.3. In the next period, we get a vertical row of three cells as seen by applying the rules for life and death to each cell. The live cell in the middle has two live neighbors, so it remains alive. The two live cells at the ends each have one live neighbor, so they die. Finally, the cells above and below the live cell in the center both come to life because each has three live neighbors. By symmetry, another update returns to the horizontal row of three cells. If we continue to iterate the rules, the pattern alternates between a horizontal and vertical line—that is, it blinks.

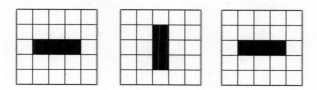

Figure 15.3: A Blinker in the Game of Life

The blinker results from the interactions of cells. It is not assumed. Complex systems scholars refer to this sort of macro-level phenomenon as *emergent*. Blinkers are among the more common and least impressive of the emergent structures produced by the Game of Life. Figure 15.4 shows three other simple configurations: a *block,* a *glider,* and the *R-pentomino.* The block is an equilibrium configuration. Each live cell has exactly three live neighbors, and each dead cell has at most two live neighbors. No live cells die, and no dead cells come to life. The middle configuration produces a cycle of size 4 that glides diagonally one cell down and to the right. More elaborate configurations, called *glider guns,* produce an endless stream of gliders. The third configuration, the R-pentomino, creates a complex sequence of patterns. If we run the model for more than a thousand steps on a large grid, it generates gliders and blinkers as well as several small, stable configurations. The Game of Life can also produce randomness.[9] Thus, the Game of Life can produce any class of outcome depending on the initial state.

These capabilities raise philosophical questions. The Game of Life consists of two-state cells arranged on a grid that update using simple rules. It can produce elaborate patterns and, with appropriate coding, it can be turned into a universal computer. The initial pattern can be thought of as the input. The rules produce an outcome that can be interpreted as a calculation. We can therefore draw a crude analogy between the model and the human brain, which also consists of spatially connected simple parts that rely on threshold-based rules, albeit more complicated ones. That is not to say that the patterns we see in the Game of Life can explain consciousness. No book exists that is titled *The Game of Life: Consciousness Explained,* though Daniel Dennett did write a book called *Consciousness Explained* in

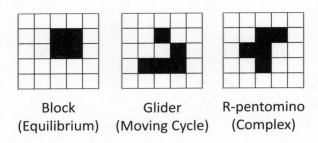

Block · Glider · R-pentomino
(Equilibrium) · (Moving Cycle) · (Complex)

Figure 15.4: Patterns in the Game of Life

which he posits that simple models like the Game of Life can provide insight into how consciousness might have evolved, an insight echoed by the physicist Stephen Hawking, who wrote, "It is possible to imagine that something like the Game of Life, with only a few basic laws, might produce highly complex features, perhaps even intelligence."[10]

Summary

In this chapter, we studied two models of interacting cells arranged on a grid. The first model, the local majority model, always goes to one of many possible equilibria, and we can interpret the model as analogous to a variety of physical and social processes. The second model, the Game of Life, can produce any class of outcome, from equilibria to randomness. That model claims no explicit connection to the real world. It provides an example of how constructing an alternative reality can produce insights—the emergence of dynamic macro-level structures from micro-level rules—that deepen our understanding of the world. As the Game of Life shows, the whole can perform functions that far exceed the capacities of its parts. If, for example, we create a slanted figure eight by connecting two 3-by-3 boxes at their corners, the Game of Life produces a cyclic pattern of length eight. It cycles through a set of patterns and then returns to the figure eight in exactly eight steps. That a pattern resembling an eight acts "as if" it counts to eight is quite amazing.

To understand how and why the Game of Life produces complexity while the local majority model inexorably moves to an equilibrium, we need additional analytic tools and frameworks. In Chapter 16, we introduce Lyapunov functions, which use difference equations to classify the state of the world. By careful construction of a Lyapunov function, we can explain why the local majority model must head to equilibrium and also why the Game of Life need not.

As a final note, the salience of the question of whether models, and by extension the real world, produce equilibria, patterns, complexity, or randomness arose naturally from our explorations with models. As we explored, we found some models go to equilibria and others do not. We think of using models to answer questions. In this chapter, we saw how models can raise questions as well.

16. *Lyapunov Functions and Equilibria*

The beauty of mathematics only shows itself to more patient followers.

—Maryam Mirzakhani

In this chapter, we learn Lyapunov functions, which provide conditions for a model to achieve equilibrium. Lyapunov functions are real-valued functions defined over the configuration system that are indexed by time. In each time step, a Lyapunov function assigns a value to the configuration. If the configuration changes—that is, if the model is not at an equilibrium—then the Lyapunov function's value decreases by a fixed amount. A Lyapunov function also has a minimum value, which means that eventually its value must stop decreasing. When that happens the model reaches an equilibrium. We can use Lyapunov functions to show, for example, why the local majority model converges.

The key takeaway from this chapter will be that if we can construct a Lyapunov function for a model, the model must go to an equilibrium. We cannot get a periodic orbit, randomness, or complexity. Even more, we can also bound the time to convergence to that equilibrium as we show when we construct a Lyapunov function for the local majority model.

The chapter has six parts. We first define Lyapunov functions and then apply them in the Race to the Bottom Game. We then construct Lyapunov functions for the local majority model and a model of ordering activities. In the fourth part, we show why we can construct Lyapunov functions for some exchange markets, and why we cannot for others. We then see why the Game of Life lacks a Lyapunov function as well. We then discuss a deceptively vexing mathematics problem that always goes to an equilibrium for which no Lyapunov function has been found. The chapter concludes by returning to the question of whether equilibria are desirable.

Lyapunov Functions

A *discrete dynamical system* consists of a space of possible *configurations*—think of these of as multidimensional states of the world such as an initial collection of live and dead cells in the game of the life—along with a *transition rule* that maps the configuration at time t into a configuration at time $t + 1$. A *Lyapunov function* maps configurations into the real numbers and satisfies two assumptions. First, if the transition function is not at an equilibrium, the value of the Lyapunov function falls by a fixed amount (more on that in a moment). And second, the Lyapunov function has a minimum value. If both assumptions hold, then the dynamical system must attain an equilibrium.

Lyapunov Theorem

Given a discrete time dynamical system consisting of the transition rule $x_{t+1} = G(x_t)$, the real-valued function $F(x_t)$ is a Lyapunov function if $F(x_t) \geq M$ for all x_t and if there exists an $A > 0$ such that

$$F(x_{t+1}) \leq F(x_t) - A \text{ if } G(x_t) \neq x_t.$$

If F is a Lyapunov Function for G, then starting from any x_0, there exists a t^*, such that $G(x_{t^*}) = x_{t^*}$, and the system attains an equilibrium in finite time.

We first construct a Lyapunov function within the *Race to the Bottom Game*, which captures strategic environments in which players choose levels of support such that each player prefers to provide just less than the average level.

The Race to the Bottom Game

Each of N players proposes a level of support in $\{0, 1, \ldots 100\}$ in each period. The player closest to $\frac{2}{3}$ of the average level of support wins a prize in that period.

The game can be used to explain reductions in state government spending for social programs such as support for the indigent. No governor or state legislature wants to appear heartless. Yet none wants to offer lavish programs that attract indigent populations from neighboring states. Each state wants to offer some funding, but less than the average. The same incentives exist for countries choosing environmental regulations or tax rates. Countries prefer to be less restrictive on environmental policies and have lower than average taxes in order to attract business.

Whether the Race to the Bottom Game attains an equilibrium depends on the behavioral rules of the players. If, for example, players choose random levels of support, then outcomes will be random. Random levels would not make sense given the game's payoff structure. Here we assume the following behavioral rule that aligns with experimental findings.[1] In the first period, we assume that every player chooses a random level of support less than 50. Thereafter, each player chooses a level at least 1 less than $\frac{2}{3}$ of the previous period's average. If that number is less than zero, then each player chooses zero.

It is straightforward to show that the maximum level of support from any player satisfies the conditions for a Lyapunov function. The maximum level of support has a minimum at zero. And in each period the maximum level of support falls by at least 1 given that levels of support take integer values. Thus, at some point, everyone proposes zero support. The players have raced to the bottom. In this example, the model produces an undesirable result. To prevent a race to the bottom requires changing the game. To increase support for the indigent, a federation could shift to federal funding or impose a floor on spending.[2]

As an aside, suppose that we allow players to choose any real number in the interval between zero and 100 rather than integer values. If in each round, players chooses a level of support equal to $\frac{2}{3}$ of the previous mean, the average level of support will decrease over time but it will never attain the equilibrium of zero. As in Xeno's paradox, the process would get closer and closer to zero but never reach it. Thus, we must assume a minimal decrease (A) to ensure an equilibrium.

Equilibrium in the Local Majority Model

We now return to the local majority model. We define our Lyapunov function to be the *total disagreement* in the population: the sum over all cells of the number of neighboring cells in the opposite state. To prove the model attains an equilibrium, we must show that if a cell changes its state, then total disagreement falls by at least a fixed amount.

The algebra is not too complicated. First, if a cell changes its state, it must have been in a minority relative to its neighbors. We know that at least five of its neighbors were in the opposite state and at most three were in the same state. Therefore, when the cell switches states, the number of cells that disagree with the cell decreases by at least 2 (see figure 16.1). To calculate the change in total disagreement, we must add in the changes to total disagreement contributed by neighboring cells. The five or more cells that now agree with the cell's state have lower disagreement (by 1 each) and the three or fewer cells that previously agreed now have higher disagreement (by 1 each). Therefore, total disagreement across all neighboring cells falls by at least 4.

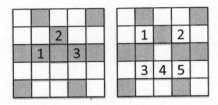

Figure 16.1: Total Disagreement Falls by 4 in Local Majority Model

We have therefore proven that even though some cells may have more disagreement, total disagreement satisfies the conditions for a Lyapunov function. The local majority model must, therefore, converge to an equilibrium—not just sometimes or most of the time, but all of the time. We also know the rate of convergence. Any time that a cell changes its state, total disagreement falls by at least 4. It follows that a configuration with total disagreement of 100 must reach an equilibrium within 25 periods. More generally, a configuration with total disagreement of D must reach equilibrium in $\frac{D}{4}$ periods. As noted in Chapter 15, the equilibrium reached will almost always be an inefficient pattern of splotches that includes frustrated cells.

Self-Organization: New York and Disney World

Our next application uses Lyapunov functions to prove the existence of an equilibrium in the *self-organizing activities model*. The model consists of a population of individuals and a set of activities that each individual can do during the day. The key assumption of the model will be that each person prefers less-crowded activities. Fewer people means less wait for the machines at the gym and a shorter line at the bakery or coffee shop. This model was motivated by a quote by Thomas Schelling from *Micromotives and Macrobehavior* in which he describes amazement at how cities self-organize—how traffic patterns, the flow of pedestrians, the number of people in parks and restaurants, and store inventories all reach appropriate levels with little central planning. How does the corner store always have four bottles of pure maple syrup from Cedarville, Michigan? How come the bakery runs out of fresh rye bread each day about twenty minutes before closing time? This order emerges even though the city's diverse actors—the tourists, store owners, residents, and delivery people—have limited information about the entire city.

Self-Organizing Activities Model

A city offers A activities. Each day consists of L time periods. Each person in large population of size M chooses a **routine**, an order to participate in a set of L activities (out of a larger set of K possibilities) across the L time periods. A person's **congestion** equals the number of other people who choose the same activities as her at the same times.

To prove the model converges, we show that the *total congestion,* the sum of the congestion levels of the entire population, satisfies the conditions for a Lyapunov function. When a person lowers her congestion, she lowers her contribution to total congestion and also lowers the congestion by 1 for each person she no longer meets and raises congestion by 1 for each new person she encounters. Given that she lowers her own congestion, it follows that more people belong to the first group than the second. For example, suppose that a person was going to a crowded gym at 8:00 a.m. and a crowded coffee shop at 4:00 p.m. If she switches time slots and finds the coffee shop to be nearly empty at 8:00 a.m. and the gym to be less crowded at 4:00 p.m., she reduces congestion for herself and for all the people she met previously. She does increase congestion for the smaller number of people she now meets, but total congestion falls (and by at least 1). Given that total congestion cannot fall below zero, the system must reach an equilibrium.

Although, in general, we have no guarantee that the system will locate an efficient equilibrium, this model almost always converges to a configuration of almost minimal total congestion. In an inefficient configuration, more people choose one activity—say, going to the gym—during a period than some other activity—say, going to the coffee shop. If the difference in congestion is high at those two activities, an individual can lower her congestion by switching the times she goes to the gym and the coffee shop. If the coffee shop and the gym had equal numbers of people in that other period, the exchange would reduce total congestion.[3]

The model explains some of the order that we see in the world. It gives insight into how cities can self-organize to near-efficient configurations without central planners. It also tells us why amusement parks, such as Disney World, may not. Each day, Disney World has new attendees, who lack the time to try new routes. Without help from central planners, some attractions at Disney World would have huge lines, while others would have no wait. Disney World tries to limit these inefficiencies by allowing people to sign up for particular attractions at specific times and by having employees steer people to less crowded areas.

Pure Exchange Economies

We can also use Lyapunov functions to explore when pure exchange economies attain equilibria and when they may not. A *pure exchange economy* consists of a set of consumers, each of whom has an endowment of goods as well as preferences. We might think of a set of people who show up at a marketplace or bazaar with something to trade with others, such as eggplants, cheese, or woven blankets. Each trade has a cost in time and effort for both parties. In order for two people to trade, each must benefit by an amount that exceeds this cost of trading.

Rather than construct a Lyapunov function that always decreases by a fixed amount and has a minimum, we do the opposite: we show that *total happiness* always increases by a fixed amount and has a maximum value. By assumption, any time two people trade, their happiness levels increase by at least the cost of trading. In addition, each person brought a fixed endowment of goods, so total happiness has a maximum value. The assumptions of a Lyapunov function are met, so the system goes to an equilibrium. At that equilibrium, the allocation need not be efficient. Of course, if it is not efficient, some people in the market might be able to identify a trade that makes them happier.

In constructing this argument, we assume that only people who participate in the trade derive happiness (or unhappiness). In other exchanges that might not be true. Imagine that Iraq promises to trade oil in return for nuclear weapons from Pakistan. The leaders of both countries may be happier, but total happiness, measured globally, would fall. The rest of the world might not be happy with Iraq building up a nuclear arsenal.

The impact felt by people in other countries is called a *negative externality*. When an exchange market includes negative externalities, trades need not raise total happiness. In our earlier example of a pure exchange market, when people trade fruits, vegetables, blankets, and tools, few externalities exist. The presence of externalities implies that we cannot say whether the system will reach an equilibrium. Trades in arms and oil such as the one described could beget other trades. Iraq's growing stockpile of nuclear weapons could lead Saudi Arabia to demand more military support from its allies. This in turn could lead to actions by other countries in the region. Global happiness, or global security for that matter, may jump up and down with each action. We cannot be sure that trades will ever stop.

Whether or not Lyapunov functions exist in a trading context depends on the size of the negative externalities, as can be seen through an example told to me by a former undergraduate student. Her employer was moving to new offices that would include a large room with open desks where analysts would sit. Her manager proposed randomly assigning analysts to desks and then allowing people to trade where they sat. He thought this would lead to a good outcome because of a belief that free exchange produces efficiency.

My former student realized that even though any two people who traded desks would be happier, their former and new neighbors need not be. A person might feel hurt if a current neighbor, especially one he may have chosen to be near, traded to a desk across the room. The former neighbor might also not like the new neighbor, who might talk loudly on the phone. The former neighbor may then himself move. Relocations might continue for a long time. And each one might chip away at morale. The plan looked risky. The organization needed its employees to trust one another and treat one another with respect. Those behaviors would be hard to maintain in employees who kept having people trade to move away from them. When her manager thought through the model, he abandoned his plan.[4]

The story does not end there. This same manager had also purchased office chairs in a variety of styles and colors and had planned to randomly assign chairs and allow people to trade. In this instance, my student (using model thinking) told her manager to allow trading of chairs. Chair trades would not produce any externalities and would be fun for the employees. Chair trading is a pure exchange market. The two cases provide a clear case of models guiding conditional actions. Exchange markets work for chairs but not for desk locations.

Models Without Lyapunov Functions

When we attempt to construct a Lyapunov function for a model and fail, we can still accumulate knowledge. Often we gain insight into why a model does not produce equilibria. In the Game of Life, some configurations attain equilibria, others do not. When one does produce an equilibrium, we might be able to write down a configuration-specific Lyapunov function. For example, any initial configuration that takes the form of a diagonal line will decrease by length 2 each period, as the two live cells at the end of the line die off and no new cells come to life. The configuration ends with all dead cells. For these configurations, the number of live cells would be a Lyapunov function. If we begin with another configuration, such as the R-pentomino, which produces a complex sequence of configurations, we cannot construct a Lyapunov function because the system does not go to equilibrium.

Our inability to construct a Lyapunov function does not imply that a model or system does not reach equilibrium. It could. Some systems go to equilibrium in every known case, yet no one has been able to construct a Lyapunov function. One famous example, the *half or triple plus one rule (HOTPO)* problem, also known as the Collatz conjecture, is deceptively simple. HOTPO starts with an integer. If the number is odd, we multiply by 3 and add 1. If the number is even, we divide by 2. The process stops when it reaches 1. If we start with the number 5 (which is odd), we triple it and add 1 to obtain 16. We divide 16 by 2 to get 8, divide 8 by 2 to get 4, and divide by 2 twice more to reach 1, the equilibrium. For every number up to 2^{64}, HOTPO stops. Nevertheless, no one has proven whether HOTPO always goes to equilibrium. The mathematician Paul Erdos reportedly said, "Mathematics is not yet ripe for such problems."[5] The inability of mathematicians to determine whether HOTPO goes to an equilibrium points to a more general lesson: models offer the possibility of proving results. We have no guarantee we can derive them. Often we write down a model only to find proving results difficult, if not impossible.

Summary

In this chapter we have seen how Lyapunov functions help us to prove whether a system or model achieves equilibrium and how quickly it does. Even our failed attempts to construct a Lyapunov function can be of use. They can provide insights into the causes of complexity. Such was the case with exchange economies with externalities and the trading of desks. In neither case could we construct a global variable that would always decrease or increase. Thus, we had no guarantee that these processes would go to equilibrium.

When we think back to the uses of models—to reason, explain, design, communicate, act, predict, and explore—we find that Lyapunov functions can help with each. As just noted, Lyapunov functions help us to reason through why systems equilibrate. They can be used to design information systems such as Disney World's sign-up times. We can use insights from the model to inform actions such as not allowing the trading of desks, to communicate how a system achieves equilibrium, to predict the time to reach equilibrium, and to explore, as we did when showing how cities self-organize.

The equation of a straight
Line models completely that line
The equation is the line
there is no need for the line
Plugging in enough numbers into the
Equation produces a translation
with is the line   Both Picture and line
=
But people work with lines all the time

17. *Markov Models*

History is a cyclic poem written by time upon the memories of man.

—Percy Bysshe Shelley

Markov models capture systems that transition probabilistically between a finite set of states. A political system might transition between democratic and dictatorial, a market between volatile and stable regimes, or a person between happy, contemplative, anxious, and sad. In a Markov model, the movements between states occur according to fixed probabilities. The probability that a country transitions from authoritarian to democratic in a given year might be 5%; the probability that a person transitions from anxious to tired within an hour may be 20%. If, in addition, the system can move from any one state to any other through a sequence of transitions and if there exists no simple cycle of back and forth, a Markov model attains a unique *statistical equilibrium*.

In a statistical equilibrium, individual entities continue to move between states but the probability distribution across the states remains fixed. A statistical equilibrium in a Markov model of ideology would allow for people to transition between liberal, conservative, and independent, but the proportions of people of each ideology would remain unchanged. When applied to a single entity, a statistical equilibrium means that long-run probability of the entity being in each state does not change. A person could be in a statistical equilibrium in which he is happy 60% of the time and sad 40% of the time. The person's mental state could change from hour to hour, but his long-run distribution across those states does not.

The unique statistical equilibrium implies that long-run distributions of outcomes cannot depend on the initial state or on the path of events. In other words, initial conditions do not matter, and history does not matter. Nor can interventions that change the state matter. As time marches on, a process that satisfies the assumptions inexorably heads to its unique statistical equilibrium and then stays there. Here again, a model reveals conditional logic: if the world fits the assumptions of a Markov model, history does not matter in the long run. The Markov model does not say that history can never matter. First, the model allows for outcome path dependence—what happens next will depend on the current state. Second, it allows for history to model in the long run as well, but for that to be the case, one of the model's assumptions must be violated.

Markov models have myriad applications. They can be used to interpret dynamic phenomena such as democratic transitions, buildups to war, and the success of drug interventions. They can also be used to rank webpages, academic journals, and sports teams. They can even be used to adjudicate authorship of books and articles. In this chapter, we cover each of these applications. We begin with two examples and then state a general theorem for a statistical equilibrium to exist. Then, in the third section, we turn to the applications. In the discussion at the end of the chapter, we reengage the question of how and when history matters in light of our knowledge of Markov models.

Two Examples

A Markov model consists of a set of *states* and *transition probabilities* between those states. In our first example, we characterize a person's mood on a given day as either mentally engaged or bored. These are formally represented as the two states of the model. The transition probabilities then characterize the probability of moving between states. For example, as shown in the diagram below, we assume that when mentally engaged, a person has a 90% chance of remaining in that state and a 10% chance of becoming bored, and that when bored, a person has a 70% chance of remaining bored and a 30% chance of becoming mentally engaged.

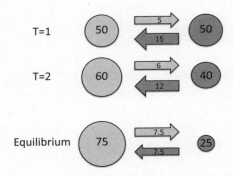

Figure 17.1: A Markov Process

Suppose that these transition probabilities hold for 100 students taking a biology class. Initially half of the students are engaged and half are bored, as shown in 17.1. Applying the transition probabilities described above, we expect that on the next day, 5 (10%) of the mentally engaged students will become bored, and 15 (30%) of the bored students will become mentally engaged. This results in 60 mentally engaged students and 40 bored students. The day following that, 6 of the 60 mentally engaged students should become bored and 12 of the bored students should become mentally engaged, resulting in 66 mentally engaged students and 34 bored students. As we continue to apply the transition rules, the process converges to a statistical equilibrium with 75 mentally engaged students and 25 bored students. In that equilibrium, students continue to move between the two states but the number of students in each state does not change.

If instead the process begins with all 100 students mentally engaged, the next day only 90 students will be mentally engaged. On the following day, that number falls to 84. If we continue to iterate the process, we find that in the long run 75 students will be mentally engaged and 25 students will be bored. The model attains the same statistical equilibrium.

In our second example, we categorize countries into three states: free, partly free, or not free. Figure 17.2 shows the percentage of countries in each of the three categories over the thirty-five-year period ending in 2010 based on Freedom House data. The figure shows a clear trend toward increased democratization. Over the past thirty-five years, the percentage of free countries has risen by 20%. If that linear trend continues, by 2040 two-thirds of all countries would be free, and by 2080 eight out of nine countries would be free.

A Markov model leads to a different prediction. To make the predictions, we set the length of a period to five years and loosely calibrate transition probabilities based on the past data (see table 17.1).[1]

| | | State in Next Period | | |
		Free	Partly Free	Not Free
	Free	95%	5%	0%
Current State	Partly	10%	80%	10%
	Not Free	5%	15%	80%

Table 17.1: Markov Transition Probabilities for Democratization

If we initialize the model using the percentages of countries in each category in 1975, the calibrated model (as we would expect) almost perfectly matches the 2010 distribution: 48% of countries are free, 31% are partly free, and 21% are not

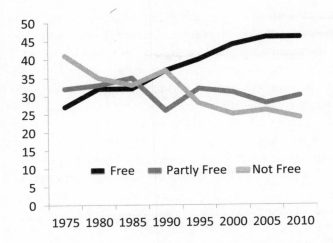

Figure 17.2: Freedom House: % Free, Partly Free, and Not Free

free. The actual percentages in 2010 were 46%, 30%, and 24%. If we continue to run the model, it predicts that in 2080, 62.5% of countries will be free, 25% partly free, and 12.5% not free.

The less rosy prediction of the Markov model stems from the fact that a linear projection fails to take into account that free countries can transition to being partly free and also to being not free. As more countries become free, the number of free countries becoming partly free increases. The reasons for this are manifold. For one, operating a democracy requires fiscal authority and institutions capable of implementing policies. To borrow the phrasing of Flores and Nooruddin, democracy may not easily take root in some places.[2] In those places, we should expect transitions from free to partly free. The Markov model captures these.

The Perron-Frobenius Theorem

Both of our examples converge to unique statistical equilibria. That was not by chance. Any Markov model with a finite set of states, fixed transition probabilities between them, the potential to move from any state to any other in a series of transitions, and no fixed cycles between states converges to a unique equilibrium.

The theorem implies that if those four assumptions are satisfied, the initial state, history, and interventions that change the state cannot change the long-run equilibrium. If nations move between dictatorships and democracies according to fixed probabilities, then interventions that impose or encourage democracies in some countries have no long-term effects. If fluctuations in dominant political ideologies satisfy the assumptions, then history cannot influence the long-run distribution over ideologies. And if a person's mental state can be represented as a Markov model, then words of encouragement or supportive gestures have no long-run impact.

Perron-Frobenius Theorem

A Markov process converges to a unique statistical equilibrium provided it satisfies four conditions:

Finite set of states: $S = \{1, 2, \ldots, K\}$.

Fixed transition rule: The probabilities of moving between states are fixed, for example, the probability of transitioning from state A to state B equals $P(A, B)$ in every period.

Ergodicity (state accessibility): The system can get from any state to any other through a series of transitions.

Noncyclic: The system does not produce a deterministic cycle through a sequence of states.

The takeaway from the theorem should not be that history cannot matter but that if history does matter, one of the model's assumptions must be violated. Two assumptions—the finite number of states and no simple cycle—almost always hold. Ergodicity can be violated, as when allies go to war and cannot transition back to an alliance. Such examples notwithstanding, ergodicity generally holds as well.

That leaves the restriction to fixed transition probabilities between states as the assumption least likely to hold. Thus, the model says that when history matters, underlying structural forces must change transition probabilities (or change the set of states). Consider the challenge of helping families to escape poverty. The forces that create social inequality have proven immune to policy interventions.[3] In Markov models interventions that change families' states—such as special programs for underperforming students or a one-day food drive—can provide temporary boosts. They cannot change the long-run equilibrium. In contrast, interventions that provide resources and training that improve people's ability to keep jobs, and therefore change their probabilities of moving from employed to unemployed, could change long-run outcomes. At a minimum, the model gives us a terminology—the distinction between states and transition probabilities—along with a logic to see the value of changing structural forces rather than the current state.

The Sales-Durability Paradox

The *sales-durability paradox* states that the prevalence of a product (or an idea) depends less on its relative sales than on its durability. Markov models can explain the paradox by letting states represent the proportion of people who own a type of good. Here, we consider two types of floor coverings: tile (the durable good) and linoleum (the good with higher sales). The paradox arises when the good with higher sales, in this case linoleum, is less prevalent.

In our model, we assume that linoleum outsells tile by a ratio of 3 to 1. To capture durability differences, we assume that each year 1 in 10 people replace their linoleum floors, while only 1 in 60 people replace their tile floors. The resulting Markov model has an equilibrium in which two-thirds of floors are tile.[4]

The same logic that underpins the sales-durability paradox explains the positive relationship between market share and brand loyalty (the likelihood that someone switches brands). If we write a Markov model, lower brand loyalty must imply lower market share in equilibrium because loyalty operates just like durability. This empirical regularity is known as the *double jeopardy law*. If you have low brand loyalty, you tend to have low sales.[5]

Markov: One-to-Many

Markov models can be applied in a variety of contexts. We can use them to model genetic drift between the four nucleic acids: adenine (A), cytosine (C), thymine (T), and guanine (G). If each nucleic acid has a small and equal probability of becoming one of the other three types, we can write a transition matrix for drift. We can use them to model health trajectories by letting states represent health categories such as excellent, moderate, and poor. The model can evaluate how interventions such as drug protocols, behavioral changes, and surgeries change the transition probabilities and equilibrium distributions over outcomes. Interventions that produce better equilibria—that is, more people in excellent health—merit pursuing.[6]

Markov models can also be used to identify patterns in international crises and distinguish between transitions that lead to war and those that produce peace and reconciliation.[7] This application requires us to estimate two different models, one using cases where crises led to war, and one using cases in which reconciliation occurred prior to war. If the transition probabilities in the two models differ significantly, then we can compare existing patterns, such as bombing, hostage taking, no exchange of prisoners, and escalated posturing, and see which process better fits the data.

Using Markov models to discriminate between patterns in this way can adjudicate authorship controversies. Given an author's known writings, we can estimate the probability that one word follows another. In the text of this book, the word "the" follows the word "for" four times as often as does the word "example." We could represent that information as transition probabilities in a large matrix. The matrix for this book would look different than the matrix for a book written by someone else. If we were to construct separate word transition matrices for Melville, Morrison, and Mao, we would see differences in their transitions between word pairs.[8]

Using a technique like this, we can use models to aid in assigning authorship of the *Federalist Papers*—eighty-five essays written in 1787 and 1788 by Alexander Hamilton, John Jay, and James Madison to convince New Yorkers to support the United States Constitution. Each essay was signed with the pen name Publius. Though the authorship of most of the essays has been settled, several remain a matter of dispute. A Markov model assigns all of the disputed essays to James Madison.[9] Hamilton or Jay could have written those essays, but if either did, he wrote in the style of Madison. A similar analysis of four discourses and twelve short unattributed essays discovered by Arlene Saxonhouse showed that at least three could be attributed to Hobbes with high probability.[10] In neither of these cases does the model necessarily give the correct answer. The model produces knowledge. We rely on our wisdom to decide how to weigh this model against other models or intuition.

For our last application, we describe how Markov models were used to create Google's original PageRank algorithm. PageRank transformed search on the World Wide Web.[11] The World Wide Web consists of a network of websites connected by links. To estimate the importance of each site we could count the links to and from a site. In the network of sites in figure 17.3, sites B, C, and E each have two links, A has one link, and D has no links. This method provides a crude estimate of importance, but it has flaws. Sites B, C, and E all have two links, but site E seems more important than site B given its position in the network.

PageRank considers each site to be a state in a Markov model. It then assigns a positive transition probability between two sites if they share a link. For the moment, we assign equal probability to any link; that is, we assume that a searcher at A would be equally likely to move to B or E. If our searcher goes to E, she then alternates between C and E forever. Alternatively, if she chooses B, she goes to C,

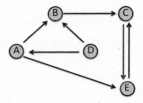

Figure 17.3: Linkages Between Sites on the World Wide Web

and again starts alternating between C and E. In fact, beginning at any site results in alternation between C and E. Again, C and E appear to be the most important sites. Unfortunately, this model does not fit two assumptions of the Perron-Frobenius theorem. The system cannot get from any site to any other: there is no way to get from C to D. In addition, the transition probabilities create a loop between C and E.

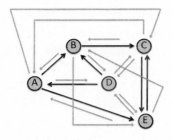

Figure 17.4: Adding in Random Movements Between Sites

To fix both problems, Google added in a small random probability of moving from any site to any other as shown in figure 17.4. The model now satisfies all assumptions of the theorem and has a unique statistical equilibrium. Sites can be ranked by their probabilities in that equilibrium. A searcher who begins at A will most likely end up at C or E within a few searches. Once there, she will bounce back and forth between those two sites until trying a random site. If she goes to A or D, the path back to C will most likely go through B or E. It follows that site B should have a higher ranking than A or D, but that all three should be unlikely. In the unique statistical equilibrium shown in figure 17.5, that happens to be the case. A, B, and D are all rarely visited, but B is the most visited of the three.

196

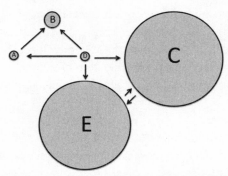

Figure 17.5: Statistical Equilibrium of PageRank Model

PageRank can be thought of as a combination of the random walk model and a Markov model. If we think of PageRank as an algorithm, we realize that we can use it to produce rankings of any network. We can let nodes represent baseball or soccer teams and transition probabilities denote the percentage of time that one team defeats another.[12] If the teams play only once, the transition probabilities can be assigned based on margin of victory. The resulting ranking, though not definitive, complements subjective expert assessments. We can also use PageRank to compute species' importance using food web data.[13]

Summary

Markov models describe dynamic systems that move between states according to fixed transition probabilities. If we additionally assume that the process can move between any two states and that the process does not produce a cycle, then a Markov model attains a unique statistical equilibrium. In the equilibrium people or entities move between states in such a way that the probability distribution across states does not change. It follows that as a process nears that equilibrium, the changes in the probabilities diminish. Represented as a graph, the slope of the curve flattens. Recall our earlier discussion of California's population growth when we learned linear models. California's population growth has slowed because as the population of California has grown, the number of people leaving California has increased. That result holds true even if the proportion of Californians leaving does not change.

When applying Markov models to explain phenomena or predict trends, a modeler's selection of the states proves critical. The choice of states determines the transition probabilities between those states. A Markov model of drug addiction could assume two states: being a user or being clean. A more elaborate model might distinguish users by frequency of use. Regardless of the choice over states, if the four assumptions hold (and in this instance, the key test would be whether transition probabilities remain fixed), then the system will produce a unique statistical equilibrium. Any one-time change in the state of a system has at most a temporary effect. Reducing drug use in equilibrium would require changing transition probabilities.

Continuing with that same logic, we can infer that a one-day event to spur interest in education may lack meaningful impact. Volunteers coming into a community and cleaning up a park may produce few long-term benefits. Any one-time influx of money, regardless of its size, will dissipate in its effect unless it changes transition probabilities. In 2010, Mark Zuckerberg donated $100 million to the Newark, New Jersey, public schools, an amount that was matched by other donors. That one-shot donation, which amounted to approximately $6,000 per student, has produced few measurable effects on test scores.[14]

Markov models guide action by distinguishing between policies that change transition probabilities, which can have long-term effects, and those change the state and can only have short-term effects. If transition probabilities cannot be changed, then we must reset the state on a regular schedule to change outcomes. A person's work life may create transition probabilities that lead toward competitive, selfish, and stressful mental states. Daily exercise, meditation, or religious practice may move a person into a more grateful, compassionate, and relaxed state to start each day. Weekends perform a similar function, as do regular date nights for married couples. Both temporarily move a person's state away from the equilibrium.

Not every dynamic system satisfies the assumptions of the Markov model. For those that do not, history, interventions, and events can have long-term consequences. In the Polya process, outcomes change the long-run equilibrium. A large intervention or shock to a system can change transition probabilities or even the set of states. Major technological improvements such as the steam engine, electricity, the telegraph, or the internet change the set of possible states of the economy. Political and social movements that define new rights or create new policies also change the set of states. We might therefore think of history as a sequence of Markov models rather than as a single process moving toward an inevitable equilibrium.

Markov Decision Models

A *Markov decision model* amends a Markov model by including actions. An action generates a reward, which is conditional on the state and also affects the transition probabilities between states. Given the effect of an action on transition probabilities, the optimal action does not always maximize the immediate reward.

As an example, we consider students who have a choice between two actions: surfing the internet and studying. Surfing the internet always produces the same payoff. Studying produces a high payoff when the student is engaged and a low payoff when the student is bored. To add in the effect of actions on transition probabilities, we assume that a bored student who surfs the internet remains in the bored state and that an engaged student who surfs the internet becomes bored half of the time. A student who studies has a 75% chance of being mentally engaged in the next period regardless of her present state:

Actions: *surf the internet (U), study (S)*

States: *bored (B), mentally engaged (E)*

Reward Structure

	Bored (B)	Engaged (E)
Surf (U)	6	6
Study (S)	4	8

Transition Mapping

	Bored (B)	Engaged (E)
Surf, Bored (U, B)	1	0
Surf, Engaged (U, E)	$\frac{1}{2}$	$\frac{1}{2}$
Study, Bored (S, B)	$\frac{1}{4}$	$\frac{3}{4}$
Study, Engaged (S, E)	$\frac{1}{4}$	$\frac{3}{4}$

A solution to a Markov decision problem consists of an action to be taken in each state. Myopic best response behavior, which we encountered earlier, selects the reward-maximizing action in each state. In the example, that corresponds to surfing when bored and studying when mentally engaged.

The myopic solution results in the student falling into the bored state. Once that occurs, she chooses to surf the internet and remains bored in all remaining periods. Thus, her long-run average reward equals 6. The always-study solution puts her in the engaged state 75% of the time and in the bored state 25% of the time, producing an average return of 7. This solution produces a higher average payoff because she is more often in the mentally engaged state.

As seen in this example, framing a choice as a Markov decision problem can produce better actions. By taking into account the consequences of an action on our state, we choose more wisely: Sleeping late produces a higher immediate reward than getting up early and exercising. Buying an expensive coffee produces a higher reward than making our own cup. Yet, in the long run we may be happier exercising and saving money on coffee. Did we need a model for this? We might have instead turned to Proverbs 21:17: "Whoever loves pleasure will be a poor man; he who loves wine and oil will not be rich." That may be true, but had we turned to Ecclesiastes 8:15: "And I commend joy, for man has nothing better under the sun but to eat and drink and be joyful," we would have found an opposite proverb. By embedding our choices within a Markov decision framework, we can use logic to decide which commonsensical advice makes sense in a given setting.

18.   *Systems Dynamics Models*

The principles governing the behavior of systems are not widely understood.

—Jay Wright Forrester

In this chapter, we cover systems dynamics models.[1] These models analyze systems with feedbacks and interdependencies. They are used to model ecologies and economies, supply chains and production processes. They improve our capacity to think through logical chains that include negative and positive feedbacks. A systems dynamics model consists of sources, sinks, stocks, flows, rates, and constants. Sources produce inputs into the system. Sinks absorb outputs. Stocks keep track of levels of variables, and flows capture feedbacks between levels of stocks. Rates and constants apply to the flows, which can be fixed or change over time.

Systems dynamics models can include both positive and negative feedbacks. Positive feedbacks, such as the Matthew effect covered in Chapter 6, occur when an increase in a variable or attribute produces an additional increase in that same variable. Success breeds success, sales lead to more sales, and, in the case of academic papers and patents, citations generate more citations.

Negative feedbacks dampen trends. We must be careful not to infer normative implications from the word *negative*. Negative feedbacks often produce desirable properties. They can prevent bubbles and crashes. When we eat, our brain receives signals to stop eating. When a company's profits increase beyond normal economic returns, competitors enter, reducing those profits and preventing the company from exploiting customers. When a species proliferates, its members compete for food, reducing population growth. In each case, negative feedbacks contribute to system-level robustness.

Using systems dynamics models, we can often identify the causes of complexity. When a system includes both positive and negative feedbacks, it can produce complexity. That was true of the Game of Life in which existing cells caused new cells to come to life but overcrowding caused cells to die.

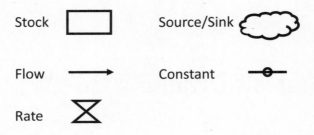

Figure 18.1: The Components of Systems Dynamics Models

Systems dynamics models that represent flows and stock levels as mathematical functions can be calibrated to explain past values of stocks, to predict future values, and to estimate the effects of an intervention. We can then use the models to explain, predict, and guide action. Systems dynamics models can also be qualitative. We can label each arrow with a plus or a minus to clarify logic.[2]

The remainder of the chapter consists of five parts. To introduce terminology, we build a qualitative model of a bakery. We then construct a predator-prey model based on the Lotka-Volterra equations. Our version assumes interacting populations of foxes and hares and embeds both negative and positive feedbacks. We next show how by using systems dynamics models we can anticipate vicious cycles. We then describe the WORLD3 model, a large model of the global economy. We conclude with a discussion of how systems dynamics models often produce counterintuitive results, which demonstrates the limits of human reasoning and the value of models as logical aids.

The Parts of a Systems Dynamics Model

A *systems dynamics model* consists of sources, sinks, stocks, and flows. A *source* produces a *stock,* the amount or level of some variable. A *flow* describes how the level of a stock changes. A *sink* catches the output of a flow from a stock. Sinks and sources are placeholders for processes not included in the model. The level of a stock changes over time based on sources and flows. In a systems dynamics model of an amusement park, for example, the number of people at the park (a stock) increases as people arrive (a source). The rate of increase could in turn depend on other parameters such as the weather, the amount of advertising, or the price of admission.

Figure 18.2: A Systems Dynamics Model of a Bakery

Systems dynamics models use the representational system shown in figure 18.1. Sources and sinks are represented by clouds. Stocks are represented by boxes, and flows by arrows identified by a plus or a minus sign. Variable flows are represented by inverted triangles and constant flows by circles bisected by the flow arrow. A positive arrow represents a positive feedback, where more begets more. A negative arrow represents a negative feedback from one variable to another.

To build familiarity, we first construct a basic systems dynamics model of a bakery that consists of a baker, bread, and customers. The baker makes bread and customers buy it. If the rate at which the baker produces bread exceeds the rate at which customers purchase bread, the stock of bread grows, and the bakery fills with bread. Alternatively, if the rate of sales exceeds the baker's production rate, the bakery will perpetually sell out. To make the model more realistic, we can allow the baker to adjust the rate of bread production as a function of the stock of the bread as shown in figure 18.3, which includes a flow (an arrow) from the stock of bread to the rate at which the baker produces bread. We place a negative sign on the arrow to denote that the rate decreases as the stock of bread increases. If the adjustment rate is set properly, the model will produce an equilibrium where the rate at which loaves are baked equals the rate at which customers buy bread so that inventories equilibrate.

To make the model even more realistic, we can add a second stock, *line,* that equals the number of people waiting outside the bakery, as well as a second source, *potentials,* which adds people to the line. A short to moderate line may attract customers, while a long line could turn customers away. To capture the variable effect of the length of the line on the rate of arrival from the source, we write (+/ −) above the arrow. We also include a plus sign above the arrow from the stock of line to the rate at which customers buy bread, assuming that with more people in line, people decide faster.

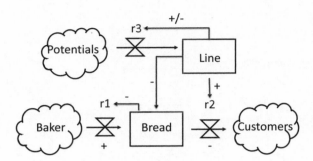

Figure 18.3: A More Elaborate Model of the Bakery

This model could be calibrated to data. We could estimate the rates at which people join the line based on its length. The baker could then determine an optimal rate of adjustment for baking as a function of the stock of bread and the length of the line. That rate would provide a starting point from which a better rate might be learned. Even without calibration, the act of writing the model adds value. The baker realizes the importance of line length to his overall sales.

The Predator-Prey Model

We now introduce the *predator-prey model,* an ecological model that captures the relationship between the number of hares (the prey) and the number of foxes (the predator). The model include two positive feedbacks: hares produce hares, and foxes produce foxes. It also includes a negative feedback: foxes eat hares. The model assumes that if the level of hares is high, foxes produce more offspring. Figure 18.4 qualitatively represents these assumptions but does not quantify the relationships. From the figure we see that as the number of foxes increases, the number of hares decreases, which in turn results in fewer foxes. As the number of foxes falls, hares should proliferate, leading to more foxes. The logic suggests the possibility of a cycle, or perhaps an equilibrium. We cannot be sure.

To gain insight into what occurs, we need to construct a quantitative version of the model. We assume linear flows that depend on the stock levels. Absent any foxes, the number of hares grows at a fixed rate, and absent any hares, the number of foxes decreases at a fixed rate owing to a lack of food. The model will assume that the probability of a hare and a fox meeting is proportional to the number of foxes times the number of hares. To capture foxes eating hares when these interactions

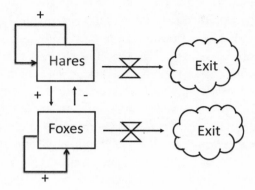

Figure 18.4: A Systems Dynamics Model of the Predator-Prey Model

occur, we assume that foxes grow at a constant rate times that product and that hares decrease at a constant rate times that product. The resulting equations are known as the *Lotka-Volterra equations*.

Lotka-Volterra Model

An ecosystem consist of H hares and F foxes. The population of hares grows at rate g and the population of foxes dies off at rate d. When hares and foxes meet, hares die off at rate a and foxes grow at rate b. These assumptions produce the following differential equations:[3]

$$\dot{H} = gH - aFH \qquad \dot{F} = bFH - dF$$

These equations have an extinction equilibrium ($F = H = 0$), as well as an interior equilibrium given by the equations $F = \frac{g}{a}$ and $H = \frac{d}{b}$.

The differential equations describe how the numbers of hares and foxes change over time. When the equations equal zero, the number of hares and foxes do not change and the system is in an equilibrium. One equilibrium, the *extinction equilibrium,* consists of no hares or foxes. Therefore, the model predicts that under some conditions, predator-prey relationships lead to the extinction of both species. That cannot occur in all cases; otherwise, no species would be left on earth.

The *interior equilibrium* contains positive numbers of foxes and hares. In that equilibrium, the number of foxes increases with the growth rate of hares and decreases if each interaction between a fox and a hare reduces the hare population at a faster rate. Both of those results are intuitive. If hares reproduce faster, the systems can support more foxes. And if each fox requires more hares to remain alive, the system can support fewer foxes. Both results align with our intuitions. We want such results: models should produce intuitive findings.

Models should also produce less intuitive findings, and this model does. It shows that the equilibrium number of foxes does not depend at all on the foxes' rate of death. If foxes die at a faster rate, the equilibrium number of hares increases, and the remaining foxes have abundant food, meaning foxes grow at a faster rate. That faster growth rate of foxes exactly cancels out the foxes' higher death rate.

Similar logic applies to the hare population. The equilibrium number of hares does not depend on the growth rate of hares or on the rate at which hares are consumed by foxes. The number of hares does depend on the rate at which foxes die and the rate at which foxes turn hares into more foxes. Our intuition fails us in these cases because we cannot think through the feedbacks. The direct effect of increasing the growth rate of hares is more hares. The indirect effect, more foxes, implies fewer hares. These two effects cancel out. Nonintuitive findings such as these are a hallmark of systems dynamics models. Our intuition fails because we latch onto direct effects and fail to think through the entire logical chain. Even if the direct effect of increasing (or decreasing) a rate or flow may be to increase (or decrease) a stock, the presence of systems effects in the form of positive and negative feedbacks means that other stocks will also change values, so the net effect of a change in a rate or flow may be reduced, canceled, or even reversed.

Using mathematics, we could show two equilibria for Lotka-Volterra equations. We do not know which, if either, of those equilibria would be realized. It is true that if the model starts at an equilibrium, then it will stay there. But until we run the model, we do not know if the equations will produce an equilibrium, a cycle, randomness, or complexity. All that we know is that an equilibrium exists.

Simulations of the equations produce *lagged cycles*. First one species becomes populous, then it reduces in number and the other species increases in number. Empirical studies show these cycles to be common. Figure 18.5 shows the number of wolves (predators) and moose (prey) on Isle Royale, a forty-five-mile-long island in Lake Superior, over a fifty-year period. Notice that species levels of predators and prey fluctuate with lagged cycles. The patterns are not as regular as those produced by the model as we should expect given that the model omits geography, other species, weather variation, and heterogeneity within the two species.

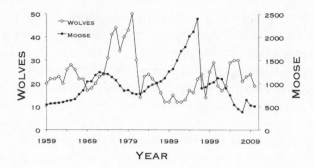

Figure 18.5: Wolves and Moose on Isle Royale, Michigan (isleroyalewolf.org)

This analysis of Lotka-Volterra equations reinforces our earlier observation that we should not conflate the existence of an equilibrium with its attainment. In this case, the system produces cycles and not the equilibrium. However, the dynamics cycle around the equilibrium. Thus, the equilibrium tells us, on average, the number of foxes and hares. It follows that our earlier counterintuitive result—that increasing the growth rate of foxes (or hares) has no effect on the equilibrium level of foxes (or hares)—still holds in the aggregate.

Using Systems Dynamics Models to Guide Action

Systems dynamics models can include both positive and negative *feedback loops*. Positive feedback loops can result in *virtuous cycles,* such as when increased trust between two countries leads to more trade, reduced military engagement, and, therefore, more trust. Positive feedback loops can also produce *vicious cycles*. A reduction in jobs in a region may result in less incentive for people to acquire skills, which may in turn induce firms to leave the region owing to a lack of qualified laborers, producing even less incentive for workers to acquire skills.

Systems dynamics models can help us anticipate vicious cycles. In 2008, many national economies were under severe financial stress. When asset prices fell, over-leveraged banks teetered on the edge of insolvency. Investors and depositors became concerned about the safety of their investments. Some countries, such as the United States, insure bank deposits up to a limit. Other countries, notably Australia, offered no deposit insurance.

To prevent a panic, Australia decided to introduce deposit insurance. The logic seems sound: insuring deposits prevents a run on banks. However, it considers only a part of the system. And in doing so, it suffers from a tragic flaw, which becomes obvious once we write down a systems dynamics model. In a model of a financial system, each bank (a stock) has some level of assets. Depositors put money into the banks and earn a return. The bank's borrowers use that money to make investments. Deposit insurance guarantees the depositors' money held in banks.

People also keep money in the stock market and in money market funds. Each type of investment is a stock. Once we start drawing the arrows—the flows between the boxes—the policy's flaw becomes evident. The direct effect of the deposit insurance increases the safety of banks, making banks more attractive (arrow #1 in figure 18.6). It also makes the other types of investments less attractive. Imagine yourself as investor with money in both banks and money markets during a turbulent time. Your bank deposits are now insured. Your money market funds are not. The prudent action would be to increase bank deposits (arrow #2) and withdraw from money markets (arrow #3).

A vicious cycle ensues: the decrease in investments in money markets makes them riskier. That increased risk produces even more withdrawals from money markets, creating a positive feedback loop (circular path 4). Withdrawals beget increased risk, and risk begets more withdrawals, which in turn begets more risk. The policy appears all but certain to create a collapse of the money market industry, and it did. Within four days of insuring bank deposits, the government froze money market accounts, saving the industry from collapse, a decision with catastrophic consequences. Millions of retirees who relied on withdrawals from those accounts to pay for food, housing, and other essentials could not afford basic necessities.[4]

Though in retrospect, the vicious cycle seems obvious, we have no guarantee that if Australian policymakers had constructed a systems dynamics model they would have seen the consequences of their policy. However, the process of constructing the model would have revealed the effects of insuring bank deposits within the broader financial system and may have also made the ensuing vicious cycle evident. This example also demonstrates the limits of data. Data from other countries would have shown that insuring deposits stabilizes financial systems. However, in those countries the deposit insurance was not created during a crisis and would have been misleading.

The WORLD3 Model

We next consider a more elaborate systems dynamics model that covers the global economy, known as the *WORLD3 model*. The model originated in the 1970s and predicted the collapse of the world economy unless governments changed their

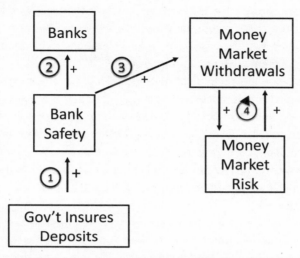

Figure 18.6: A Systems Dynamics Model of a Financial System

growth and environmental policies.[5] The WORLD3 model includes multiple inter-acting processes growing at different rates within a common framework, allowing policymakers to see interdependencies.[6] Mainstream economists often dismiss the WORLD3 model as too complicated and as failing to take into account rational re-sponses by economic actors.

The model assumes that population and economic output grow at fixed per-centages each year and that economic output creates pollution. Over time, land becomes less productive, population levels exceed the economy's ability to produce sufficient goods, and the world economy collapses. This prediction is reminiscent of the dire warnings of Malthus nearly two centuries earlier.

The model contains approximately 150 variables, 300 equations, and 500 pa-rameters, including fertility rates, rates of economic growth, and rate of land use. To calibrate the model, these parameters' rates of increase must be estimated from data. The WORLD3 model includes interactions among variables, implying that changes in multiple parameters will often produce nonlinear effects. Testing the robustness of the model therefore requires changing pairs and triplets of variables simultaneously. Five hundred parameters create more than 12,000 pairs and over 20 million triplets of parameters—far too many for anyone to analyze.

The model predicts population will fall to 4 billion in 2100. John Miller finds that by tweaking just two parameters, the fraction of industrial output allocated to consumption and the reproductive lifetime of females, nearly doubles the model's predicted world population to 7.4 billion. The huge increase results from positive feedbacks. Longer reproductive life spans imply more children, who require more food. Increasing the share of output that goes to food allows more children to survive. The women who survive have longer reproductive life spans and more children. The result is massive population growth.[7]

The finding of a doubling in population from small parameter changes is troublesome. However, the fact that outcomes depend on parameters is not a weakness. To the contrary, the model was built to guide action, to help identify effective policies. For example, the model shows that reducing fertility rates, which in fact did occur, would reduce population growth. Moreover, in that the model was calibrated, it provided an estimate of how much population growth would be reduced. The model could then be included in an ensemble of models to produce more accurate predictions.

Over time the model's original predictions have become less accurate, in part because rates of population growth have slowed as population has increased. They no longer match the model's assumptions. This is the sort of adaptive response anticipated by economists.[8] While proponents of the WORLD3 model accept this criticism, they hasten to point out that many of the model's predictions, including those relating to economic growth and total world population, have been quite accurate. As to the reductions in fertility, advocates note that if the WORLD3 model played any part in its own undoing—if the WORLD3 model created an awareness of overpopulation and of the importance of the environment—they are happy to be wrong.

Summary

When constructing a systems dynamics model, we choose the key parts (the stocks), describe the relationships between those parts (the flows), and then simulate the model to discover the implications. These models differ from Markov models in that the rates (which play the role of transition probabilities) adjust. Therefore, the model does not necessarily go to an equilibrium. We have to run the model to see what will happen. In addition, because we do not have to solve for the outcome, we need not worry about the tractability of our assumptions.

Systems dynamics models can have many variables and can include any type of feedbacks between those variables. One can write models without them, but once the boxes defining the stocks have been drawn, a modeler almost cannot help but draw arrows between them. The modeler feels obliged to ask, "What other variables might be affected and how might changes in those variables feed back into the current model?" resulting in more elaborate models.

This flexibility can come at a cost: the more stocks and flows created, the less understandable the model becomes. The art of constructing a useful systems dynamics model lies in including just enough detail to reveal where our intuition fails but not so much detail that we create a morass as confusing as the real world. The most useful systems dynamics reside in that boundary. Those models can reveal unintended effects and contribute to better policy actions. As we just saw, even the best-intended policies, such as the Australian deposit insurance policy, can produce undesirable outcomes.

Systems dynamics models also show how negative feedback loops can limit the effect of interventions. Laws that mandate safety features on cars, such as antilock brakes or airbags, may cause people to drive more recklessly. Widening roads may cause more people to move to the suburbs, thus increasing congestion. Decreasing the nicotine in cigarettes may cause smokers to consume more cigarettes. Developing better treatments for sexually transmitted diseases such as HIV may make people more likely to engage in unsafe sex. The list goes on and on.[9] Many of these negative feedbacks seem obvious in retrospect, but anticipating them ahead of time may not be. The act of writing down a qualitative systems dynamics model brings these feedback loops to light and makes us better thinkers.

The fact that systems dynamics models encourage us to include feedbacks is a strength of the approach. In 1696, England's King William III introduced a homestead tax with a base rate of two shillings per house plus an additional fee based on the number of windows: houses with more than ten windows paid an additional four shillings, and houses with more than twenty paid an additional eight shillings. The king taxed windows because they are observable, can be objectively measured, and correlate with housing values. Had the king relied on assessments of property values, he would have invited favoritism and bribery. His window tax was such a good idea that over the next century, it spread to France, Spain, and Scotland. France did not get rid of its window tax (*impôt sur les portes et fenêtres*) until 1926.

As model thinkers, we would expect people, who are purposive and adaptive, to respond to the tax. They chose a variety of routes. Some people bricked up the windows on their existing houses. The window tax led architects to alter housing designs. Many middle-class homes built during the period of the tax lack second-story bedroom windows. One row of houses in Edinburgh featured no bedroom windows at all.[10] Tax revenues fell. Campbell's law held again: politicians created a measure and people found a way to skirt it. More elaborate systems dynamics model would include implications of the reduction in windows by adding arrows from the stock variable called windows to attributes such as the health of citizens, which would be reduced from a lack of fresh air and light.

The great value of systems dynamics models resides in part in their ability to help us reason through the effects of our actions. We can often think through the direct effects of policies. Taxing windows will raise revenue. Requiring antilock brakes will save lives. Though we cannot always anticipate every indirect effect—the positive and negative feedbacks, with models we can think more clearly and deeply through the implications of the feedbacks we do identify.

19. *Threshold Models with Feedbacks*

Integration of racial, ethnic, and other groups that mark significant lines of social inequality is a vital ideal for a democratic society.

—Elizabeth Anderson

In this chapter, we cover models of threshold-based behaviors. Threshold-based behavior occurs when people's actions change when an external variable exceeds or falls below a threshold. Threshold-based behavior occurs when a person buys a coat when the price falls below $100 or joins a social movement when its membership reaches 1,000 people. Threshold-based behavior is intuitive and easy to analyze, and it often produces counterintuitive results, such as when tolerant behavior produces segregation. Threshold-based models can often produce tipping points. For example, when a person's decision to join a social movement depends on the number of people already in the movement, as more people join the movement the total number of participants is more likely to exceed other people's thresholds to join, creating a tip.

The models in this chapter can be classified as *agent-based models*—computer programs that model each agent individually. Agent-based models allow for more granularity than systems dynamics models, which represent an entire population by a single stock variable. Agent-based models position agents in space and can include behavioral diversity. Those added degrees of freedom offer advantages, but we must keep in mind that too much detail undermines some of our reasons for modeling. We should not, for example, model each neuron in a person's head when building a model of how people choose whether to join a social movement. The optimal level of granularity will depend on the purpose of the model.

The chapter begins with Granovetter's model of riots, followed by a double riot version that models the growth process for start-ups. We then cover two segregation models. The first considers people moving between rooms at a party. The second considers segregation at the scale of cities. Next we introduce the ping-pong model, which includes negative feedbacks and can produce a cycle or an equilibrium. In the discussion at the end of the chapter, we again return to the Game of Life and see how the double-threshold rule creates a combination of positive and negative feedbacks. Positive feedbacks create correlated behaviors that negative feedbacks dampen. We also return to the topic of model granularity.

Granovetter's Riot Model

In a threshold-based model, an individual takes one of two actions, depending on whether an aggregate variable exceeds a threshold. If the variable's value exceeds the threshold, the individual takes one action. Otherwise, she takes the other action. Our first model captures riots and social movements. In it, each person makes a choice to either join the riot or hang back. The decision hinges on the number of people involved in the riot. The model does not take a normative position. The social movement or uprising could be a justifiable revolt against a despot or soccer hooligans destroying property. The model applies to both cases.

The *riot model* assigns each person a threshold. A person joins the riot when the number of joiners exceeds that threshold.[1] Initially, only those who have a threshold of zero join. For the purposes of this discussion, we assume a social movement rather than a riot, so in this case joining may involve standing in a central square or marching. Suppose that on the first day, 200 people with a threshold of zero start a movement. On day two, those 200 people continue to protest, and they are joined by those whose participation threshold lies below 200. If this consists of an additional 500 people, then on day three, people with thresholds above 700 join. This could be several thousand people.

Riot Model

Each of N individuals, indexed by i, has a **riot threshold**, $T(i) \in \{0, 1, ...N\}$. Initially, any individual with a riot threshold of zero, $T(i) = 0$, joins the riot. $R(t)$ equals the number of people rioting at time t. Individual i participates at time t if $T(i) < R(t-1)$.

Analysis of the model reveals that the diversity of thresholds matters at least as much as the average threshold. We can see why by comparing three scenarios involving a population of 1,000 people. In the first scenario, everyone has a threshold of 10, so no social movement occurs. In the second scenario, 5 people have a threshold of zero, 10 people have a threshold of 1, and everyone else has a threshold of

20. In this scenario, 5 people join initially. The next day, 10 more join. Thereafter, no one else joins. In the third scenario, each person has a unique threshold ranging from 0 to 999. For convenience, we can number people from 0 to 999 according to their thresholds, where person i has a threshold of i. In the first period, person 0 joins. The second day, person 1 joins. On the third day, person 2 joins. Each day one more person joins, until all 1,000 people participate. The first scenario has the lowest average threshold, yet it produces no social movement because no one has a threshold of zero. In the second scenario, some people have thresholds of zero, but not a sufficient number to create a widespread movement. Only in the last scenario does the social movement take hold.

The model reveals the importance of the entire distribution of thresholds, not just the mean. It therefore shows the difficulty of predicting which social movements will be successful. The model can also guide action by informing revolutionaries who wish to start an uprising against a despot that in addition to having a core group of people to begin the movement, they also need to create a population of others who will join them. Variants of the riot model can be applied to standing ovations, to changes in political views (the acceptance of gay rights), to fashion changes (wearing bow ties), and to market dynamics (joining in a stock market or real estate bubble). In each case, people's behavior may be approximated by a threshold-based rule, and that threshold varies across people. In each of these contexts, the likelihood of a large event—be it a mass movement or a fad of thicker-rimmed eyeglasses—may depend as much or more on the distribution of thresholds than on the mean value.

Market Creation and Double Riots

The riot model can be extended to cover internet start-ups that create new markets of buyers and sellers. To create a new market, a start-up must create a population of buyers and a population of sellers. A site that matches dog owners and dog sitters needs to sign up dog sitters as well as dog owners. Similar incentives exist for sites that offer package delivery, transportation, or housecleaning. Each must create two populations to succeed, and the populations must grow at approximately the same rate. Otherwise, either the sellers or the buyers will be unable to find a match and they will leave disappointed. In other words, the start-up must create a simultaneous *double riot*.

The successful start-up Airbnb provides a mini case study of a double riot. Airbnb matches people willing to rent a house, room, or apartment with people looking for a place to stay for a short period of time. Airbnb needed to build two populations: renters and people letting out their apartments. People looking for a place to rent would visit the site only if the site had a sufficient number of places available for rent. Therefore, Airbnb needed to sign up people willing to list apartments. The first two launches of Airbnb failed. Listing an apartment on the site required effort—downloading pictures and including other information. No one had an incentive to list until Airbnb had a large population of renters.

Thus, Airbnb needed enough listings to create a riot among renters—that is, to get renters to come to the site. They also needed enough renters to create a riot among those who wanted to list rooms and apartments. Whether Airbnb would take off would depend on the thresholds for the two groups. The bigger problem was getting people to list, as that required more effort. Airbnb overcame this problem by going door-to-door and helping people list their apartments. Once that happened, the renting riot began and the listing riot followed.[2] The business succeeded because the founders were able to bootstrap a sufficient number of initial renters so that a double riot ensued. They constructed the tail, and the tail wagged the dog.

Two Models of Segregation

Our next two models, both developed by Thomas Schelling, explore segregation. People segregate by race and ethnicity at multiple scales. We segregate by nation and by regions within nations. The United States is racially segregated by neighborhoods within cities, and even by tables within school cafeterias. These observations can be read as evidence of intolerance. That inference contradicts the increasing number of interracial and interethnic marriages. How can the same people choose not to live near or even eat lunch with others of different races, yet they choose to marry across racial groups?

We could account for these facts if the people in multiracial marriages belong to different social classes than those who sit at segregated tables. But that is not true. Interracial marriage occurs at all income levels, and segregated lunch tables can be found even at the most elite colleges and universities. Schelling's models can accommodate both sets of facts. They show how tolerant people can produce segregation.

The first of these models, *Schelling's party model,* can be thought of as a mashup of the random walk model and the riot model. The model describes a party that takes place in a house with two rooms. The hosts of the party have invited guests that visibly sort into two types. The types could be men and women, blacks and whites, Spaniards and Australians, or goths and jocks. The key assumption is that each person be able to distinguish everyone else's type.

Schelling's Party Model

Each of N individuals has an observable type A or B. Each person randomly chooses one of two rooms. At each moment a person moves to the other room with probability p. Person i has a **tolerance threshold** T_i and leaves her room if the percentage of people in the room of her type falls below that threshold.

To see how segregation arises despite tolerance, imagine a party with 20 Australians and 20 Brazilians. Each person is tolerant and will remain in a room so long as 25% of the people in the room have the same ethnicity. Suppose that initially one room contains 12 Australians and 9 Brazilians and the other contains 8 Australians and 11 Brazilians. No one feels compelled to move, but there will be random movements between rooms, and these will alter the percentages in each room. In figure 19.1, one room contains 11 Australians and only 4 Brazilians. This configuration hovers at a tipping point: if any of the Brazilians leaves, the percentage of Brazilians will fall below 25%, causing the 3 remaining Brazilians to leave as well. If that happens, Brazilians will never move into that room.

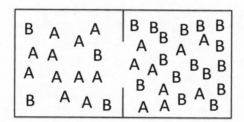

Figure 19.1: A Configuration Poised to Produce Segregation

Recall from the random walk model that a one-dimensional random walk will cross any threshold. The number of Australians in a room is a one-dimensional random walk. Therefore, if the party lasts long enough, segregation is inevitable. Even the best parties only last so long, so not all parties end in segregation. We also know that larger parties are less likely to segregate because the random walk has to cross a much higher threshold when the party has more people. At a party of 1,000 people in two equal sized rooms, we would not expect the proportion of either type to fall below 25% of either type in a room. We would expect that to happen at a party of 12 people. We should therefore expect more segregation at small parties.

We should also expect more segregation when people have diverse tolerance thresholds. To see why, assume 10 Brazilians and 10 Australians and assign each person a tolerance threshold between 5% and 45% in such a way that each group has an average tolerance threshold of 25%, as shown in figure 19.2.

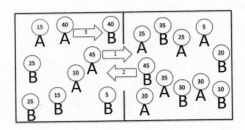

Figure 19.2: Relocations Caused by Heterogeneous Thresholds

The room on the left contains 5 Brazilians and 4 Australians, so the proportion of Australians is less than 45%, causing the least tolerant Australian to move to the room on the right (denoted by the arrow #1). When this Australian moves to the other room, she reduces the proportion of Brazilians in that room, causing the least tolerant Brazilian to move (denoted by arrow #2). These two relocations lower the percentage of Australians in the room on the left below 40%, causing the second-least-tolerant Australian to follow her into the room on the right (denoted by arrow #2). A cascade ensues. However, as shown in figure 19.3, the result need not be full segregation, as the most tolerant individuals are comfortable in either room. The model produces two effects of diverse thresholds: they make the tip to segregation more likely and make complete segregation less likely, as very tolerant people are content in either room.

This model can help explain variations in gender ratios across profession—why more women work as nurses and more men work in sales. Those differences could be due to preferences, but they could also arise if some people prefer to work with people of the same gender. This can be made more formal in a *revolving-door model*, which makes two empirically based assumptions: (i) women who exit a profession choose a new profession with more women, and (ii) women leave professions at a higher rate then men.[3] If women in the life sciences leave biomedical research at a faster rate than men and take jobs in professions such as health care that employ more women, their actions increase gender segregation in both professions.

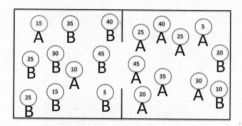

Figure 19.3: Segregation Resulting from Heterogeneous Thresholds

Schelling's Segregation Model

Schelling's segregation model places agents in distinct locations in geographic space, a checkerboard. Otherwise, it is identical to the party model. It assumes two types of people and makes the same behavioral assumption as Schelling's party model.

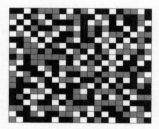

Figure 19.4: Initial Configuration in Schelling's Segregation Model

> # Schelling's Segregation Model
>
> N individuals, each of whom has a type A or B, are randomly arranged on an M-by-M checkerboard with room for open spaces. Each person i has a **tolerance threshold**, T_i, and relocates to a random new location if the percentage of the people of her same type on the eight neighboring squares falls below her threshold.

If individuals have an average threshold near 50%, the model produces segregation, as shown in figure 19.5. Segregation arises because the individuals consider only their local neighborhoods, which have at most eight occupants. Almost any random initial configuration includes some people surrounded by others of the opposite type. If individuals move into regions with more individuals of their same type, they can cause relocations by people of the other type. As the relocations accumulate, segregation occurs. We need not walk through the logic again as to why threshold diversity exacerbates these effects.

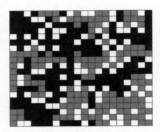

Figure 19.5: Configuration After Relocations in Schelling's Segregation Model

That tolerant people can produce segregated residential patterns serves as the foundational insight of Thomas Schelling's seminal book *Micromotives and Macrobehavior*: what occurs at the macro level need not align directly to the micro level motivations of individuals.

Threshold Models with Negative Feedbacks

Our next model, the *ping-pong model*, assumes threshold-based behavior that produces negative feedbacks. Recall that negative feedbacks can stabilize a system or, as we saw in the predator-prey model, produce cycles. The model assumes a finite number of entities, which could be people or mechanical, biological, or chemical devices. In each period, each entity takes either a positive (+1) or a negative (-1) action. In the first period, each agent chooses a random action. The initial state equals the sum of the actions. All subsequent states of the system equal the previous period's state plus the average of all actions and a random term. Each entity has a threshold, drawn from a distribution, and chooses the action that reduces the absolute value of the state if the absolute value exceeds its threshold. Put simply, if the state's value exceeds an entity's threshold (either negatively or positively), the entity does what it can to reduce the magnitude of the state.

Ping-Pong Model

Each entity in a population of size N randomly takes an initial positive (+1) or a negative (-1) action. The initial **state of the system**, S_0, is set equal to zero. All future states of the system, S_t, equal the average action plus a random variable:

$$S_t = \frac{\sum_{i=1}^{N} A_i(t)}{N} + S_{t-1} + \epsilon_t$$

Each entity i has a **response threshold** $T_i > 0$ drawn uniformly from the interval $[0, \text{RANGE}]$. An entity takes the same action as before if the magnitude of the state, $|S_t|$, is less than its threshold and takes an action to reduce the magnitude of the state otherwise.

If $|S_t| \leq T_i$, $A_i(t+1) = A_i(t)$, otherwise $A_i(t+1) = -\text{sign}S_i(t)$ where ϵ_t is randomly drawn from $\{-1, +1\}$.

Multiple applications of the ping-pong model should come to mind. Here are two. People allocate time and resources to multiple charitable causes. If cause receives too much attention or money, people may donate money to other charities to equalize their revenues. Or a country may have two UNESCO World Heritage sites that rely on volunteers. If one site has an abundance of volunteers, then some may reallocate their energies to the other site.

As foreshadowed by the name, the ping-pong model can produce cyclic behavior around the equilibrium. In one period too many people choose one action, and in the next too many choose the other. When all entities have a threshold of zero, all entities choose action one (+1) in one period and action minus one (-1) in the next.

To explore how threshold diversity contributes to whether people behave like ping-pong balls or find an equilibrium, we consider three cases, each involving 100 people. In the first case, we assume thresholds are uniformly distributed between zero and 10. If the state in the first period equals -6, it will exceed the threshold of approximately 60 people. Approximately 30 of these 60 will have taken action one and will switch actions. The sum of the actions will now exceed 50, so the new state of the system (the average of the previous two periods) will exceed 20. This value exceeds all thresholds, producing the ping-pong effect shown in the top graph in figure 19.6.

If we increase the threshold diversity making them uniformly distributed between zero and 100, the ping-pong effect all but disappears. To see why, assume

221

that the state equals -6 in the first period. On average, only six people's thresholds will be met. If three change actions, the state will move toward zero. This dampening of the deviation can be seen in the bottom graph in figure 19.6, which corresponds to thresholds between zero and 100. As we might expect, if we consider a moderate case, with thresholds uniformly distributed between zero and 60, we see a more moderate cycle, as visible in the middle graph. Thus, in systems with negative feedbacks, threshold diversity produces stability, but it has the opposite effect in the models with positive feedbacks.

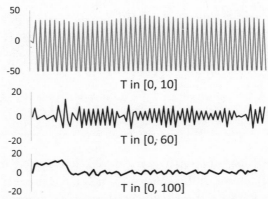

Figure 19.6: Time Series for Ping-Pong Model for Three Threshold Ranges

Summary: Model Granularity

The basic logic of feedbacks is straightforward: positive feedbacks reinforce actions, negative feedbacks dampen them. A system with only positive feedbacks will either blow up or collapse. A system with only negative feedbacks will either stabilize or cycle. A system with both positive feedbacks and negative feedbacks has the potential to produce complexity.

In systems dynamics models, feedbacks operate on stock variables (the number of hares) and rates (increasing the rate of purchases at the bakery). Agent-based models, such as the threshold models we covered here, produce feedbacks as a result of individual actions. These more granular models can include a diversity of thresholds, which exacerbates the effects of positive feedbacks and tempers negative feedbacks. Diversity in the tail of the distribution made riots much more likely. Diversity in tolerance thresholds increased segregation. Diversity of responses in systems with negative feedbacks stopped the wild fluctuations that existed with homogeneous feedbacks. In models of economic competition among firms, hetero-geneous production costs can play a similar function. As prices rise or fall, firms vary in their responses owing to their differences in costs.[4]

This difference between the two types of models raises the issue of model gran-ularity. Does one create a variable (or box) labeled "hares" and describe how the number of hares increases and decreases as a function of other variables, or does one model each hare individually? Disaggregating variables increases descriptive accuracy, but models are not graded on that criterion. Remember Box's dictum that all models are wrong, and Borges's map that was the same scale as the real world. Many modelers, including Einstein, have taken the position that we should seek the correct level of granularity. If writing a model to explain the force that is exerted by a human arm, for example, we need not include DNA.

For studying social systems, no ideal level may exist. We may need to explore them at multiple levels of granularity. By constructing multiple models, each at a distinct granularity, we can engage in a dialogue across those models. If we are attempting to make sense of trade patterns between Sweden and Finland, we could start with the two countries as variables and identify broad macro patterns. We might then disaggregate each country's imports and exports into industries, and then into firms within those industries. Industry-level data would enable us to better explain past patterns and make more precise predictions of future patterns. The deeper dive into firms within those industries, including information such as their cost structures and growth trajectories, could produce even better results, but we would need a lot of information to construct a useful model with that many moving parts. We might even model the leadership within those firms. Most likely, this last level of unpacking would yield few benefits, but it could be that some leaders are known to pursue expansionist strategies.

To summarize, finer granularity need not be better. Models can have too much detail. Even when we can understand more granular models, we benefit from con-structing coarser models as a point of comparison. By comparing the differences in the models' predictions, explanations, and policy prescriptions, we see how as-sumptions affect results. We see the conditionality of our assumptions. Thus, our many models should not only differ in the variables they include, they should also differ in the extent to which they disaggregate the same variables.

Algorithmic Riots

The riot model and the ping-pong model can provide insights into stock market crashes and price rebounds. We consider two illustrative cases here. The first occurred on Monday, October 19, 1987, when the Dow Jones Industrial Average fell by over 22%. The next day, this collapse reverberated around the globe. The cause of the crash remains a topic of analysis. At the time, the United States economy was in its fourth year of expansion. In the first eight months of the year, the Dow Jones Industrial Average had risen over 40%. Despite or perhaps because of this run-up in values, many people believed stock prices were overvalued. On the Sunday prior to the crash, the United States secretary of the treasury, James Baker, threatened to weaken the dollar if Germany did not lower tariffs, a comment that at the time did not seem of great import. The next day, the market crashed. Fourteen months later, the market returned to its previous value.

To apply the models, we assume a single financial asset representing the entire market. We assume that each person holding this asset has a *crash threshold*. If the price of the asset falls more than the crash threshold in a given day, the investor sells the asset, taking her money out of the market. This rule captures the behavior of trend or *noise traders* and creates a version of the riot model. If some percentage of investors woke up on October 19 and decided to sell a substantial amount of assets, they would have caused a drop in the market. If that drop exceeded other investors' crash thresholds, those others would sell assets as well, causing a downward spiral. The result is a classic positive feedback loop. Selling begets lower prices, which causes more selling.

We now add insights from the ping-pong model to the analysis. If prices falls too low, some people will apply a second rule, a *bargain threshold rule*. According to this rule, a person buys if the price falls below this value. Here, our investor acts based on value, not trends. When prices fall dramatically, the bargain thresholds create a negative feedback. Buyers rush in to buy at a bargain price, halting the price fall.

Actual markets are more complex than this simple account of sellers with thresholds and buyers waiting in the wings. The stock market contains many types of traders, including large institutions, pension funds, foreign governments, speculators, portfolio insurers, and speculators, as well as small investors.[5] As a result of this diversity, someone is almost always willing to buy as prices fall, producing the negative feedback necessary to stabilize the market.

Early analyses of the crash emphasized the prevalence of (computer) program trading. These are threshold-based rules encoded in computer programs. Rules such as *sell all stocks if the market index falls below some set price* were carried out automatically, with no human supervision. Most analysts now believe that program trading contributed to the 1987 crash but was not the primary cause. More detailed analyses of the 1987 crash reveal that a large number of portfolio insurers—traders who guarantee a rate of return to the portfolios of their investors—produced strong positive feedbacks that were not tempered by negative feedbacks. As the market fell, these portfolio insurers sold off stock to prevent losses. As the crash unfolded, these insurers sold more and more stock. In effect, insurers acted as if they were populations of individuals with diverse thresholds. One portfolio insurer sold over $1 billion in stock. To put that in perspective, only $20 billion in stock was sold that entire day.

The second crash, the May 6, 2010, "flash crash" dropped the Dow Jones Industrial Average by 5% in three minutes. It was the result of algorithmic trades. Owing to the complexity and speed of modern financial markets, no one knows for certain what exactly caused the flash crash. We know that a large mutual fund made a huge sell order, dumping over $4 billion of stock futures into a market containing high-speed trading algorithms that try to exploit beneficial trades. The algorithms sensed a price trend and starting executing trades at breakneck speed. Imagine the riot model at high speed. This produced a toxic market, in which traders worry that large institutional investors know something that they do not know and so they exit the market.[6] Many of the algorithms stopped trading given the abnormal market behavior; other algorithms kept selling, and a crash ensued, all in the span of a couple of minutes. Within twenty minutes, the bargain rules went into effect and, as predicted by the ping-pong model, brought prices back up (nearly) to the original price.

20. *Spatial and Hedonic Choice Models*

Our theory is, if you need the user to tell you what you're selling, then you don't know what you're selling, and it's probably not going to be a good experience.

—Marissa Mayer

In this chapter, we cover models of individual choice over alternatives represented by their attributes. These models were developed to capture consumer choices. A person buying a house takes into account its square footage, the number of bedrooms and bathrooms, and the quality of construction. These can also be applied to college deans of admissions or hiring directors as they select among applicants or to voters as they decide between candidates. An admissions dean considers an applicant's SAT scores, grade point average, and extracurriculars. A voter evaluates candidates based on their positions on education, infrastructure, crime, and taxes. In addition to helping us understand individual choices, these models provide insights into why we have the choices we do—for example, why we have so many choices of breakfast cereals.

In the models we cover, we characterize some attributes as *spatial* and others as *hedonic*. A spatial attribute, such as the color of a jacket or the thickness of a slice of bread, has no best value. Each individual prefers a particular amount of the attribute: a consumer of baby back ribs has a preferred level of spiciness, and an amateur downhill skier has a preferred angle of descent on a slope. The model assumes that the closer a product's attributes are to a person's ideal point, the more the person values the product. These ideal points vary across people: one person may prefer spicier ribs than another.

On hedonic attributes, more (or in some cases less) is always better. People prefer longer battery life in a smartphone, more square footage in a house, more durability in the soles of shoes, and better gas mileage in their cars. Most product choices are hybrids—they contain both spatial hedonic attributes. A car's color will be a spatial attribute. Its gas mileage is a hedonic attribute.

Throughout the chapter, we assume that people choose the alternative that they value the most. We do this for reasons mentioned in Chapter 4, on modeling human behavior: rational behavior provides a benchmark, is analytically tractable, makes a unique prediction, and fits empirically when the situation is repeated and the stakes are large.

Models of spatial and hedonic competition are widely used in economics and political science in part because they can be taken to data.[1] In this chapter, we get a hint of their applicability. We begin with a spatial model of product competition. We then apply the model to politics and show how it can be used to analyze status quo effects, agenda power, and the influence of veto players. We then cover the hedonic attribute model and a hybrid model to reveal insights into price competition. Along the way, we show how to take data to the models to infer the positions of candidates and judges based on their votes on bills and legal cases and to infer implicit prices for unpriced attributes such as cleaner air or a shorter commute.[2]

The Spatial Competition Model

The *spatial competition model* assumes alternatives defined by a set of attributes and consumers defined by ideal points. The simplest version of the model considers products with a single attribute. In Hotelling's original spatial model, that attribute was geographic location.[3]

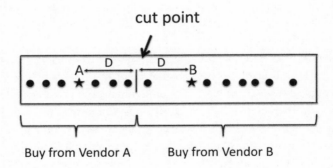

Figure 20.1: Hotelling's Geographic Model of Ice Cream Vendors on a Beach

Hotelling's model assumes a collection of consumers spread along a beach, represented by circles in figure 20.1, along with two ice cream vendors, denoted by *A* and *B*. Each customer buys one ice cream from the nearer vendor. The *cut point* is an equal distance between the two vendors and determines who buys from whom. The seven consumers to the left of the cut point buy from vendor *A*, and the six consumers to the right of the cut point buy from vendor *B*.

Given the idea of consumers preferring closer goods, we can reinterpret distance more abstractly. For example, we could imagine the ice cream vendors being in the same location but offering different levels of butterfat in their ice cream. The same figure can represent vendor *B* offering a creamier product than *A*. In the reinterpretation, the consumers' locations are not physical position on the beach but preferred levels of butterfat.

We can again apply one-to-many model thinking and use this same model to analyze political competition. The *Downsian model* reinvents Hotelling's geographic space as an ideological continuum from left to right. We can reinterpret figure 20.1 as follows: vendor *A* represents a liberal political candidate, vendor *B* a more conservative candidate, and the circles represent the ideological ideal points of voters. To extend the analogy, we assume voters prefer nearer candidates.

The shift from geographic locations of firms and product attributes to political ideologies involves a transition from physical attributes such as location and butterfat level to the more abstract concept of ideology. While we have clear measures of physical attributes, assigning ideologies requires a method for translating the actions of candidates into numbers. If the candidates have voting records, we can assign ideologies by first gathering all of the votes the candidate has cast. We should ignore all votes that lack an ideological component—unanimous proclamations establishing National Milk Day and the like. On all other votes, we can rely on expert opinion to assign a liberal position and a conservative position. A candidate's spatial location on the interval can be set equal to the percentage of the time she votes the conservative position.[4] A candidate who always takes the conservative position is placed to the far right. A candidate who votes conservative half of the time and liberal half of the time is placed in the center.

With this model, we can adjudicate claims that American political parties have become more ideologically distinct by empirical tests. One analysis, shown in figure 20.2, reveals a marked and increasing polarization of the average ideal points in each party. This does not prove that polarization has increased, but it provides evidence. The analysis also reveals that the polarization is mostly due to a Republican shift to the right.

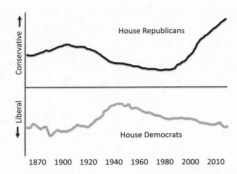

Figure 20.2: Increasing Ideological Polarization in Congress

Increasing the Number of Attributes

The general spatial competition model includes an arbitrary number of attributes. A couch can be represented by physical dimensions: length, width, and depth, type of construction, and type of upholstery. The value (or utility) that a consumer obtains from a product depends on the product's distance to her ideal point across these same dimensions. We can write this value function as a constant term minus the distance between the alternative and her ideal point.[5]

Spatial Competition Model

An **alternative** consists of N **spatial attributes**: $\vec{a} = (a_1, a_2, \ldots, a_N)$.

An **individual** is represented by an **ideal point**: $\vec{x} = (x_1, x_2, \ldots, x_N)$.

The **payoff** (utility) to an individual from an alternative equals
$$\pi(\vec{x}, \vec{a}) = C - (x_1 - a_1)^2 - (x_2 - a_2)^2 - \cdots - (x_N - a_N)^2$$
where $C >$ is a constant.

Example: $\vec{x} = (3, 4, 6)$, $\vec{a} = (2, 1, 8)$, $C = 20$:
$$\pi(\vec{x}, \vec{a}) = 20 - (3-2)^2 - (4-1)^2 - (6-8)^2 = 6$$

In the general spatial competition model, two chocolate bars might be represented by the percentages of cocoa and amounts of sugar they contain, as shown in figure 20.3. The cut line will be the perpendicular bisector of the line connecting the two products. Consumers with ideal points to the left of the cut line prefer A, and consumers to the right prefer B.[6]

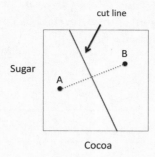

Figure 20.3: Product Attributes of Two Chocolate Bars and the Cut Line

The model can also accommodate any number of products. To add a third choco-late bar, we add another point in attribute space. To determine which bar consumers buy, we then draw additional cut lines, as shown in figure 20.4. The multiple cut lines carve up the space of ideal points into three regions, known as *Voronoi neigh-borhoods,* that partition the space of ideal points based on their distances to the products.

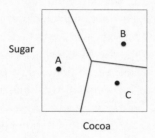

Figure 20.4: A Spatial Model with Three Products (Voronoi Neighborhoods)

The Downsian Model of Spatial Competition

We next apply the model to explain candidate positioning. To do so, we assume vote-seeking candidates, who place primary emphasis on winning elections. We begin with an example to think through the incentives of candidates. Figure 20.5 shows two scenarios with thirteen voters and two candidates. Recall that voters prefer the nearer candidate. In the top diagram, the liberal candidate, denoted by L, receives five votes, while the conservative candidate, denoted by R, receives eight votes. In the lower diagram, the liberal candidate moves to the center, attracts seven votes, and wins the election.

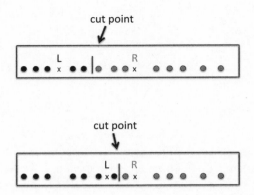

Figure 20.5: Party Moving to Center to Win Election

The liberal candidate has an incentive to move to the center. By the same logic, so does the conservative candidate. The conservative candidate could move to the left, though remaining right of L, and also win. Continuing with this reasoning, the liberal candidate, L, could take a position even closer to the median voter's ideal point. If we continue to apply this logic, we see that candidates should converge on that point. This result is known as the *median voter theorem*.

The median voter theorem can be interpreted in strong or weak form. In strong form, it implies candidates adopt identical positions at the median voter's ideal point. That clearly does not hold empirically. In weak form, it implies that candidates have an incentive to move toward moderate positions. Empirical evidence suggests that this does happen. During the course of an election, candidates move toward the center. That movement need not be a mad rush. Candidates who possess ideological convictions or benefit from core supporters at the extremes will reposition with caution.

This model reduces each candidate and each voter to a single ideological point, a strong assumption. Czech writer and politician Václav Havel takes exception to one-dimensional ideological projections: "To wandering humankind, it [ideology] offers an immediately available home: all one has to do is accept it, and suddenly everything becomes clear once more, life takes on new meaning, and all mysteries, unanswered questions, anxiety, and loneliness vanish. Of course, one pays dearly for this low-rent home: the price is abdication of one's own reason, conscience, and responsibility, for an essential aspect of this ideology is the consignment of reason and conscience to a higher authority.[7]" Havel makes a good point. We should not abdicate reason for ideology. Models provide us with tools to reason better. This particular model helps us to understand why politicians act as they do. Using data, we can determine the confidence of each politician's placement on the left-right interval. A politician who always takes a moderate position can be placed in the middle of the interval with high confidence.

Incidentally, Havel's denial that he can be reduced to a point could be tested with data. If his criticism holds, if we cannot pin down his ideology based on votes, we need not abandon the model. We could represent the uncertainty over Havel's ideology by assigning him an interval rather than a point. Or, we could construct a *time series* of his measured ideology to see if he remains consistent. Studies of the ideological positions of Supreme Court justices show that some justices become more liberal as they spend more time on the bench.[8]

Last, we could increase the dimensionality of the model. A two-dimensional model could distinguish between social policies and fiscal policies. The model can then capture a politician who advocates liberal positions on social policies and conservative positions on fiscal policies. For the United States Congress, one dimension explains approximately 83% of the variation in votes. Adding a second dimension adds only another 4%.[9]

In addition to allowing us to model preferences with greater accuracy, adding dimensions also changes our theoretical findings. We start with a two-dimensional model. From the one-dimensional model, we know that if a candidate is not located at the median on either issue, then the other candidate could win an election by matching the first candidate's position on one issue (thus making that issue irrelevant) and taking the median position on the other issue. Similarly, if one candidate takes the median position on one issue but does not on the second issue, then the other candidate could take the median position on both issues and win the election. It follows that the only position that has the possibility of not being beatable is the *two-dimensional median*. Figure 20.6 shows how the two-dimensional median, represented by a circle, can be defeated. If the square candidate positions to the left on issue one and to the right on issue two, she produces a cut line in which she wins three votes.

Building intuition from this example, we see that the two-dimensional median will be unbeatable only if voter ideal points are arranged such that fewer than half lie in every direction from the median, a condition called *radial symmetry*. Radial symmetry would be satisfied if voters' ideal points were uniformly distributed across a disc or a square, a very strong assumption. This result, that any position can be

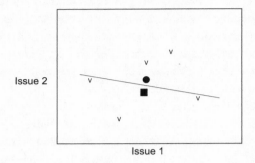

Figure 20.6: The Two-Dimensional Median Loses to a Challenger

defeated, is known as *Plott's no-winner result*. It holds in two or more dimensions.[10]

The difference in results is stark. A one-dimensional model implies that candidates position at the median. Higher-dimensional models imply that they should not. Which type of model do we believe? We should place complete faith in neither model, but instead gain insights from both. The one-dimensional model reveals a powerful incentive for vote-seeking candidates to move toward moderate policies. The higher-dimensional models demonstrate the limits to those incentives. No position guarantees a win, so we should not expect an equilibrium. We should instead expect complexity, an endless dance of competition for votes through coalition building.[11]

Status Quo Effects, Agenda Control, and Veto Players

We can also apply the Downsian model as a lens for interpreting the ideological dimensions of bills passed by committees, councils, legislatures, and presidential systems. Here again, the key will be to map pieces of legislation onto the same single ideological dimension as committee members. We consider three strategic effects here: the influence of current policy (*status quo effects*), the power of agenda control, and the effect of adding veto players.

Throughout, we rely on an example involving a committee of three people with ideal points at 40, 60, and 80, in which the committee member with an ideal point at 40 gets to propose a policy that must be approved by a majority. Figure 20.7 shows the effect of the status quo on final policy. If the status quo is at 80, she needs legislation that the median voter, the committee member at 60, prefers to the status quo. In this case, the median voter would accept a proposal at 40, the proposer's ideal point, as it is just as good as the status quo.[12] If we move the status quo to 70, the median voter would reject a proposal of 40. The proposer has to offer a policy at 50. Finally, if the status quo sits at the median voter's ideal point of 60, the proposer has no power. We can thus draw the following inference: the proposer has the most power when the status quo has an extreme value.

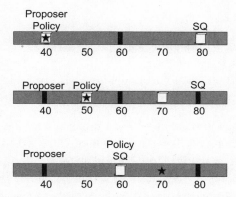

Figure 20.7: The Effect of the Status Quo on Final Policy

This insight applies in any context where people vote and opinions can be mapped to one dimension. The head of a nonprofit making a proposal to her board of directors to increase spending on affordable housing efforts has little agenda power if current spending levels align with the wishes of her median board member. She can have power if current policies do not align with that board member's ideal.

To show the power of proposing, we consider the case where the status quo sits at 70, shown in figure 20.8. The top diagram shows the previous case, where the proposer has an ideal point at 40 and proposes a policy at 50. The middle diagram shows the case in which the median voter can propose his ideal point of 60 and also obtain the vote of the committee member at 40. The bottom panel shows the case where the proposer has an ideal point of 80. She cannot offer a policy that both she and the median voter prefer to the status quo, so she accepts the status quo.

This exercise reveals the limit of proposer power. Legislation may move in the direction of the ideal point of the person in power, but as we have learned, the extent of that power will be mitigated by the representativeness of the status quo.[13]

Last, we can use this same model to consider multiple levels of government and an increased number of veto players. Here, we interpret the three committee members as the median voters in the House and the Senate, and the President. For

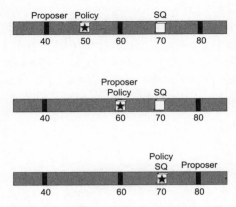

Figure 20.8: The Effect of the Proposer on Final Policy

legislation to pass, each of these three must prefer it over the status quo. In this scenario, each has veto power. Refer again to figure 20.8, where the status quo sits at 70. If all three voters must approve any change, then no proposal can defeat the status quo. Any policy to the left of 70 will be vetoed by the voter at 80. Any policy to the right of 80 will be vetoed by both of the other voters.[14] If all three voters can veto legislation, there will be no new legislation unless the status quo lies outside the interval from 40 to 80—that is, if the current status quo is to the left or right of anyone's ideal point. The model reveals a tight link between the number and ideological diversity of veto players and the extent of gridlock. That insight holds more generally. Organizations with diverse veto players will be unable to take action.

The Hedonic Competition Model

The *hedonic model of competition* also represents the alternatives (typically products) by attributes. However, in this model, the attributes consist of quality, efficiency, or price, in which more (or, in the case of price, less) of the attribute is always preferred. To capture heterogeneity, the model allows individuals to attach different weights to the dimensions.

Using *linear regression,* we can infer implicit values for the attributes of goods using the hedonic competition model (also known as the *Lancaster model*). The application is straightforward. The model assumes the payoff to be a linear function of the product's attributes and the person's weights. If we have data on the selling prices of thousands of homes as well as the relevant attributes of those homes (square footage, number of bedrooms and bathrooms, quality of local schools, size of yard, and quality of construction), a regression will produce the average weight (in dollars) of each attribute for those people who bought the houses. This is known as *hedonic regression*.

Hedonic Competition Model

An **alternative** consists of N **valence attributes**: $\vec{a} = (v_1, v_2, \ldots, v_N)$.

Individual preferences are captured by **weights** $\vec{x} = (w_1, w_2, \ldots, w_N)$ assigned to the attributes.

The **payoff** (utility) to an individual from an alternative equals
$$\pi(\vec{x}, \vec{a}) = w_1 \cdot v_1 + w_2 \cdot v_2 + \cdots + w_N \cdot v_N$$
.

Example: $\vec{v} = (3, 1, 2)$, $\vec{w} = (4, 2, 5)$:
$$\pi(\vec{v}, \vec{w}) = 4 \cdot 3 + 2 \cdot 1 + 5 \cdot 2 = 24$$
.

Some of those attributes, such as a swimming pool or a new kitchen, have market prices. As a check on the model, we can compare the estimated prices to the market costs. If the regression shows that a swimming pool adds $150,000 to the price of a house and swimming pools cost $15,000, we know the model is missing attributes. If the regression shows the added value to be only $8,000, then this likely means that people do not recover the full costs of adding pools.

Other attributes, such as the commute time from the house to the city center, do not have market prices. In those cases, the regression produces an implicit price for that attribute, and that implicit price can be informative. The table below shows hypothetical price data for six houses.

House	Square Ft.	# Bedrooms	Commute Time	Price
204 Pine	2000	3	30 min	$200,000
312 Maple	3000	5	60 min	$380,000
211 Geddes	2500	4	10 min	$310,000
342 Martin	1500	2	20 min	$150,000
125 Clark	2000	4	30 min	$220,000
918 Brown	4000	6	50 min	$360,000

If we assume the houses are identical on all other attributes and we run a hedonic regression, we obtain the following equation:

$$\text{Price (\$)} = 100(\text{Square Ft}) + 20{,}000(\text{\# Bedrooms}) - 2000(\text{Commute Time})$$

The regression equation estimates that people value each additional square foot at \$100, each bedroom at \$20,000, and each minute saved commuting at \$2,000 over the period of home ownership. A person who lives in a house for twenty years spends 4,000 to 5,000 days commuting. If we take the lower number, each extra minute of daily commute time results in 4,000 minutes, or over 60 hours, commuting. The \$2,000 estimate equates to around \$30 an hour. In other words, people pay for proximity "as if" they are paying themselves \$30 an hour to sit in traffic.[15]

A Hybrid Model of Product Competition

The spatial and hedonic models differ in how they represent preferences over attributes. In the *spatial model of competition,* each person has a preferred level of each attribute, and his value for an alternative increases as it gets nearer his ideal point on those dimensions. In the hedonic model, people prefer either more or less of each attribute.

Many of the choices we model—over consumer goods and services, ideal life partners, public policies, religions, and job applicants—include both spatial and hedonic attributes. We may each have our own preferred level of crispiness for french fries, yet we all prefer to pay less per serving. Crispiness is spatial. Price is hedonic. Employers likely differ in the personality characteristics they look for in potential employees. Some firms prefer extroverts. Others may prefer introverts. All firms prefer more honesty and integrity. Thus, personality type is a spatial attribute, while honesty is a hedonic attribute.

We can thus create a hybrid model in which the alternatives contain both spatial and hedonic attributes. This model can be used to analyze market entry, product differentiation, and the extent of price competition. If we return to our example of the chocolate bars, before choosing a new product's attributes, an entrant might first place the three existing products in attribute space and then survey consumers to learn about the distribution of their ideal points. The entrant could then estimate the Voronoi neighborhoods for her proposed product. If that neighborhood contains few consumers, she should not expect substantial sales. Any entrepreneur considering entering a market can take this approach. A boot designer can plot existing designs of insulated boots, of which there may be dozens, and find that none come in shiny patent leather. Someone designing a smartphone app for making to-do lists can map the features of existing apps, measure total market demand, and project possible sales.

We can visualize a price reduction in the spatial model of competition as a movement of the cut line. Refer back to figure 20.3 showing the two chocolate bars. The cut line corresponds to the ideal points of consumers who are indifferent between *A* and *B*. If the firm producing *B* lowers its price, and if consumers prefer to pay less for candy bars, then this will shift the cut line toward *A* and increase *B*'s market share. We do not need the model to know that *B* lowering its price should increase its market share. We do need the model to estimate the magnitude of that effect. The key will be to distinguish between *crowded markets,* with a large number of products in a low-dimensional attribute space, and a *sparse market,* where there are few competitors. In a crowded market, each product has a small Voronoi neighborhood. In a sparse market, the Voronoi neighborhoods are huge.

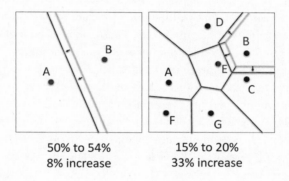

50% to 54%   15% to 20%
8% increase   33% increase

Figure 20.9: Price Competition in Sparse and Crowded Markets

Price changes have different effects in the two types of markets. Figure 20.9 shows the effects of a hypothetical 10% price reduction in candy bar *B*, from $2.00 to $1.80. The diagram on the left shows a sparse market. Lowering the price for *B* shifts the cut line between product *A* and *B* and increases *B*'s market share from 50% to 54%, an 8% increase in *B*'s market share. The 10% price drop and 8% sales increase reduces revenue by 3%. Lowering prices would be a bad idea. The diagram on the right shows a crowded market with seven types of candy bars. Here, the price drop has a smaller effect on *B*'s absolute market share, an increase of 5% from 15% to 20%, but this 5% represents a larger proportional increase (33%) in *B*'s market share. The overall effect is a 20% increase in revenue.[16] Thus, the model predicts stronger price competition in crowded markets than in sparse markets, and extreme competition for *commodities*: products that are indistinguishable, like crude oil, pork bellies, and red wheat #2. It predicts less price competition for high-end fashion goods, where designers can sustain substantial price markups because product dimensionality creates a sparse market.

This relationship between the number of attributes and the extent of price competition suggests that a good strategy would be to add new attributes. This would make the market more sparse, reduce price competition, and lead to higher profits. Even if that inference is correct, the strategy may be easier stated than accomplished. People must value the new attribute. For each successful attribute—cordless stereo speakers—one can find multiple failed attempts—Bic's ill-fated disposable underwear.

Summary

The spatial model of competition, the hedonic model of competition, and a hybrid of the two provide a framework within which we can represent different products, political candidates, or even job applicants. These models can measure ideological positions, price implicit attributes, and evaluate potential market entry positions. They generate insights into how market competition creates an incentive for differentiation, how political competition creates an incentive for convergence, and how price competition should be more intense for products with fewer attributes.

In the models, we make rather strong and empirically dubious assumptions. For example, we assume that people do not change their preferences, and that they do not succumb to social pressures. If that were so, why do firms and politicians spend enormous amounts of money trying to change preferences? We could shrug off this criticism by referring again to Box's dictum that all models are wrong.

We can also construct a more nuanced response that distinguishes between *fundamental preferences*, the outcomes that a person desires, and her *instrumental preferences*, the person's preferences over the attributes that produce the fundamental outcomes. A student's fundamental preferences may strike a balance between being popular, healthy, and scholarly. She may pursue these fundamental ends through instrumental actions—waking early, going to the juice bar, and completing her homework so as to have time to be social in the evenings. Her choice of the fruit shake helps her to achieve a fundamental preference for good health. It is an instrumental preference. If she comes across a scientific paper revealing the high sugar content in fruit shakes, she may switch to drinking water. If so, her instrumental preferences change even though her fundamental preferences do not. Once again, we see how a model is not an end in itself but provides an architecture to structure our thinking.

Many Models of Value

In a market economy, we can measure an individual's value for a good by the price she is willing to pay. An individual might value a pastrami sandwich at $7, a painting by Goya at $3 million, and a one-acre lot in Ocala, Florida, at $75,000. Many economic models ignore the source of these valuations. George Stigler famously quoted the sensualist Mitya from Dostoyevsky's *The Brothers Karamazov,* who said, *"De gustibus non est disputandum"* (In matters of taste, there can be no disputes). The models we cover say less about tastes than about how tastes translate to the monetary values people assign to goods.

Hedonic attribute model: This model explains a good's value based on intrinsic attributes. Differences in valuations depend on different underlying preferences over the good's attributes. These attributes could be the physical components of the good. A pastrami sandwich consists of rye bread, six ounces of pastrami, Swiss cheese, mustard, pickles, and onions. Its value can then be written as a weighted linear combination of those components. More elaborate hedonic models can include interaction terms. The pastrami may be even more valuable if served on grilled rye bread.

Coordination model: This model explains prices as socially constructed. The value of a Goya painting depends on what other people believe its value to be. Initially, people have beliefs or opinions about the value of the painting. They then interact with other people in their social network and update their beliefs. Two people could both set their values equal to the mean of their two values; one person could change her value to match the other's value; or each person could move their valuation in the direction of the others. Given any of these three assumptions, valuations converge locally. People connected to one another will have similar valuations. The ultimate value assigned to a good will depend on the initial distribution of values, the social network, and the order in which pairings occurs.

Predictive models: This model explains prices as forecasts of future value. The value of the one-acre lot in Ocala, Bitcoin, or a stock depends on how much people will pay for them in the future. These valuations depend on predictive models, which in turn depend on attributes and categories. We might categorize Ocala as warm, low-tax, and inland. Variation in people's valuations arises from different predictive models. Investors use multiple predictive models. These models may rely on attributes or, as in the case of valuing Bitcoin, also make assumptions about coordination.

These three models provide three distinct explanations for the value of a good. No one model will be best in all cases. Each will have cases where it works best. The models function as arrows in our quiver. The value of a pastrami sandwich most likely derives from its intrinsic properties. The value of the Goya painting may be largely socially constructed: a painting has value if people believe it has value. The price of the land in Florida likely depends on valuations derived from predictions of future real estate values.

21. *Game Theory Models Times Three*

Deductive reasoning travels from the most abstract to the least abstract. It begins with a set of axioms and uses the laws of logic and mathematics to manipulate them to form predictions about the world.

—Rachel Croson

Game theory models strategic interactions. Many of the models that follow, including our models of cooperation, signaling, mechanisms, and collection action, involve games. We do not take up the analysis of games in much depth because entire books are devoted to the subject. Our goal will be to provide a gentle introduction. To that end, we present examples of the three main classes of games: normal-form games, in which players choose from a discrete set of actions (typically two); sequential games, in which players choose actions sequentially; and continuous-action games, in which players can choose actions of any magnitude or effect size. These examples introduce the main concepts, help us to understand later models, and add value in their own right.

The remainder of the chapter has four parts. We begin by covering 2-by-2 zero-sum games. In a zero-sum game, each of two players chooses among two actions. No matter what actions the players choose, the amount won by one player is exactly offset by the losses of the other. We use zero-sum games to define the basic terminology of game theory, to distinguish between strategies and actions, and to introduce the concept of iterated elimination of dominated strategies. We then study the Market Entry Game, a sequential game, in which an entrant competes against an existing firm, and we replicate that game many times to create what is known as the chain store paradox. In the third part, we consider an effort game in which individuals choose effort levels to win a prize of a fixed amount. Increasing effort improves a player's chances of winning the prize. The chapter concludes with a brief discussion of the value of game theory models generally.

Normal-Form Zero-Sum Games

In this section we analyze two-player *normal-form zero-sum games*. In both games, each player chooses an action and receives a payoff that depends on the player's own action and the other player's action. In addition, the players' payoffs sum to zero. In the first game, *Matching Pennies* shown in figure 21.1, each player chooses one of two actions: heads or tails. The row player wants to match the other player's choice, and the column player wants to mismatch. Payoffs are shown in the matrix below:

	Heads	Tails
Heads	1, -1	-1, 1
Tails	-1, 1	1, -1

Figure 21.1: The Matching Pennies Game

A strategy is a rule for how to play the game. It could be a choice of a single action, a randomization between actions, or, as we see in the next section, a sequence of actions. A *Nash equilibrium* of a game is a pair of strategies such that each player's strategy is optimal given the strategy of the other player. In Matching Pennies, in the unique equilibrium strategy both players randomize equally between the two actions. To prove randomization is an equilibrium, we need to show that if each player randomizes, the other player cannot do better than randomizing. This is straightforward. If the row player (actions and payoffs in bold) plays heads with probability $\frac{1}{2}$ and tails with probability $\frac{1}{2}$, then the column player earns zero regardless of her action. Therefore, randomizing is also an optimal strategy for the column player. By symmetry, randomization is optimal for the row player as well.

This optimality of randomization has implications for behavior in strategic settings. Sports are zero-sum: one team (or player) wins and one loses. During penalty kicks, a striker wants to randomize between aiming for the left or right corner. In tennis, a server wants to randomize serving to the inside or outside. On fourth and goal in football, an offense wants to randomize between run and pass. In each case, the opponent also wants to randomize their planned responses. Any non-randomness can be exploited. The same holds in card games such as poker. A good poker player bluffs randomly. If she always bluffed, her opponents would learn her strategy and stay in the game. She would then lose every time she bluffed. Similarly, if she never bluffed, her opponents would learn to fold. Optimal bluffing makes her opponents uncertain whether to stay or fold.

In our second game, the *Minimize Risk Game* shown in figure 21.2 each player can take a risky action or a safe action. This is an asymmetric zero-sum game. The payoffs depend not just on the actions but also on which player takes which action. In this game, the row player has a *dominant strategy* to play safe. No matter what action the column player chooses, the row player earns a higher payoff by choosing safe. The column player does not have a dominant strategy. If the row player chooses risky, the column player should choose risky. If the row player chooses safe, the column player should choose safe.

	Risky	Safe
Risky	-10, 10	0, 0
Safe	10, -10	0, 0

Figure 21.2: The Minimize Risk Game

By thinking through the incentives for the row player, the column player can deduce that the row player will never choose risky because risky is dominated by safe. Therefore, the column player knows that the row player will choose safe. Given that, the column player should also choose safe. This type of reasoning in which one player rules out dominated strategies for the other player is known as *iterative elimination of dominated strategies*. In this game, using iterative elimination of dominated strategies shows that both players choosing safe is the unique Nash equilibrium.

Sequential Games

In a *sequential game,* players take actions in a specific order, as shown on a *game tree,* which consists of nodes and edges. Each node corresponds to a moment when a player must take an action. Each edge from that node denotes one of the possible actions. At the end branches of a game tree, we write the payoffs associated with following that path of actions. The game tree in figure 21.3 shows the Market Entry Game.

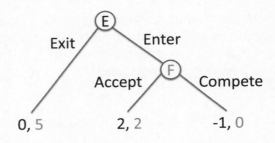

Figure 21.3: The Market Entry Game

The *Market Entry Game* involves two players: a potential entrant and an existing firm. If the entrant chooses not to enter the market (the left branch of the tree), it earns no payoff and the existing firm earns a profit of 5. If the entrant enters the market, the existing firm must choose between accepting the new entrant and seeing its profits fall from 5 to 2 or competing with the new firm and driving its profits to zero and the entrant's profits negative. We assume the entrant's profits to be negative because it has to pay for the cost of entering.

In a sequential game, a strategy corresponds to an action choice at each node. Suppose that the existing firm chooses to compete if the entrant enters. If the entrant knows this, the entrant would not enter, as doing so would produce negative profits. This set of actions, the entrant choosing to not enter and the existing firm planning to compete if the entrant did enter, are a Nash equilibrium. However, this is not the only Nash equilibrium, nor is it the most likely outcome. There is a second equilibrium in which the entrant chooses to enter the market and the existing firm accepts the entrants move and does not compete.

To select among these two equilibria, we apply a refinement criterion. In sequential games, a common refinement chooses the *subgame perfect equilibrium*. We solve for the subgame equilibrium using *backward induction*: we start at the end nodes and choose the optimal action at each. We then work backward up the game tree assuming that each player chooses the best action given the actions of the other player at subsequent nodes. In the Market Entry Game, we start at the end node for the existing firm. It has an optimal action: to accept. We then move up the game tree and see that the entrant has an optimal strategy to enter.

This game becomes even more interesting when replicated. Imagine that the firm exists in many markets. Perhaps it is a chain store with franchises in dozens of cities. Suppose also that there exists a sequence of potential entrants. The firm is going to play one Market Entry Game after another in sequence.

If the firm reasons using backward induction starting from the last market, it will accept the entrant in that last market, as that is the payoff maximizing action. Continuing with the same logic, the firm will accept the second-to-last entrant. It will also accept every other entrant. It follows that in the unique subgame perfect equilibrium of the sequence of games, all of the potential entrants choose to enter, and the firm accepts all of them.

Though entrance and acceptance in every market is the unique subgame perfect equilibrium. In practice, it may not occur. Imagine we are on the board of directors of the existing firm and we are confronting the first entrant, who (having studied game theory) enters the market. We may want to compete to try to deter entry in the other markets. Competing would be an intelligent strategy if it is credible, that is, if we can build a reputation as willing to compete. The outcome we hope to create differs from the subgame perfect equilibrium.

Game theorists refer to the disconnect in this game between what game theory predicts and what actual players would try to produce as the *chain store paradox*. It is one example where what game theory considers to be optimal behavior may not be the behavior chosen by a sophisticated player when the stakes are large. The example does not disprove game theory or undermine the rational choice assumption, so much as it reveals why we must always challenge assumptions.

Continuous Action Games

We now study a game in which players choose from a continuum of possible actions. In the game, actions correspond to effort levels. By choosing greater effort, a player increases her probability of winning a prize. This game allows us to model any number of players.

The derived expression for equilibrium effort reveals a number of insights. As we would expect, individual effort increases with the size of the prize. Also, in equilibrium, total effort will be less than the value of the prize. That too would be expected given that we assume players optimize. Players should put forth effort to win, but not an unreasonable amount.

The Effort Game

Each of N players chooses an effort level expressible in monetary terms to win a prize of value M. The probability that a player wins the prize equals her effort divided by the total effort of all players. If E_i equals the effort level of player i, her probability of winning is given by the following equation:[1]

$$P(i \text{ wins}) = \frac{E_i}{(E_1 + E_2 + \ldots + E_N)}$$

Equilibrium: $E_i = \frac{M}{N} - \frac{M}{N^2}$

We can see the effects on individual and total effort by increasing the number of players. Here, the findings are less intuitive. According to the model, individual players' effort levels decrease but the total effort by all players increases. Thus, the model implies that efforts by organizers of research grant opportunities, architectural competitions, and essay contests to attract large numbers of entrants may, paradoxically, produce lower-quality winners because in the larger contests, participants have less incentive to put in effort.

Summary

We began this chapter by covering zero-sum games. These games assume no mutually beneficial combinations of actions. Any action that proves good for one player necessarily hurts another player. In a zero-sum decision, any action harms someone as much as it helps someone else. Taking money from one person and giving it to another is a zero-sum action. Many personal actions and policy choices are zero-sum in at least one dimension. We have only so many hours in the day, so much money to spend, and so many resources to allocate. That said, a zero-sum action in one dimension may not be zero-sum in another. A budget relocation could be zero-sum in monetary terms but positive or negative sum in terms of human happiness or fulfillment.

We should always explore whether a proposed policy change creates a zero sum game. For example, many people argue for school choice—giving parents the ability to choose the school their child will attend—because it increases competition. Market logic suggests that by being forced to compete, schools have incentives to improve quality.

However, schools only have an incentive to improve quality if excess capacity exists. Otherwise, school choice can create a zero-sum game among the students. Imagine a city with 10,000 students and 10 schools each with a capacity for 1,000 students. If the students rank the schools in the same order, spots in the best schools will have to be allocated by lottery. Those who win the lottery will go to better schools. Those who lose the lottery will go to worse schools, who remain in operation because of a lack of excess supply. Students are playing a zero-sum game. If new schools open, or if existing schools improve, the game is no longer zero-sum. Everyone can win.

Both the market and the zero-sum models provide insights. The market model reveals incentives for quality improvements and for the creation of new schools. The zero-sum model shows that school choice alone means that some students will gain while others lose. The relative weight we should place on each model depends on the context: Does sufficient excess capacity exist in the better schools so that they can absorb the additional students? Do schools have the resources and expertise to improve their quality? Will entrepreneurs create new schools? Does the transportation system enable students to get to multiple schools in order to create competition?

Our takeaway should be that neither of the two models gives us a correct answer, but each produces useful insights. School choice will create competition. It also creates a massive sorting problem with features of a zero-sum game. Whether the positive aspects of competition will outweigh the negative sorting costs depends on the context. We must array our lattice of models on the set of facts to make a good policy choice.

The Identification Problem

Data on people's actions often reveals clustered behavior. Good students are more likely to be friends with other good students than with students who struggle. People who engage in criminal behavior are more likely to interact with other people who commit crimes than are people who do not commit crimes. Any number of social goods and ills—smoking, physical fitness, obesity, and even happiness—cluster in social networks. People also cluster by beliefs: Democrats cluster. Republicans cluster. Libertarians cluster.

We have two models that can explain clustering: *peer effect models* and *sorting models*. Peer effect models explain clustering with game theory. Individuals play a coordination game with their friends. In sorting models, people move to be near others who are like them. A cluster of good students could result from either students coordinating on a common behavior (peer effects) or arise because good students choose to hang out with one another (sorting). Given a snapshot of data, the two are indistinguishable.

Data: Students earn either high H or moderate M scores, with each being equally likely. Students belong to friendship cliques of size 4 with the following distribution: $p(\{H,H,H,H\}) = P(\{M,M,M,M\}) = \frac{5}{16}$, $p(\{H,H,H,M\}) = P(\{M,M,M,H\}) = \frac{3}{16}$, and $p(\{H,H,M,M\}) = 0$.

Peer effect model: Students originally form random groups of size 4: $p(\{H,H,H,H\}) = P(\{M,M,M,M\}) = \frac{1}{16}$, $p(\{H,H,H,M\}) = P(\{M,M,M,H\}) = \frac{1}{4}$, and $p(\{H,H,M,M\}) = \frac{6}{16}$. People who belong to groups consisting of only one type remain unchanged. A person who has the opposite type from everyone else switches type, so an $\{H,H,H,M\}$ group becomes $\{H,H,H,H\}$. In groups with equal numbers of each type, one member switches type. The group $\{H,H,M,M\}$ is equally likely to become $\{H,H,H,M\}$ or $\{M,M,M,H\}$.

Sorting model: Students originally form random groups of size 4. In any group with two types, a person who has the opposite type as at least two other people switches groups with someone of the other type. It follows that $\{H,H,H,M\}$ becomes $\{H,H,H,H\}$ and $\{M,M,M,H\}$ becomes $\{M,M,M,M\}$, and that any group of the form $\{H,H,M,M\}$ is equally likely to become $\{H,H,H,M\}$ or $\{M,M,M,H\}$.

Both models are consistent with the data, creating an **identification problem**. With only a snapshot of data, we cannot determine whether smoking, reading manga, or longboard skateboarding are peer effects or sorting. In some instances, we can reason through which model applies. The tendency for people to say "pop" in the Midwest and "soda" on the coasts is something that we can safely assume to be driven by peer effects—few people move to Boston so that they can refer to Coke as "soda." On more important behaviors such as educational performance, drug use, obesity, and happiness, we need time series data to discern which model applies. By looking across time, we can discern if people change their behaviors to fit in with their friends (peer effect), or if they change their friends and retain their behaviors (sorting). In many cases of interest such as school performance, both effects may be in play.[2]

22. *Models of Cooperation*

No one has ever become poor by giving.

—Anne Frank

Experts asked to name the most important scientific questions produce a limited set of responses: How did the universe form? How does consciousness emerge? Can we find a cure for cancer? One question experts identify spans the social and biological sciences: How does cooperation arise?[1] Cooperation entails taking an action that is not in one's self-interest, which suggests that we should not expect to see much of it. And yet we see cooperation in myriad domains and at multiple scales. Cells cooperate through adhesion, where one cell produces extracellular material to which others can attach. We see cooperation among ants, bees, humans, organizations of humans, and even nations, which cooperate in the creation of treaties and international laws.

In this chapter, we use models to take up the questions of how cooperation emerges, how it is maintained, and how we might create more of it. These models cannot explain in full the variety of cooperation that exists in the world—why ravens share their discoveries of carrion, why naked mole rats collectively defend against predators, why climbing vines lay down fewer roots when planted adjacent to kin, why termites and bees build elaborate nests, and why ants lock appendages to form bridges for the carrying of food—but they will produce insights.[2]

Although we see many examples of cooperation within and across species, we also see failures. The extent of cooperation depends on the circumstances. Federations gain and then lose members; Britain participated in the creation of the European Union and then exited from it. The same people who volunteer labor for a school fundraiser may cut in line at the supermarket or cheat on their taxes. A lion who hunts water buffalo in a pack may secrete away a warthog kill. And not every species cooperates. The roots of black walnut trees release juglone, an herbicide, into the soil, to inhibit the growth of nearby plants.

The diversity of behaviors of cooperating entities—cells, roots, ravens, people, business firms, and nations—obliges a many-model approach. We might best model cells and plants as following fixed rules; ravens, ants, and lions as using more rules that condition on the environment or on past outcomes; and people, business firms, and nations as looking ahead and performing cost-benefit calculations.

The first key takeaway from this chapter will be that cooperation can emerge and be maintained through a variety of mechanisms. We highlight four mechanisms that enable cooperation: *repetition, reputation, local clustering,* and *group selection.* These mechanisms all enable cooperation without external intervention or management. They can apply to cooperating mole rats, bees, and humans alike. Humans also have other more formal ways to induce and maintain cooperation. In the discussion at the end of the chapter, we describe institutional solutions such as paying people to cooperate, punishing them if they do not, and legally mandating cooperative behavior.

The second takeaway will be that the efficacy of any one of these mechanisms depends on the behavioral repertoires of those cooperating. Some mechanisms, notably repetition, work for almost any behavior. Reputations and norms require forward-looking behavior and information sharing. They will be most effective for more sophisticated actors.

The effect of clustering, on the other hand, depends on the model. Cooperation among actors who are selected for or against by evolutionary forces emerges most often on sparse networks. Cooperation through norms requires dense networks. The efficacy of group selection depends in a nuanced way on the ability of the actors to be forward-looking and on their speed of adaptation. Making actors more forward-looking enhances the power of group selection. Allowing them to adapt more quickly can hinder it. To explore these questions and to unpack the interplay between behavioral assumptions and cooperation, we rely on the familiar Prisoners' Dilemma game as well as a cooperative action model. The second model allows us to capture actions that benefit multiple players as well as to model cooperation on networks.

The remainder of the chapter takes the following form. We begin with a description of the Prisoners' Dilemma and show how cooperation can be maintained among rational actors. We then show how repetition also can induce cooperation between rule-based actors and why evolving cooperation is more difficult than maintaining it. We then consider less sophisticated biological actors and show how kin selection and local clustering can promote cooperation. The last two sections cover group selection and the question of how we use these models to produce more cooperation.

The Prisoners' Dilemma

The name *Prisoners' Dilemma* derives from a story of two people accused of jointly committing a crime. The authorities have circumstantial evidence so they offer each person a chance to confess. The accused confront a dilemma. If neither confesses, each receives a minor sentence based on the evidence. If only one confesses, then that person receives no punishment while the other is punished severely. If both confess, both receive harsh punishment, though not as extreme as in the case where only one confesses. Figure 22.1 represents this story as a two-player game. Each player can either *cooperate* (C) or *defect* (D). The gray numbers represent the payoff to the column player and the black numbers the payoff to the row player. Each player has a dominant strategy to defect: whatever the action of the other player, defecting produces a higher payoff. However, if both players defect, each receives a lower payoff than if both cooperated. Thus, self-interest leads to actions that are collectively worse.

	Cooperate	Defect
Cooperate	3, 3	1, 4
Defect	4, 1	2, 2

Figure 22.1: An Example of a Prisoners' Dilemma Game

The Prisoners' Dilemma captures the core incentives of many real-world contexts. It can model the arms race between the United States and the former Soviet Union, where defecting corresponds to spending money on weapons and cooperating to economic development. It can model political campaigning and whether to go negative (defect) or to run positive campaign ads (cooperate). It can even explain why male peacocks have such long tails: each peacock has an incentive to appear stronger and more robust than the others.

Some instances of the Prisoners' Dilemma can only be recognized after the fact. The first adopters of many technologies, such as banks that moved early into ATM

	Cooperate	Defect
Cooperate	R, R	-S, T
Defect	T, -S	0, 0

$$T > R > 0 > -S$$
$$2R > T - S$$

Figure 22.2: The Prisoners' Dilemma

machines, saw their profits increase. When others followed, profits fell from increased competition. Choosing to put in ATM machines proved to be an analog of the choice to defect.[3]

The general form of the Prisoners' Dilemma, shown in figure 22.2, assumes a baseline payoff of zero if both players defect. The game can then be expressed with three variables: a *reward*, R, from cooperating, a *temptation*, T, to defect, and a *sucker's payoff*, S, from being exploited (see box). The inequalities shown in the box ensure that choosing defect is a dominant strategy and cooperating produces the efficient outcome.

Cooperation Through Repetition and Reputation

We first show how repetition of the game and the building of reputations can maintain cooperation among rational actors. The fact that cooperation can be maintained does not guarantee that it will be realized; it says only that if cooperation happens to emerge, rational players can sustain it. To prove that repetition maintains cooperation, we construct a *repeated game model* in which after each play of the game, with probability P, the game will be played again. In theory, play could go on forever.

The players apply repeated game strategies, which give an action based on the history of past play. Here we consider a repeated game strategy known as *Grim Trigger*, which cooperates in the first play of the game and cooperates in any future play of the game so long as the other player has never defected. If the other player ever defects, Grim Trigger defects forever. It is unforgiving. If both players use the Grim Trigger strategy, both cooperate forever.

To prove Grim Trigger maintains cooperation in the repeated game, we need only show that if one player chooses Grim Trigger, then the other player receives the highest possible payoff by also playing Grim Trigger. Given that a deviation by the second player produces endless defection by the first player, the second player need only compare the expected payoff from always cooperating to the expected payoff from the one-time benefit of defecting plus the payoff when both players defect thereafter.[4] Whether of not Grim Trigger produces the higher payoff depends on the extent of temptation, the reward from cooperating, and the probability that the game repeats.

Repetition Maintains Cooperation

In the repeated Prisoners' Dilemma, Grim Trigger maintains cooperation if the probability of continued play, P, exceeds the ratio of the difference in the temptation payoff, T, and the reward payoff, R, to the temptation payoff:[5]

$$P > \frac{(T - R)}{T}$$

The result tells us that if the temptation payoff exceeds triple the reward payoff, $T > 3R$, the game must be repeated with a probability in excess of two-thirds. The inequality also tells us that cooperation becomes easier to maintain if the reward increases, the probability of continued play increases, or the temptation to defect decreases. Each of these implications reveals an intuitive route to more cooperation: increase the reward, make continued interaction more probable, and reduce the temptation to defect. Though these are quite straightforward inferences, they might not have been at top of mind prior to writing the model.

In pondering the necessary condition for cooperation, we can also infer less straightforward insights. The expression implies that if players thought that the probability of continuation would fall below the threshold in the future, then rational players would stop cooperating before the probability change occurs, not when the change occurs.[6]

The logic that repetition supports cooperation among rational players hinges on a particular feature of the model: a probability of continuing play. If, instead, we had assumed a fixed number of repetitions—say, that the game was to be played three times—rational players would not cooperate, which we can prove by backward induction. Suppose that the game is only played three times and that the first player announces that she will play Grim Trigger. Assume that $T = 3$, $R = 2$, and $S = 1$. Given these payoffs, if the second player cooperates in all three rounds, she earns a total payoff of 6. We need to check that no other strategy generates a higher payoff. Defecting in the first round produces a payoff of only 2, because after her defection the first player will defect in the last two rounds. Defecting in the second round produces a payoff of 5. Neither would be rational. Defecting in the third round, though, produces a payoff of 7: 2 in each of the first two periods, and 3 in the last period. Therefore, a rational player defects in the last round.

The first player, who played Grim Trigger, should recognize the defection will occur in the third round and also defect. It then follows that the other player would realize that both players are going to defect in the third round and so would defect in the second round of the game. By the same logic, the first player would also defect. This unraveling would continue to the first round. The same reasoning applies if we repeat the game any finite number of times. In the last round played, rational players defect. As result, both have an incentive to defect in the second-to-last round, and so on and so on. The only rational strategy is to always defect.

Our analysis so far considers two players in isolation. It does not take into account how a person's defection might influence how others treat that person in future interactions. In effect, we drew a boundary around the two people playing the game. We can extend the model to include a community of people who monitor the behavior of one another and punish people who deviate.

To do this formally, we assume that each day people randomly form pairs and play Prisoners' Dilemma games. The members of the community believe that these games will go on forever, so the probability of future play equals 1. Under these assumptions, an individual will not be likely to play against the same person the next day, so the incentive to defect will be higher. However, we allow for the possibility that a defection can be recognized by the community. If so, the person earns a bad reputation and, by agreement, no one in the community will cooperate with that individual in the future. If we let P_D denote the probability that a person gets caught defecting, earns a reputation as a defector, and is punished in all future games, then the condition for cooperation to be maintained through reputations, $(P_D > \frac{(T-R)}{T})$, is identical to the condition for repetition to maintain cooperation, except that P_D, the probability that a person has been caught defecting, replaces P, the probability of repeated play.

In the reputation model, the community enforces cooperation. Someone who has defected and has been caught will be defected against by all future players. Here again, individuals calculate the benefits and costs of defecting. They must also believe that others will adhere to the punishment, which in this case means that all others will defect. For that to be true, individuals must either know one another or have some method of identifying or tagging past defenders. It follows that, all else equal, people in small communities should be better able to enforce cooperation through repetition. In small northern towns, people leave their cars running in store parking lots during the winter. They do not fear the car being stolen (a defection) because they know everyone in the town. Anyone who stole a car, even as a prank, would incur a reputation loss.

Physical tags can make reputations public information in order to maintain cooperation. In Nathaniel Hawthorne's novel *The Scarlet Letter,* Hester Prynne is forced to wear a scarlet *A* for committing adultery. Some cultures amputate the hands of convicted thieves, a rather costly tag. Tagging of defectors even occurs in other species. The cleaner fish, *Labroides dimidiatus,* can clean parasites from other fish (cooperate) or consume tastier alternatives (defect). If a fish cooperates, its neighbors will be free of parasites. The lack of parasites is observable to other fish. The cleanliness of neighboring fish becomes a tag, a visual reputation.[7]

Connectedness and Reputation

To support cooperation through a reputation mechanism requires that an individual's neighbors know of a deviation. To assess the likelihood of word of a deviation spreading, we can apply three insights we learned when adding networks to the contagion model. First, the greater the degree of the network, the more likely it is that word of deviation would spread. Second, variation in the distribution of degrees, in particular the existence of superspreaders, would amplify the likelihood. Third, if an individual defects against someone who is not connected to any of the individual's other neighbors, then the neighbors will not be likely to hear of the defection. Therefore, for reputations to spread, the network must have a high clustering coefficient. The clustering coefficient is a proxy for social capital.

Cooperation Among Rule-Playing Behaviors

We now relax the assumption of rationality and assume that players apply rule-based strategies such as Grim Trigger. We will use our model to understand whether and how cooperation can emerge. Our model assumes a population of individuals who play repeated rounds of a Prisoners' Dilemma game against one another. We assume that each interaction continues with some probability as above. That construction could induce rational players to cooperate if the probability of continuation is sufficiently high.

Unlike above, here we assume that players apply behavioral rules. Some may play Grim Trigger. Others may always cooperate, and others may always defect. Variants of these strategies may be played by other species. Warbler males adopt a "dear enemy" strategy in which they do not engage in loud singing or fighting to extend their property at the expense of their neighbors. We can think of this as a cooperative action.[8]

For ease of explanation, we assume that each individual plays with every other individual. After every individual has played all her games, each announces a performance equal to her average payoff in a play of the game. We use average per-game payoff rather than total payoff because some players may, by chance, play many more games than others given probabilistic continuation. In this model setup, a strategy's performance depends on the distribution of strategies. It follows that the winning strategy can then also depend on the initial distribution. If cooperative strategies perform best initially, cooperation will likely grow in the population.

For our analysis, we randomly assign to each player one of five behavioral rule strategies: always cooperate (All C), always defect (All D), Grim Trigger (GRIM), Tit for Tat (TFT), or TROLL. GRIM cooperates initially and continues cooperating until the opposing player defects, after which it defects forever. All C and All D do what their names imply: they blindly cooperate or defect regardless of the other player's action. TFT cooperates in the first period and thereafter copies the action of the other player from the previous period; two players who both use TFT will always cooperate. TROLL seeks to exploit players who always cooperate. It defects in the first two periods, and if the other player does not defect in either of those periods, TROLL defects forever. If the other player does defect, TROLL switches to cooperate for two periods and thereafter plays GRIM.

We first calculate the payoffs for each behavioral rule strategy playing against every other strategy using the payoffs from the Prisoners' Dilemma in figure 22.1. We start with the strategy All D. If it plays against All C, it receives a payoff of 4 in every play of the game. All C, on the other hand, receives an average payoff of only 1 in those interactions. If All D plays against either TFT or GRIM, it receives a payoff of 4 in the first play and 2 thereafter. If we assume the game is repeated many times, this will average out to a little more than 2, so we write it as 2^+. When All D plays TROLL, both defect in the first two periods, and TROLL cooperates in periods three and four but defects thereafter. All D again earns an average payoff of 2^+. TROLL earns an average payoff of a little less than 2, which we write as 2^-.

We can perform similar exercises and compute the expected payoffs for every pair of strategies.[9] Table 22.1 shows the payoff for each strategy against each of the other strategies.

Player's Strategy	Opponent's Strategy				
	All C	All D	TFT	GRIM	TROLL
All C	3	1	3	3	1
All D	4	2	2^+	2^+	2^+
TFT	3	2	3	3	3^-
GRIM	3	2	3	3	2^+
TROLL	4	2	3^-	2	3^-

Table 22.1: Average Payoffs for Row Strategies Against Column Strategies

The table reveals a mix of mutual cooperation, mutual defection, and strategies taking advantage of flaws in other strategies. A careful examination of the table reveals that four of the five strategies cooperate with themselves. We will refer to these as the potentially cooperative strategies. Only TFT cooperates with all four of these potentially cooperative strategies. So if any combination of those four accounted for the bulk of a population, TFT would perform well, if not best.[10]

The thousands of human experiments run on the Prisoners' Dilemma reveal tremendous heterogeneity in the strategies people choose. We will therefore use the payoffs in the table to think through the outcomes given different distributions. Based on the diversity of payoffs for the different combinations of strategies, the best strategy will depend on the composition of the population. In a population that consists mostly of All C, the strategy All D performs best. If individuals choose to adopt the best strategy, or if selection operates quickly, then the population might never manage to cooperate. If learning or selection happens at a moderate rate, players should move away from All C. Once the population contains few All C, All D will perform less well than GRIM, TROLL, and TFT. One of these strategies should take hold. This pattern of defectors performing well initially and then cooperation taking hold can be found in many experiments with human subjects as well as in simulations with computer-based artificial agents. We might describe what happens in those cases as the emergence or evolution of cooperation.

One can imagine any distribution across these five strategies or any other ensemble of strategies, compute average payoffs, and then think through what might occur through learning or selection. In a later chapter, we construct formal models of learning and selection. We rely here on informal arguments, as we only wish to make the point that whether cooperation emerges depends on the initial strategies in population and how people learn or evolve new strategies.

A necessary condition for cooperation to emerge or evolve is that the payoff from cooperating exceeds the payoff from defecting given the population. Otherwise, both selection and learning would lead the population toward defection. To simplify the analysis, we can imagine a population that consists of cooperative strategies, such as GRIM, All C, and TFT, and defecting strategies, such as All D. We can then calculate what would have to be true for the cooperative strategies to perform better on average. That calculation reveals that evolving cooperation is more difficult than maintaining cooperation, and that cooperation cannot bootstrap itself—a small population of cooperators cannot cause cooperation to emerge.[11]

This distinction between maintaining coordination, emerging or evolving coordination, and bootstrapping coordination merits revisiting. Cooperation can be maintained if, when all players cooperate, cooperation performs best. Maintenance corresponds to cooperation through GRIM being a Nash equilibrium of the repeated game. Cooperation can emerge or evolve if the strategies that cooperate when paired in a population outperform, on average, those that do not. As just argued, the conditions for emergence of cooperation are harder to satisfy than the conditions for maintenance of cooperation. In fact, the mathematics shows us that bootstrapping is all but impossible. If the proportion of cooperators is near zero, then cooperators earn lower payoffs than defectors. The takeaway should not be that bootstrapping coordination can never occur, only that it cannot happen in this model. To obtain cooperation, we need a proportion of people to cooperate initially. That could happen with people who reflect on the game, but it seems less likely for bees and tree roots. To understand how bootstrapping could occur, we need more elaborate models that allow for local learning, evolution, and group selection. We turn to those now.

Cooperative Action Model

To study how cooperation can emerge, we introduce a *cooperative action model* in which individuals can either take a cooperative action or refrain from doing so.[12] The cooperative action imposes a cost on the individual and produces a benefit to others. Refraining from action imposes no cost and produces no benefit.

There are several differences between the cooperation action model and the repeated Prisoners' Dilemma. First, the individuals in the cooperative action model are not playing a repeated pairwise game in which they apply strategies and earn payoffs. Instead, individuals are either cooperators or non-cooperators. Second, the model does not assume rational actors or individuals who apply more sophisticated rules. Third, the individuals belong to an interaction network. Their cooperative actions impact only those with whom they are connected, their neighbors. Last, because the individuals have fixed types, they take the same action with all of their neighbors. A cooperating individual with five neighbors pays the cost of cooperating five times and produces a benefit to five others.

Cooperative Action Model

A population of N individuals consists of cooperators and defectors connected in a network. Cooperation incurs a cost C and produces a benefit B to the other player for each interaction. Defecting produces no cost or benefit. The *ratio of cooperative advantage*, $\frac{B}{C}$, captures the potential gains from cooperation.

The network plays a key role in allowing cooperation to emerge and even bootstrap itself. A small cluster or group of cooperators who interact mostly with one another perform well and then spread in the population. In an ecosystem, offspring often locate adjacent to parents. If the offspring of cooperators are more likely to be cooperators, then bootstrapping cooperation becomes even easier.

To show that clustering can bootstrap cooperation, we start with a partially filled network. Each node on that network is a potential location for an individual. In the biological context, the locations would be feasible habitats. We then populate a portion of the network with individuals who are either cooperators or defectors. We might, for example, first draw a random network with an average degree of 10, then roll a die at each node. If the die comes up six, we place an individual at that node. If not, we leave the node empty. If we do place an individual on a node, we roll the die again. If we roll a five, we place a cooperator on the node. Otherwise, we place a defector. This procedure populates one-sixth of the nodes on our network, and one-sixth of occupied nodes contain cooperators.

Given this construction, individuals will differ in their number of neighbors. Some will have no neighbors. Some will have four or five neighbors. To enable the growth or demise of cooperation, we populate the remainder of the network by iteratively filling in the nodes adjacent to occupied nodes. We assume that an empty node takes the type (i.e., is either a cooperator or a defector) of the highest-performing type among its neighbors. Figure 22.3 shows two segments of linear networks. Cooperators are represented by dark lines, defectors by gray lines, and empty nodes by dashed dark lines. Each segment contains an empty node in the center with two neighbors, one defector and one cooperator. In this figure, cooperating creates a benefit of 2 and imposes a cost of 1.

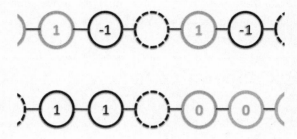

Figure 22.3: Payoffs to Neighbors of an Empty Node in Two Linear Networks

In the top segment of figure 22.3, the defector to the right of the open node has a cooperating neighbor, so it earns a payoff of 1. The cooperating neighbor to the left of the open node has a defector as a neighbor, so it earns a payoff of -1. Given our rules for node filling, because the defecting neighbor has a higher payoff, the empty node will become a defector. In the bottom segment, the defecting neighbor of the empty cell has a defecting neighbor, while the cooperating neighbor of the empty cell is connected to another cooperator. In the lower segment, therefore, we get the opposite outcome. Here, the cooperating neighbor has the higher payoff, so the empty node will become a cooperator.

In this example, a lone cooperator cannot spawn an additional cooperator, but two adjacent cooperators can. It follows that a small cluster of cooperative nodes surrounded by empty cells could expand into open nodes. Therefore, regions of cooperation can emerge from a handful of cooperators.

We can write more general conditions about whether an empty cell becomes a cooperator or defector based on the proportions of neighboring cooperators and defectors and the ratio of cooperative advantage. It follows that cooperation becomes easier to bootstrap in networks with lower degree. This finding is the opposite of what we found when analyzing how reputation maintains cooperation, where a more connected network increases the likelihood of a defection ruining someone's reputation, so more connections help to maintain cooperation. This provides another example of many-model thinking producing conditional knowledge. The question of whether connected networks produce more or less cooperation has no single answer. If cooperation is maintained by sophisticated actors using reputation, more connected networks will be more cooperative. If cooperation is bootstrapped or evolved among unsophisticated actors, like trees or ants, less connected networks should promote more cooperation.

Clustering Bootstraps Cooperation

If the neighbors of an open node include a cooperator of degree D with K cooperating neighbors and all non-cooperators of the empty node have no cooperating neighbors, then the open node becomes a cooperator if and only if the ratio of cooperative advantage exceeds the ratio of the degree to the number of cooperators:[13]

$$\frac{B}{C} \geq \frac{D}{K}$$

Group Selection

Our final mechanism for bootstrapping, evolving, and maintaining cooperation, *group selection*, relies on competition or selection among groups.[14] To model group selection, we divide the population into subgroups. Within each subgroup, individuals engage in a version of the cooperative action model where each individual either cooperates or defects. As before, we can assign a performance to each individual. We also assign a performance to each group equal to the average performance of its members. The model assumes selection among groups in which copies of the highest-performing groups will replace lower-performing groups. This advantages groups of cooperators, which will perform better.

The intuition that cooperative groups should take over given group selection has a catch: within any group, the defectors outperform the cooperators. Consider two groups of size ten. The first group contains two cooperators and eight defectors. The second group contains two defectors and eight cooperators. Assume benefits equal 2 and costs equal 1, as above. In the first group, each defector's performance equals 4, as it receives a benefit of 2 from each cooperator. Each cooperator incurs a cost of 9 and receives a benefit of only 2, so its performance equals -7. The average performance of a group member equals 1.8. In the second group, each defector receives 2 from each of eight cooperators, so its performance equals 16. Each cooperator's performance equals 5: it receives 14 from the other seven cooperators but pays out 9 in costs. The average performance in the second group equals 7.2.

These calculations reveal a paradoxical insight: within each group, defectors perform better than cooperators, yet the higher-performing group contains more cooperators. A tension should be apparent: *individual selection favors defection but group selection favors cooperation*. This tension operates across a variety of ecological, social, political, and economic contexts. Trees whose roots cooperate with other trees fare worse individually, but they contribute to stronger ecosystems better able to spread into open spaces. Cooperative individuals within communities may reap fewer benefits than defectors, but cooperative communities will grow in size. Politicians who support their party may be less likely to be reelected than those who focus on themselves, but cohesive parties will be more likely to grow. And an individual working at a firm may fare worse by building talents useful only to her current employer, yet if she does, her firm will be able to outcompete others.

The cooperative action model helped us to identify and quantify this tension. To see whether group selection can bootstrap, evolve, or maintain cooperation, we need to add more detail to our model. Traulson and Nowak propose an elegant model in which populations grow and new members resemble the high performers. This construction builds in both individual and group selection. Selection occurs at the level of individuals, but higher performers are more likely to come from cooperative groups. When a group becomes sufficiently large, it divides in two, creating a new group. To prevent the population from becoming too large, the formation of a new group causes the eradication of a randomly chosen existing group. This last feature builds in a weak form of group selection.[15]

These models show that group selection increases cooperation provided that the benefit from cooperative action is relatively large and the maximal group size is small relative to the number of groups. The finding that the efficacy of group selection depends in part on the ratio of maximal group size to the number of groups reveals the necessity of competition. Having more groups implies a greater likelihood of a group of all cooperators. It also implicitly assumes more competition. The more unexpected result is that smaller maximal group size enables more cooperation. A smaller maximal group size prevents groups of cooperators from becoming dominated by defectors; it limits the effects of individual selection. Think back to our group of eight cooperators and two defectors. The defectors perform better. If the group were allowed to grow to size eighty, it would contain a much larger proportion of defectors before the split occurs. If the group splits once it has twelve members, in the worst-case scenario the group consists of two-thirds cooperators when splitting.

The potential for group selection to increase cooperation can be applied within organizations. Most organizations assign compensation based primarily on individual performance. Splitting employees into teams that compete against one another and allocating bonuses and opportunities based on team performance creates the possibility for inducing cooperative behavior. If resources go to teams, individuals have incentives to work well within those teams, to cooperate.[16] These incentives should increase cooperation within teams if the benefits from cooperation are high and if the size of teams is small relative to the number of teams.

When evaluating the potential for group selection, we must think carefully about the sophistication of the individual actors. Trees adapt slowly, so group selection will not have to operate very fast. People adapt quickly, so if the individual incentives to defect are high, group selection will have to operate at a correspondingly fast rate. People, though, may also recognize the group selection effect. They may take into account the competition among groups and see their self-interest in creating a strong group. This makes cooperation more likely. All this is to suggest that we should be careful not to place too much confidence on a specific constraint that shows cooperation will increase in a particular model. Instead, we should apply our judgment across many models and ask if the qualitative insights hold.

Summary

The puzzle of how cooperation takes hold, grows, and is maintained has been studied by thousands of scholars across a range of disciplines. That inquiry has been aided by models, most prominent among them the Prisoners' Dilemma. If we assume rational actors in a repeated game, the puzzle goes away. Cooperation can be maintained through threat of punishment. Punishment can occur directly through repeated play or indirectly because of reputation. Those mechanisms may explain how cooperation arises in high-stakes settings with sophisticated people, but it will not explain why ants, bees, trees, and naked mole rats are so cooperative. When we took up cooperation among rule players, we found that evolving cooperation is not as easy. Rational actors can maintain cooperation in environments in which rule-playing actors cannot evolve it.

We also found that unsophisticated rules, such as Tit for Tat, though not optimal, can cooperate with one another and not be exploited. Subsequent work has shown that Tit for Tat performs less well if we assume random mistakes in play. If a mistake occurs and a player defects, two players each using Tit for Tat will produce a cycle of defecting and cooperating actions. If both players accidentally defect, Tit for Tat will result in mutual defection until another mistake occurs.

In real Prisoners' Dilemma games, mistakes happen. On September 1, 1983, Korean Airlines flight 007 drifted into Soviet airspace en route to Seoul, Korea, from

Anchorage, Alaska. A Soviet SU-15 shot down the plane, killing all 269 people on board. The United States saw this as a defection by the Soviet Union. The Soviet Union, thinking the plane was on a spy mission, thought this was a defection by the United States.

To avoid endless punishments following a mistake, other strategies—such as Win Stay, Lose Shift—are more forgiving. Under this strategy, the mutual cooperation payoff (R) and the temptation payoff (T) are coded as wins. The other two payoffs are coded as losses. Win Stay, Lose Shift starts by cooperating; thereafter, if it won, it sticks with whatever it did last period. If it lost, it switches to the other action. By working through some examples, you can see how Win Stay, Lose Shift returns to cooperative behavior.[17]

We also described two other mechanisms. Clustering can enable cooperation to bootstrap itself. This mechanism relies on cooperators playing one another and having cooperation spread through selection. Group selection operates through a similar logic. Groups of cooperators perform well and replace groups of defectors. If we construct models, we find that cooperation that arises through clustering and group selection requires more stringent conditions than does cooperation through repetition or reputation. We also learned how the success of the various mechanisms depends on how we model the individuals. We should not expect these mechanisms to operate identically for people, ants, and trees. More sophisticated actors may be better able to sustain cooperation by being forward-looking, yet they might also be more likely to see the benefits of defecting when surrounded by cooperators.

Most of our discussion has framed cooperation as beneficial. Entities can also cooperate in order to exploit others. Firms form cartels to keep prices artificially high, and countries form coalitions to restrict the supply of a resource, such as oil, for their own benefit, not for the benefit of humanity writ large. Cancer cells cooperate to fight off our immune systems.[18] So as we study cooperation, we should keep in mind that it need not be for the common good. The water buffalo do not benefit from the lions' cooperative actions.

23. *Collective Action Problems*

Managing environmental resources sustainably has always been difficult, ever since Homo sapiens *developed modern inventiveness, efficiency, and hunting skills by around 50,000 years ago.*

—Jared Diamond

In this chapter, we cover *collective action problems*: situations in which self-interest does not align with the collective interest. These problems arise in matters small and large. At airports, travelers individually benefit by standing as near as possible to the baggage carousel, but collectively everyone would be better off if people stood back a few feet. In a democracy, people have little incentive to become informed voters given the very low probability of a single vote turning an election, yet a democracy performs better with an informed citizenry. Collective action problems can be thought of as many-player Prisoners' Dilemmas: each person has an incentive to defect, but collectively, everyone does better by cooperating.

People often study collective action models in the context of historical examples such as the management of the Scottish commons or the lobster habitats along the coasts of Newfoundland and Maine.[1] History also contains dramatic failures. Among the most famous is the collapse of the Polynesians on Easter Island, described by Jared Diamond.[2] Easter Island lies over two thousand miles west of Chile in the South Pacific, with no other inhabitable island within a thousand miles in any direction. Given that location, Easter Islanders have always had to manage for themselves. For over a thousand years they lived well. Some estimate that by the early seventeenth century, Easter Island's population exceeded fifteen thousand people. In the sixteenth century, the Easter Islanders marshaled sufficient resources to free up labor to build giant stone heads, called *maoi,* that weigh up to eighty tons. While the Easter Islanders were busy constructing maoi, they were not cooperating in the management of their forests. By 1722, when Europeans first landed on the island, food was relatively scarce and the population had dropped to around two thousand. Few trees over ten feet tall remained. Many species of birds and animals had gone extinct. To use Diamond's phrasing, the civilization collapsed. The collapse became complete when viruses carried by the Europeans killed nearly all of the remaining population.

According to Diamond's account, the collapse of the civilization on Easter Island, as well as the collapses of the Mayans in Central America, the Anasazi in the American Southwest, and the Vinlanders on Greenland, resulted from a combination of overharvesting of natural resources (caused by institutional and cultural failures) and climatic changes. The Vinlanders grazed animals on marginal land and tore up fragile sod to make houses. In short order, the land became barren from overuse, and the Vinlanders starved. Like the Easter Islanders, the Vinlanders had failed to manage a common pool resource. By chopping down too many trees and using up too much turf, they produced a collapse.

Though evocative and compelling, these examples lead many to see collective action problems as something of relevance only in the past. That framing is unfortunate. As the world becomes more interconnected and complex, collective action problems are far more relevant today. We confront collective action problems at almost every scale of human organization. The provision of public education, physical and mental health care, infrastructure, public safety, a justice system, and national defense are all collective action problems, as are managing global fisheries, combating climate change, and in particular reducing the amount of carbon in the atmosphere. In addition, as work becomes more team-based, it necessarily produces collective action problems. Workers have incentives to free ride on the work of others. They also have incentives to overdemand shared workspaces to ensure space for their teams to work.

The chapter is organized as follows: We first define a generic collective action problem, and then analyze three specific types. We start with public goods provision problems, in which individuals contribute money to fund roads, schools, and social services or time and effort to clean a park or watershed. We then study congestion problems, where individuals must restrict use of a resource such as a road system, beach, or park. We finish with renewable resource extraction problems, where individuals consume a resource that can regrow, such as fish, lobster, and trees. Congestion problems reset each day. If too many cars clog London's streets, the city can increase the fee on cars entering the city and solve the problem, and so past overuse has no long-term effects. However, an overharvested forest or fishery takes decades to regrow. We pay consequences for past failures to cooperate.

In each of the three specific models, the nature of the misalignment between individual incentives and collective goals differs, so solutions differ as well. We can solve public goods problems through taxes, and in some cases through sorting. Congestion problems can be solved with fees or usage restrictions. Solving renewable resource problems requires more elaborate monitoring and sanctions as well as conflict resolution mechanisms.

The solutions we offer here provide foundational insights that must be tailored to a local context. Any real-world situation includes layers of complexity that our models leave out. Balinese water temples solve a water allocation problem that is a sequential congestion problem, with upriver people drawing the resource first. International fishing rights that limit access solve a common pool resource problem with a moveable resource, as Norway's solution to coastal fishing could be undermined by overfishing in the nearby waters of Sweden, Russia, and Denmark.[3] Real-world solutions rely in part on the mechanisms discussed in Chapter 22 for building cooperation in the Prisoners' Dilemma: repetition, reputation, network structure, and group selection. Group selection enters indirectly: communities and nations that succeed in solving these problems will thrive, and their successes will be copied by others.

Collective Action Problems

In a *collective action problem,* each person has a choice between contributing and free riding. Free riding is in the individual's best interest. It earns her a higher payoff. Yet when everyone contributes, people receive greater benefits.

A Collective Action Problem

In a **collective action problem**, each of N individuals chooses to free ride (f) or contribute (c) to a collective action. An individual's payoff depends on her own action and the total number of cooperators. Individuals receive a higher payoff from free riding, $\text{Payoff}(f, C) > \text{Payoff}(c, C+1)$, but the sum of payoffs is maximized when everyone contributes.

A collective action problem can be represented as a many-player version of the Prisoners' Dilemma. We can thus refer back to the solutions proposed in Chapter 22 to get insights into how to create and maintain cooperation. However, those approaches will be incomplete for two reasons: collective action problems involve groups and communities, not just pairs, and many collective action problems take particular forms that make solutions more effective than others.

Public Goods

Our first specific type of collective action problem involves public good provision. Public goods satisfy *non-rivalry* (one person's use in no way interferes with any other person's use) and *non-excludability* (individuals cannot be prevented from use). Public goods include clean air, the national defense, early tornado warning signals, and the production of knowledge. The United States Constitution lists establishing justice, ensuring domestic tranquility, and providing for the common defense as responsibilities of the government. These too are public goods.

Private goods—bicycles, oatmeal cookies, and protractors—are neither non-rival nor non-excludable. Knowledge is both. Juxtaposing an oatmeal cookie and knowledge of trigonometry highlights the difference. A teacher might say, "Carla ate the last oatmeal cookie, so no one else can have one." She would never say, "Melissa, I am so sorry, but Carla just used the Pythagorean theorem, and now no one else can use it."

The non-excludability and non-rivalry of public goods produce a collective action problem not because people do not want to contribute. They do. The problem arises because people undervalue their contributions. For every dollar a person contributes, she adds to everyone's utility. In the formal model we describe here, each person allocates her income between a public good and a representative private good. Think of the private good as money that can be spent on anything else. Extending the model to include multiple public and private goods would only complicate the analysis.

A Public Good Provision Problem

N people each allocate an income $I > N$ between a public good (PUBLIC) and a private good (PRIVATE) that each cost \$1 per unit. Each person has the following utility function:

$$\text{Utility(PUBLIC, PRIVATE)} = 2\sqrt{\text{PUBLIC}} + \text{PRIVATE}$$

Socially optimal allocation: PUBLIC $= N$ (if $N = 100$, each person contributes \$100).

Equilibrium allocation: PUBLIC $= \frac{1}{N}$ (if $N = 100$, each person contributes \$0.01).[4]

In the model, we assume that utility is concave in the public good and linear in the private good. Those two assumptions require motivation. Recall that concavity corresponds to diminishing returns: as a person consumes more, she values it less. Concavity in the amount of the public good implies diminishing marginal returns to the public good. This is a standard assumption. People benefit more from adding a third lane to the highway than from adding a fourth lane. People benefit more from cleaning heavily polluted air than from removing the last few particles per billion. We assume linearity in the private good because it represents a composite of all private goods. While utility may be concave in any one good, be it chocolate, televisions, or denim jackets, it is probably closer to linear for all goods. The assumption has the added advantage of making the model easier to analyze.

We first solve for the socially optimal allocation, which we define as the allocation that maximizes the sum of the utilities of the population: the greatest happiness of the greatest number.[5] The socially optimal allocation calls for each individual to allocate $1 to the public good for each member of the population. Notice that the amount each person contributes to the public good increases with population size. This result does not depend on our particular function. It follows from the fact in a larger population, the non-rival public good can be enjoyed by more people. More people enjoy clean air or national defense, so more should be provided.

The equilibrium contributions equal 1 divided by the population size. As the population increases, people have a greater incentive to free ride off the contributions of others. We can see why by increasing the population size by one. This new person receives the same utility from the public good as everyone else had previously. If other people's contributions stayed the same, this new person has weaker incentives to contribute to the public good than the others had previously. Therefore, he will contribute less than the others contributed. Moreover, whatever amount he does contribute increases the total amount of the public good and creates an incentive for everyone else to contribute less than they had previously.

The model thus reveals that as populations become larger, public good provision problems exacerbate. The optimal level of the good increases, while the incentives to give fall. The formulae for amounts derived in our model (the N and $\frac{1}{N}$) do depend on the functions assumed, but the phenomenon of underprovision holds more generally.

This analysis assumes self-interested people, a common assumption in economic models. Evidence from surveys, experiments, as well as casual observations, reveals that people often have other-regarding preferences. People want good schools and roads for others as well as for themselves. We can include other regarding preferences by adding an *altruism parameter* to our model. A value of zero for that parameter corresponds to the self-interested rational actor from economics, and a value of 1 corresponds to everyone caring as much about others as they care about themselves. As shown in the box, pure altruists, people who care about everyone equally, contribute the socially optimal level. Anything less than pure altruism leads to underprovision.

The calculations show that in large populations, people contribute a portion of the optimal level that is (approximately) equal to the square of the altruism parameter. Though the extent of underprovision depends on the utility function, the example demonstrates the limits of altruism. People who care about others half as much as themselves contribute one-fourth of the optimal level. People who care about others a third as much as themselves contribute a mere one-ninth of the optimal level.

Public Good Provision Among Altruists

N people have altruistic preferences with weight α on aggregate utility:

$$(1 - \alpha) \cdot \text{Utility}_j(\text{PUBLIC}, \text{PRIVATE}) + \alpha \cdot \sum_{i=1}^{N} \text{Utility}_i(\text{PUBLIC}, \text{PRIVATE})$$

Equilibrium pure altruists ($\alpha = 1$): PUBLIC $= N$

Equilibrium general solution:[6] PUBLIC $= \frac{[(1-\alpha)+\alpha N]^2}{N}$

Example: $\alpha = \frac{1}{2}$: PUBLIC $\approx \frac{N}{4}$

Given that we do not live in a world of pure altruists, we must look to other mechanisms, such as taxation. Governments impose taxes to pay for roads, national defense, education, criminal justice systems, and other public goods. Determining the amount of the tax requires a more elaborate model that includes income and preference heterogeneity. People could vote on an amount and a common tax rate. The spatial voting model predicts a tax rate equal to the preferred level of the public good for the median voter. That level may not be socially optimal if people have heterogeneous incomes and preferences.

Many public goods such as schools, roads, and recycling programs can be classified as *local*. A local community can exclude others, but within the community the public good is non-rival and non-excludable. For local public goods, allowing people to sort into communities based on their preferences, what is called *Tiebout sorting*, offers a possible solution to the public good provision problem. People who want better schools, public parks, pools, and police protection can vote for higher taxes to pay for those public goods. Those people who do not want the local public goods can live in a separate community and pay lower taxes. Tiebout sorting is not a cure-all. It brings attendant costs, including lower social cohesion. Moreover, when high-income people isolate themselves, they reduce public good provision in poorer communities and reduce network interactions that can transfer information and knowledge.[7]

The Congestion Model

In a second type of collective action model, involving resources such as roadways, beaches, and water systems, the value to an individual decreases with the number of users. Anyone who has spent time stuck in traffic has experienced a congestion problem. A wide-open road brings more pleasure and utility than one clogged with cars. Estimates place the costs of traffic delays in the United States in the neighborhood of $100 billion a year. In some cities, notably Los Angeles and Washington, DC, commuters spend, on average, more than sixty hours per year in traffic.

Our *congestion model* assumes a resource of fixed capacity. Each day, people can use the resource or abstain. An individual's benefit from using the resource decreases linearly with the number of other users.[8] The slope of the line, the *congestion parameter*, captures the magnitude of the congestion effect.

A Congestion Model

M of N people choose to use a resource. Their utility can be written as follows:

$$\text{Utility}(M) = B - \theta \cdot M$$

where B denotes the *maximal benefit*, and θ is a *congestion parameter*. The remaining $(N - M)$ people abstain and receive utility of zero.[9]

Socially optimal: $M = \frac{B}{2\theta}$ Utility$(\frac{B}{2}) = \frac{B}{2}$

Nash equilibrium: $M = \frac{B}{\theta}$ Utility$(\frac{B}{\theta}) = 0$

In the socially optimal solution, the number of people who use the resource equals the maximal possible benefit divided by twice the congestion parameter. Those findings align with our intuition. The number of people who use the resource should increase with the maximal benefit and decrease with greater congestion effects. In the Nash equilibrium solution, exactly double the socially optimal number of people use the resource. Congestion becomes so severe that no one receives any benefit. That result is an artifact of the assumption that not using the resource gives a utility of zero. This finding has the counterintuitive implication that a community that builds a beautiful park may not produce much utility for its citizens. In equilibrium, the park will be sufficiently crowded such that being at the park is no more enjoyable than staying home.

When a model produces a result that runs counter to common sense, we need to reason through the result. People must be happier having a park, so the model must be wrong. It is wrong, because we assumed identical preferences. If people vary in how much they enjoy the park, then some people may be receiving positive utility, while others get no benefit. Second, the model assumes the park is always crowded. That will not be true. Third, the alternative option might be going to the beach, not staying home. The new park may make the beach less crowded. Finally, people enjoy diverse experiences. If the city has a skateboard park, a dog park, and a water park, then people may get benefits from the diversity of experience over a period of weeks.

Those challenges notwithstanding, the main result still has some teeth. During busy times, congestion will rise to the point where the park produces no more benefit than any other activity. Crowding will still occur, though not as much as when there was a single park. In addition, as shown in the box below, the creation of multiple parks offers no guarantee that people allocate themselves across those parks optimally. In the example shown in the box on the next page, in equilibrium, too many people go to the larger park.

In addition to creating more parks, a community could try other solutions such as rationing, rotating access, running a lottery, fees, and enlarging capacity. Rationing gives each person or household a fixed amount of the resource. This solution works for divisible resources like water. It is less practical for roads. Rotation schemes divide use of the resource by time. During air pollution alerts, a city can restrict roads to cars with even- (or odd-) numbered license plates on certain days. Other resources, such as placements in popular public schools, cannot be rationed or rotated. In those cases, lotteries can be held.

Multiple Congestible Goods

M people go to Park 1 and $(N-M)$ go to Park 2. To account for Park 2 being larger, utilities are as given below:[10]

Park 1: Utility$(M) = N - M$

Park 2: Utility$(N - M) = 3N - 3 \cdot (N - M)$

Socially optimal: $M = \frac{N}{2}$ creating total utility N^2

Nash equilibrium: $M = \frac{N}{4}$ creating total utility $\frac{9}{16}N^2$

For access to roads, fees are a popular solution. The city of London charges a fee to enter the central city. Limited-access toll roads around the world do the same. Usage fees allocate the resource to those willing to pay the most. These may not be the people who would get the most utility. Singapore uses a combination of fees and limited access. Singapore auctions a fixed number of motor vehicle permits each year. The permits, which last for ten years, often sell for more than the price of a typical car. To reduce congestion during peak times, Singapore, like London, also charges fees for driving into the central business district. Singapore's traffic moves smoothly for a city its size, and the government raises substantial amounts of money, which can then be used for public transportation.

Enlarging the capacity of roads has had mixed success. When a city adds lanes to highways to increase traffic flow, it makes housing near the highways more desirable, a positive feedback. A resulting increase in housing creates more traffic, requiring even wider roads, producing a positive feedback loop similar to those described in the systems dynamic models covered in Chapter 18.

Renewable Resource Extraction

Last, we consider *renewable resource extraction,* in which individuals share a resource that regenerates itself. This model applies to forests, watersheds, grasslands, and fisheries. In each case, the amount of the resource available in the future depends on how much we use now. If too much was used, the resource may not regenerate fast enough. The need for regeneration of the resource makes these problems more fragile than public good problems or congestion problems. A city that underfunds public lighting in one year can increase spending the next year without enduring long term effects from its mistake. If a community overfishes a stream or overharvests a forest, it pays lasting costs because to make fish, you need fish. You do not need streetlights to make streetlights. Furthermore, the renewable resource may be a necessity: food to eat, water to drink, and fuel to keep warm. People need to extract the resource to live.

Renewable Resource Extraction Model

Let $R(t)$ denote the amount of a renewable resource at the start of period t. Let $C(t)$ equal the total amount consumed in period t, and g denote the growth rate of the resource. The amount of the resource in period $t + 1$ is given by the following difference equation: [11]

$$R(t + 1) = (1 + g)[R(t) - C(t)]$$

The *equilibrium consumption level*: $C^* = \frac{g}{(1+g)}R$

Renewable resource extraction problems exhibit a tipping point in the level of consumption. Any rate of consumption above the equilibrium extraction rate will produce a collapse, as can be shown in a formal model. We can think of the amount of the resource as a circular field, a pie. Consumption takes a bite out of that pie. Growth regenerates an amount of resource proportional to the amount remaining. For low levels of consumption, the resource will increase in size. But regrowth will not be able to compensate for high levels of consumption. In between lies an equilibrium level of consumption exactly balanced by regrowth.

If consumption exceeds the equilibrium level, the model predicts accelerating declines that become steep collapses. The slow decline followed by a steep collapse sends a warning for those who manage resources that are difficult to measure accurately, such as fish stocks. Annual catches give a clue, but they are not exact. We should not be surprised that cod fishing in the North Atlantic produced a modern collapse to rival that suffered by the Vinlanders, as described by Jared Diamond in his book on collapsing societies. Cod had been fished in the North Atlantic for over five hundred years. British explorers who first visited the Canadian coast told tales of catching cod in baskets and having difficulty rowing through the prodigious shoals of cod. By 1992, Canada had imposed a moratorium on cod fishing.[12]

Our model of resource extraction assumes a constant growth rate, allowing us to solve for the equilibrium consumption level. In reality, growth rates vary from year to year. In the case of a pasture, growth depends on temperature and rainfall. For a fish population, the growth rate depends on the amount of available food, which in turn depends on variations in weather or on climatic changes.

In the other two models, variation would not have long-term consequences. Some years we might have too much of a public good or a little less congestion. Those would affect utility, but perhaps no more than the inevitable variation in the weather. But in renewable resource extraction problems, variation leads to either collapse or abundance, provided that behavior does not change. Figure 23.1 assumes an average regeneration rate of 25% and 100 units of a resource. Given those assumptions, the equilibrium consumption level equals 20 units per year. The figure assumes a variable growth rate randomly drawn between 20% and 30%. The model also builds in a maximal level of the resource set to 150.

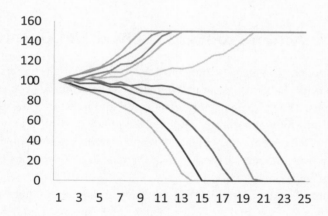

Figure 23.1: Ten Possible Paths Assuming Variation in Growth Rate of Resource

In approximately half of the paths, the level of the resource collapses. In the other half, the level of the resource increases to the maximum possible level. The variation does not cancel out. To the contrary, the effect of a reduction or increase in growth accumulates over time.[13] From this simulation, we see that optimal consumption policy would call for less consumption following lean years to prevent collapse.

Given that variation in the growth rates of a renewable resource requires consumption to vary with the resource level, we know that communities that manage a renewable resource must be able to adjust extraction levels. The method or mechanism used to make those adjustments depends on features of the resource. To borrow a phrase, "There are no panaceas."[14] No single solution will work in all cases. How local populations solve these types of problems depends on characteristics of the resource and the community.

Fish differ from cattle. A community managing multiple herds of cattle sharing a commons can monitor the behavior of individuals and the level of the resource (the amount of grass). Overgrazing can be solved by rotation schemes that allocate to each individual set times or zones for grazing. These rotations can adjust grazing based on the height of the grass. But for communities that fish, managing the resource requires more elaborate institutions that can precisely monitor individual behavior. The number of fish in the sea cannot be counted. It can only be estimated based on catches. The resource extraction problem has much more uncertainty than the cattle situation. Managing a common aquatic resource requires conservatism and more monitoring.

Collective Action Problems: Solved and Unsolved

In collective action problems, the outcomes that result from self-interested behavior do not align with the goals of the individuals. As noted, these problems occur in myriad settings. They arise in paying for non-rival and non-excludable goods. They arise in decisions for when to drive on highways. They even arise in how we drive on highways. Drivers on a busy highway who tailgate or talk on their cellphones may not take into account the costs of those actions to all of the cars behind them if an accident occurs.

Such problems exist at multiple scales. They arise within family interactions: keeping the house clean, making dinner, going shopping, and saving for vacations can all misalign individual incentives and collective well-being. They exist with communities, regions, and countries in the provision of public goods and the use and management of limited resources. They also exist at a global scale in the form of carbon emissions. Most countries would prefer to produce more energy themselves (which emits more carbon) but have lower global emission levels: individually rational actions do not align with the common good.

Collective action problems occur in the natural world as well. Trees in a forest compete for light and water. If a tree species evolves a higher canopy or deeper roots, it will increase its own chances of survival but impose a cost on other species of trees. The trees cannot pass laws to prevent growing too tall or tapping into deeper water. They do not achieve the socially optimal solution.[15]

Collective action problems tend to be easier to solve the smaller and more homogeneous the group of individuals or actors involved and the better the information (i.e., actions are easier and the state of the system can be monitored). While families generally solve collective action problems, international organizations find cooperation much harder. Efforts to reduce carbon emissions require coordination among a large group of diverse actors using imprecise monitoring mechanisms. Solving such a problem requires coordination as well as an enforcement mechanism. History teaches us that overfishing waters or overgrazing meadows invites the risk of collapse. We can apply that same reasoning to the collective action problems we face today. Elinor Ostrom, who spent decades studying real-world efforts to solve collective action problems, found that in addition to monitoring deviations, communities that solve collective action problems agree on clear boundaries, agree on well defined rules, grant the authority to impose graduated sanctions, and establish mechanisms to resolve disputes.[16]

24. *Mechanism Design*

Institutions are designed to alter human behavior. To remain effective over time, institutions need to adapt to changes in the environment or the society the institution is meant to regulate.

—Jenna Bednar

In this chapter, we show how to use models to design political and economic institutions. An institution consists of a means through which people communicate information as well as a procedure for making decisions, reallocating resources, or producing outputs based on the information revealed. In markets, people and firms communicate through prices to execute trades and make production decisions. In hierarchies, people communicate through written language to organize work plans. And in democracies, people communicate preferences through votes. Voting rules then decide policies. Well-designed institutions induce communications and actions that produce desirable outcomes. Ineffective institutions do not.

In this chapter, we present a framework for modeling institutions known as *mechanism design*. This framework highlights four aspects of real institutions: *information*, what the participants know and should be revealed to them; *incentives*, the benefits and costs of taking particular actions; *aggregation*, how the individual actions translate into collective outcomes; and *computational costs*, the cognitive demand placed on participants.

The origins of mechanism design lie in the analysis of general questions about the allocation of goods, and in particular whether market mechanisms or central planning best allocates goods. Early models posited behavioral rules such as price-taking in a market or voting truthfully. The modeler then worked through the implications of those behaviors, for example, how they aggregated. That approach was abandoned in favor of one that assumed optimizing behavior, making the constructions amenable to game theoretic reasoning. Mechanism designers then solve for Nash equilibria and compare institutions based on rational behavior.

The framework has proved useful. It can be used to find flaws in existing rules and procedures, to explain why institutions succeed or fail, and to predict outcomes. It has also been used to design a variety of institutions, including the spectrum auctions described in Chapter 2, as well as many online markets, governmental voting systems, and even the procedures that allocate space for projects on space shuttle voyages.[1]

Our treatment consists of six parts. We first describe the mechanism design framework using the Mount-Reiter diagram. In the second part, we study the problem of three people choosing between two alternatives. In the third part, we analyze three auction mechanisms and find that all yield identical results. In the fourth part, we show that this was not a coincidence and describe a foundational result, the revenue equivalence theorem, which shows that any auction mechanism that satisfies certain assumptions produces the same outcome. In the fifth part, we compare a majority rule voting mechanism with a pivot mechanism as ways to decide on whether to undertake a public project. We conclude by broadening our discussion of mechanisms along the lines introduced in our criticism of Nash equilibria.

The Mount-Reiter Diagram

A mechanism consists of six parts: an *environment* (the relevant features of the world), a set of *outcomes,* a set of *actions* (called the *message space*), a *behavioral rule* that people follow to produce actions, an *outcome function* that maps the actions into outcomes, and a *social choice correspondence* that maps the environment into a set of hoped-for outcomes. The social choice correspondence commonly consists of either the outcome that maximizes the sum of the participants' utilities or of the set of *Pareto efficient* allocations. An outcome is Pareto efficient if and only if no other outcome exists that everyone prefers. Pareto efficiency is a low bar.

Pareto Efficiency

Within a set of outcomes, an outcome is **Pareto dominated** if there exists an alternative that everyone prefers. All other outcomes are **Pareto efficient**.[2]

The Mount-Reiter diagram captures these essential parts of a mechanism graphically (figure 24.1). The diagram juxtaposes what we desire and what exists. Across the top, the social choice correspondence describes the outcomes that we normatively desire. Along the bottom, we have the mangle of reality. People apply their behavioral rules to send messages or take actions. An outcome function maps those actions into outcomes. Ideally, the lower, more complicated path on the bottom produces the same outcome as the top path, that is, the desired outcome.

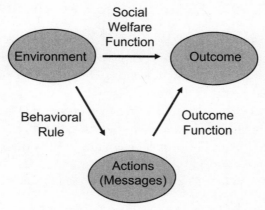

Figure 24.1: The Mount-Reiter Diagram

Not all mechanisms succeed. For example, if the environment consisted of people with preferences for a public good, the social choice correspondence maps their preferences to the optimal level of that good. However, as we saw in Chapter 23, a voluntary contribution mechanism, in which people pay for as much of the public good as they desire, results in each person providing $\frac{1}{N}$ units of the public good rather than the optimal N units. When the outcome produced by the mechanism does not align with our objective, we say the mechanism fails to implement the social choice correspondence.

The list of properties that we would like a mechanism to satisfy varies by context. We describe five here. First, we would like the equilibrium outcome of the mechanism to agree with our social choice correspondence (*Pareto efficiency*). Second, ideally participants would apply dominant strategies, that is, their best actions would not depend on the actions of others. If so, we say that the efficient outcome is *dominant strategy implementable*. Third, we would not want to have to force people to participate in the mechanism (*voluntary participation*). Fourth, if the mechanism involves a transfer or payment of resources, we do not want to have to put in additional money or destroy resources (*budget balance*). Later in the chapter when we analyze mechanisms for deciding on a public project, we see that this may be difficult to satisfy. Last, in many cases, we desire *truth-telling*. We would like the messages that people send to reveal their true information or their true type. Game theorists call this *incentive compatibility*. In most cases of interest, no mechanism can satisfy all of these desiderata. Thus, one contribution of mechanism design has been in demonstrating what is possible and what is not.

Majority Rule and the Kingmaker Mechanism

The first class of environments that we consider consists of people voting on a joint action or piece of legislation. We consider three people, whom we call Uma, Vera, and Will, who want to see a movie together and must decide between an action movie, a drama, and a comedy. The same environment would apply to three members of the military deciding whether to attack their opponent, defend their position, or cede the land. In either interpretation, the environment consists of three people with preferences defined over three alternatives. We write preferences using *orderings*. The ordering *action* $\succ$ *comedy* $\succ$ *drama* corresponds to the action movie being most preferred, followed by the comedy and then the drama. We assume the following preference orderings:

Uma: action $\succ$ comedy $\succ$ drama

Vera: comedy $\succ$ drama $\succ$ action

Will: comedy $\succ$ drama $\succ$ action

In this example, we take the social choice correspondence to be the set of Pareto efficient choices. Given the assumed preferences, the comedy and the action movie are Pareto efficient. The drama is Pareto dominated by the comedy.

We first evaluate *majority rule* as a mechanism. In the case of a tie, we assume the choice is made randomly. If people vote sincerely, the comedy receives two votes. However, suppose that Vera and Will both believe that the other two people will be split between the drama and the action movie and each votes for the drama. Suppose also that voting is sequential. Vera votes first and selects the drama. Will votes second and does the same. Uma's vote no longer matters, but suppose that to avoid conflict, she also votes for the drama. The three votes constitute a Nash equilibrium. No person has any incentive to change his or her vote. In this case, majority rule does not always implement a Pareto efficient outcome.

We next consider the *kingmaker mechanism*.[3] In this mechanism, one person is randomly selected to be the kingmaker. The kingmaker then selects a "king," who determines the group's choice. If Will is the kingmaker, he must pick between Uma and Vera. Whomever he chooses becomes king, and that person then selects the movie.

If the person selected as king acts rationally, she will select her favorite movie. Therefore, the outcome will be Pareto efficient. For this reason, the kingmaker mechanism implements Pareto efficient outcomes. The mechanism has the added advantage that if any two people have the same favorite movie, the mechanism selects that outcome. To see the logic, once again, assume that Will is the kingmaker. If Uma and Vera prefer the same movie, then that movie will be selected regardless of Will's choice of king. If, on the other hand, Will and Uma prefer the same movie, then Will should pick Uma.

Three Auctions

Now that we have a basic understanding of mechanisms, we turn to the study of auctions. Most of us have some familiarity with auctions owing to the prevalence of online marketplaces like eBay. Auctions are used in other settings as well, including government contracts, used car markets, and most web advertising. We restrict attention here to a single seller and many bidders. The object could be a house, a car, tickets to a soccer game, or a piece of art. We also assume that each bidder assigns a unique value to the object to rule out ties. The Pareto efficient outcomes are those in which the object goes to the bidder with the highest value. Any other outcome will be Pareto dominated by that outcome. We now compare three types of auctions: ascending-bid, first-price, and second-price auctions.

Ascending-Bid Auctions

In an *ascending-bid auction*, the auctioneer calls out a price. Any bidder willing to pay that price raises her hand. The auctioneer raises the price until only one bidder remains. That bidder then pays the price at which the second-to-last bidder lowered her hand. In an ascending-bid auction, a rational bidder remains in the auction until the price reaches her value. Dropping out before the price reaches her value creates the possibility of not winning the object at a price at a good price. Remaining in the auction after the price exceeds the bidder's value means the bidder could win the object but pay more than her value, resulting in a net loss.

If all of the bidders act rationally, then the bidder with the highest value wins the object and pays a price equal to the second-highest bidder's value. As an example, suppose that there exist three bidders with values $30, $60, and $80. When the price called out by the auctioneer exceeds $30, the first bidder exits the auction. When the price gets to $60, the second bidder exits. Therefore, the third bidder wins the auction and pays $60.[4]

In a *second-price auction,* each bidder submits a sealed bid. None of the other bidders see the amount. The object goes to the bidder who bids the largest amount. However, the bidder only pays an amount equal to the second-highest bid. The construction of the second-price auction makes telling the truth optimal. Imagine a bidder who values an object at $80 deciding how to bid in a second-price auction. We can assume that the other bidders have already submitted their bids. The bidder must consider three possible cases: the highest other bid could be less than $80, equal to $80, or more than $80. In each case, the bidder does best by reporting her true value for the object.

The logic becomes clearer when we work through an example. We will assume that the bidder's value for the object is $80. We consider four cases for the highest submitted bids of the others: $70 (lower), $80 (equal), $85 (just above), or $90 (higher). Table 24.1 shows payoffs for five bid values ranging from $65 to $95.

Highest Other Bid	Bidder with Value 80			
	65	80	85	95
70 (Low)	0	+10	+10	+10
80 (Equal)	0	0	0	0
82 (Above)	0	0	-2	-2
90 (High)	0	0	0	-10

Table 24.1: Net Payoff as a Function of Various Bids Given a Value of 80

As can be seen from the table, bidding 80 always gives at least as high a payoff as any other bid. Bidding her true value is always a best action (a *dominant strategy*). The same logic applies to all bidders, so all should bid their true values (the mechanism is *incentive compatible*). It follows that in a second-price auction, the bidder with the highest value wins the auction, and the amount paid equals the second-highest bidder's value.

In a *first-price auction,* each participant submits a bid, and the highest bid wins, with the bidder paying an amount equal to that bid. As in a second-price auction, the bids are submitted simultaneously, so no one knows the others' bids. A participant's optimal bidding strategy in a first-price auction depends on the participant's belief about the values (and therefore the likely bids) of the other bidders. We will assume that bidders do not know other bidders' values but that they do have correct beliefs about the distribution over those values. To be specific, we assume that the bidders' values are uniformly distributed between zero and $100 and that all of the bidders know this distribution. Bidders also know that all of the other bidders know this information as well.

Using calculus, we can show that if the values of bids are uniformly distributed and if all bidders bid optimally, then with two bidders each should bid half her true value, and with N bidders, each bidder should bid $\frac{N-1}{N}$ of her value. A person in an auction with nineteen other people should therefore bid 95% of her true value. Given this bidding rule, the bidder with the highest value always wins the object. We can also show that the amount she pays equals the expected value of the second-highest bidder. Thus, the ascending-bid auction also produces an efficient outcome and the price corresponds to the expected value of the second-highest bidder.[5]

Prior to writing down the model, many of us would have had the insight that the more bidders in the auction, the more a person should bid. Without the math, we would not have known the equilibrium bidding rule. The model gives us an exact expression for how much a person should bid. The amount increases in the bidder's value, which implies that the bidder with the highest value will win the auction, just as in the other two auction formats.

Revenue Equivalence Theorem

In each of the three auction formats, the bidder with the highest value wins. Therefore, all three mechanisms produce an efficient outcome. In addition, the expected amount paid by the winning bidder equals the value of the second-highest bidder. In other words, all three auctions produce the same expected revenue and allocate the object to the same person. That is remarkable. Even more remarkable, it can be shown that the winner and expected revenue are the same for any auction in which bidders act optimally, the highest bid wins the object, and a bidder with a value of zero receives no payoff. In other words, all auctions that satisfy those two conditions produce the same expected outcome, a result known as the *revenue equivalence theorem*.[6]

Revenue Equivalence Theorem

Any auction in which the bidders have independent private values drawn from a known, common distribution produces the same revenue to the seller and the same expected payoffs to the buyers if each bidder makes a bid that maximizes her expected payoff, the bidder with the highest bid always wins the object, and a bidder who has a value of zero has an expected payoff of zero.

The revenue equivalence theorem implies that an *all-pay auction*—in which every bidder, even the losing bidders, pays the amount of her bid—produces the same outcome as the second-price auction.[7] Even a crazy design such as a *third-price auction,* where the highest bid wins and pays an amount equal to the third-highest bid, produces the same winner and same revenue. The revenue equivalence theorem does not imply that the auction rules do not matter. In an actual auction, bidders may not use optimal strategies, or, in a first-price auction, bidders may have different beliefs about the value distributions of other bidders. If either condition holds—non-optimizing bidders or diverse beliefs—then revenue could vary across the types of auctions. Empirical and experimental tests do show some differences in how auctions perform.

As would be expected from our discussion of when to expect rationality, the higher the stakes and the more sophisticated the bidders, the more likely it is that people act rationally. In online auctions for consumer goods, we might expect some people to follow rules of thumb or suffer from biases (such as bidding in increments of $10). In a multimillion-dollar oil lease auction, bidders probably have access to full information and the requisite skills.

Also, the type of auction could influence the number of bidders. In timber auctions, first-price auctions attract more small bidders than do ascending-bid auctions because small bidders have some chance of winning if the bigger bidders submit low bids. The small bidders have no chance in an ascending-bid auction, as the bigger firms can see the smaller firms' bids and outbid them.[8]

Auctions also differ in the cognitive demands they place on participants. In some auctions, optimal behavior is easy to learn. In an ascending-bid auction, a bidder should remain in the auction until the price reaches her value. Other bidders not following optimal strategies could cause a bidder to have a higher or lower expected payoff, but they do not change the optimal strategy: a bidder should stay in as long as the price is less than her value. Similarly, in a second-price auction, a bidder should always follow the same strategy of bidding her true value. However, figuring out that truthful bidding is optimal requires multiple steps of logic.

Recall that dominant strategies are optimal regardless of the strategies of others. Both ascending-bid and second-price auctions have dominant strategies. First-price auctions do not. In a first-price auction, changes in the bidding strategy of one bidder can change the optimal strategy for another bidder. If one bidder always bids either zero or 50, then the other bidder should always bid either 1 or 51. There would be no reason to bid 60 or 70 as the winner would pay more for the object than necessary. Given the behavior of the other bidder, whenever a bid of 60 wins the auction, so does a bid of 51.

Even if an auction has a dominant strategy, not all dominant strategies are equally easy to deduce. In an ascending-bid auction, the strategy of staying in as long as the price is less than a bidder's value requires a single step of reasoning: *if the price is less than your value, buy it at this price.* In the second-price auction, a bidder has to think through multiple contingencies to see that truthful revelation is optimal. Of course, once someone bids in several second-price auctions, she should learn that the optimal bid is to tell the truth.

A last feature to consider is whether the auction encourages non-optimal behavior. In first- and second-price auctions, bidders submit their bids without knowing others' bids. In an ascending-bid auction, bidders can see the price rise and are aware of who remains in the bidding. This could cause a bidder to attach some value to winning and to overbid. Auctioneers in charity auctions try to raise bids by emotional appeals, perhaps by showing a video of children frolicking on new playground equipment that your bids will support.

The success of strategies depends on the sophistication of the bidders. It is difficult to imagine bidders in a timber auction being persuaded to bid more than their forecasted valuations. It is less difficult to imagine a person at a charity auction overbidding in light of the cause. Whether or not bidders change their values during the bidding process is a matter of conjecture. We need only recognize that it *could* happen. In first- and second-price auctions, bidders make a single bid, which allows no opportunity for emotional appeals during the auction.

Finally, in the first-price auction and the ascending-bid auction, the price equals the highest bid. In the second-price auction, it equals the second-highest bid. This leaves the appearance that the seller could have received a higher price and, in part, explains why governments do not use second-price auctions. Imagine the headline if a government received three bids for oil rights, one at $6 million, one at $8 million, and one at $12 million: "Government Gets $12 Million Bid but Sells Land for $8 Million." Anyone who knows auction theory would know that had the government run a first-price auction or an ascending-bid auction, then the top bid would not have been $12 million. It would have been $8 million.

As has been highlighted throughout the book, formal models reveal conditions necessary for a result to hold. The revenue equivalence theorem does not say that all auction mechanisms produce the same outcome. It states that all auctions with optimizing bidders in which the highest bidder wins the object and a bidder with zero value has an expected payoff of zero are equivalent. A seller could raise more money by relaxing one of those three assumptions. A seller would have difficulty making people act against their self-interest, and she would also probably not be able to extract money from someone who does not value the good. This leaves as the only possibility not selling the object to the highest bidder. One way to do this is to not sell the good at all. If the seller knows the distribution of values, she could set a *reserve price,* a minimum bid. Under some conditions, this can increase her expected revenue. Suppose that seller is certain that the three bidders for an object have values of 5, 10, and 60. Using any of the three auctions above, the winner bids $60 and pays $10. The seller could earn higher revenue by imposing a reserve price of $60 and running a first-price auction.

Mechanisms for Deciding on a Public Project

We next compare two mechanisms for deciding whether to build a public project such as a school, a new highway, or a sports arena. We assume each person has an individual value from the project and that the project has a collective cost.

A Public Project Decision Problem

Let $(V_1, V_2, \ldots, V_N)$ denote the monetary values that N people attach to a public project with cost C. The project should be undertaken if and only if

$$C < V_1 + V_2 + \cdots + V_N$$

We first consider the *majority-vote equal sharing mechanism.* In this mechanism, individuals vote whether to undertake the project. If a majority vote yes, the cost of the project is divided equally among the population.

Majority-Vote Equal Sharing

Individuals vote for or against undertaking the project. If a majority vote for the project, the project is undertaken and each pays a cost $\frac{C}{N}$. As the following example shows, this mechanism can violate efficiency and voluntary participation.

We know from the spatial voting model that whether or not the project is undertaken depends on the preferences of the median voter. In this case, that will be the person with the median value for the public project. By construction, the mechanism satisfies the budget balance condition and incentive compatibility. However, the mechanism need not satisfy either efficiency or voluntary participation as can be seen in an example. Suppose that there exist three people with values $0, $120, and $150 for a public project that costs $300. The efficient outcome is that the project should not be undertaken because the total cost of $300 exceeds the sum of the individual values. However, given that costs will be split equally, each person votes on whether to undertake the project at a cost of $100 each. It follows that two of the three individuals will vote for the project, and it will be undertaken, the inefficient outcome. Moreover, the individual with value $0 receives a payoff of -$100, so the example also violates voluntary participation.

In our second mechanism, the *pivot mechanism,* each individual submits a valuation for the project. If the sum of the valuations exceeds the cost of the project, the project is undertaken. Otherwise it is not. The amount that the individual will be taxed equals the cost of the project minus the sum of all of the other individuals' valuations. If the valuations of the other individuals exceed the cost of the project, the individual pays nothing.

The Pivot Mechanism

Individual i submits a valuation $\hat{V}_i$ for a project of cost C. If the sum of the individual valuations exceeds the cost, then the project is undertaken.

$$\hat{V}_\Sigma = \hat{V}_1 + \hat{V}_2 + \cdots + \hat{V}_N \geq C$$

Individual i pays no tax if $C - (\hat{V}_\Sigma - \hat{V}_1) < 0$ and $C - (\hat{V}_\Sigma - \hat{V}_1)$ otherwise. The mechanism is **incentive compatible** ($\hat{V}_i = V_i$), **efficient,** and **individually rational.** It also implements the efficient outcome in dominant strategies. As the following example shows, this mechanism can violate **budget balance**:

Example: $(V_1, V_2, V_3) = (60, 120, 150)$ and $C = 300$.

The project should be undertaken given that $300 < 60 + 120 + 150$. Individual 1 pays taxes of 30, the cost minus the sum of the other valuations $(300 - 270)$; individual 2 pays taxes of 90; and individual 3 pays taxes of 120. The total taxes generated the sum of 240, less than the cost of the project.

This mechanism satisfies incentive compatibility by a logic similar to that of a second-price auction. Suppose that the project has a cost of $300 and that an individual values the project at $80. There are three cases to consider. If the other valuations sum to less than $220, the individual has no incentive to submit a value more than $80, as he would have to pay that amount. If, at the other extreme, the sum of the others' valuations exceeds $300, then he pays nothing, and he could submit any valuation. If, though, the sum of the valuations of the others lies between $220 and $300, then if the individual submits a valuation of $80, he will pay $300 minus that sum and the project will be undertaken (the efficient outcome). He would not want to submit a valuation of, say, $70 because the sum of the other values could be $225 and his low valuation would prevent the project from being undertaken. Had he submitted a valuation of $80, it would have been undertaken at a cost to him of only $75.

Given that the pivot mechanism satisfies incentive compatibility, it follows that it also satisfies efficiency. The project is undertaken only if the sum of the valuations exceeds the cost. Note that because reporting one's true value is a dominant strategy, the efficient outcome is also dominant-strategy implementable. Also, because each individual pays at most her value for the project, the mechanism satisfies voluntary participation. However, as shown in the box below, the mechanism need not produce a balanced budget. In fact, only in rare cases will it do so.

For the problem of deciding on a public project, no mechanism will satisfy every criterion we might desire. The fact that we can use models to prove that can save us a lot of time trying to do the impossible. Just as engineers do not waste time trying to build perpetual motion machines, mechanism designers do not seek incentive-compatible, individually rational, efficient, budget-balanced mechanisms for the public project problem. No such mechanism exists.

The pivot mechanism is about as good as we have, but it fails to satisfy budget balance. That problem cannot be fixed by raising the amount of taxes that people pay for the project, as that would make the mechanism no longer either incentive compatible or individually rational. Individuals would have an incentive to lie, and some might be asked to contribute more than their value for the project. One possible workaround is to raise taxes in some other way and to have a pool of money available for projects. That would itself create incentive issues, though not as directly. A better solution is to have some other source of money. For example, a university that both has a large central endowment and consists of colleges that have separate endowments could use this mechanism to decide whether to construct a new student union. Each college dean would have an incentive to truthfully reveal her value for the union and the university chancellor could make up for any shortfall. A business composed of subunits that have budget authority could do the same. A project to switch to a cloud-based system could be decided upon with the pivot mechanism, and any shortcoming could be covered by upper management.

Summary

The mechanism design framework enables us to compare mechanisms across a variety of criteria. Does a mechanism produce efficient outcomes? Do people tell the truth? Would people voluntarily participate? Does the mechanism produce a surplus or loss? Using the mechanism design framework, we can also derive what is possible. It may not be possible to satisfy all of the desired criteria within the same mechanism. In these cases, modelers become engineers. We use models to try to construct workable solutions.

As technology changes, so too can our mechanisms. Take, for example, the auctions used by internet search sites such as Google. Originally, Google charged a fixed price per thousand clicks. That mechanism was not optimal given changes in information technology that allowed Google to run millions of auctions simultaneously. By using auctions, Google increased its revenue, and it allocated ad space more efficiently. Google now uses a generalized second-price auction. Each bidder submits a per-click bid to advertise for a keyword—say *mesothelioma*, a cancer caused by exposure to asbestos. The highest bidder receives the first ad slot, the second-highest bidder receives the second slot, and the third-highest bidder receives the third slot. The prices they pay are determined as in the second-price auction.

Suppose that the top four bids are \$10, \$7, \$6, and \$3 per click. The third-highest bidder will pay a price equal to the bid of the fourth-highest bidder, \$3. The second-highest bidder will pay a price equal to the bid of the third-highest bidder, \$6. And the highest bidder will pay \$7.[9] After learning the valuations of advertisers, Google could have set a reserve price and raised even more money. But that outcome would not necessarily hold if the bidders knew this was Google's plan. A bidder who thinks he is likely to be the high bidder would not want Google to know his valuation. Placing reserve prices would also harm Google's reputation. A reserve price would be seen as non-cooperative behavior because Google cannot claim to have a reserve value for spots on a webpage. The top advertising slot on a keyword search has little value to Google unless it is sold. That is not true for someone selling a collection of vintage albums or a used car. Those items have values to their sellers, so a reserve price is justified. However, Google values its reputation, and setting a reserve price to extract more revenue might anger advertisers.

To summarize, mechanism design models can aid in the design of and choice among institutions. With them, we can deduce what is possible and not possible to implement. It may not be possible to construct a mechanism that produces efficient outcomes, induces people to tell the truth, and results in a balanced budget. If so, we should not waste time and effort trying to design the impossible. Better that we devote our energies to thinking through how best to trade off efficiency, truthful revelation, and a balanced budget

We can also use mechanism design to explore bigger questions, such as when we should use a market, when we should vote, when we should rely on a hierarchical mechanism, and when we should turn to a voluntary collective to allocate a resource or decide on an action.[10] Each of these four institutions—markets, democracies, hierarchies, and collectives—functions well in some settings and not so well in others. For example, we would not want to vote on what goods people buy, nor would we want to use a market to decide on our political leaders.

Within society writ large as well as within an organization, we see each of these institutional forms. A university confronts a market for professors, relies on a democracy to hire faculty, assigns course assignments through a hierarchy, and develops strategic plans using collectives. Nonprofits, for-profits, and government agencies are also a mashup of these various institutional forms. Using the tools of mechanism design, we can formally compare how each of these institutions functions, and in doing so, make better assignments of institutions to tasks.

25. *Signaling Models*

Honest people don't hide their deeds.

—Emily Brontë

In this chapter, we study *signaling models*. These models identify conditions under which people send costly signals to reveal information or their type. A person might signal wealth by purchasing an expensive piece of art, physical stamina by climbing a mountain, or empathy by posting support for causes on social media. Signaling to reveal status has always been a part of human nature. In the nineteenth century, Thorstein Veblen refined our understanding of signaling with his development of the concept of conspicuous consumption: he observed that rather than buy goods that bring direct enjoyment or practical utility, people often make choices to signal their social status. Veblen would take delight in our modern icons of conspicuous consumption, such as the Maybach Landaulet, an automobile that retailed for nearly $1.5 million, ten-year-old bottles of Cristal that sell for over $1,500 a bottle, and Leica cameras that sell for tens of thousands of dollars.

Conspicuous consumption endures because we care what others think of us, and what we consume signals our status to others.[1] We do not see others in their entirety, so we rely on what they wear, drive, and consume to make inferences about their hidden attributes. If we see someone driving an expensive car, we can infer that she has wealth. A person donating to a charitable organization signals generosity—no selfish person would take such an action. A person announcing her PhD in theoretical biology on social media signals her intelligence and work ethic. Almost all actions have some degree of signaling. When politicians vote whether to go to war or to impose sanctions on another country, they signal ideologies. Politicians with longer-term goals, such as running for president, may cast votes that send the best signals rather than those that produce the best policy.

In this chapter, we first study a discrete signaling model, in which an individual can either send a signal or not. Individuals differ in their costs of sending the signal.

For signals to function, they must be costly or verifiable. That will be a central take-away from this chapter. For example, an employer may have a plum assignment in Barcelona, Spain, for the summer and want to select from a collection of new hires, all of whom list Spanish-language skills on their resumes. Claiming to speak Spanish is a *costless signal*. Instead, the employer could create a language badging program where earning the badge entails giving an hour-long presentation in Spanish. For the employees who can speak Spanish, this signal (the presentation) has low cost. For those employees not fluent in Spanish, the costs of preparing an hour-long talk would be prohibitive. In the formal language of signaling models, the badge *separates* the people who speak Spanish from those who do not.

We then cover a continuous signaling model, in which signals can vary in their magnitude. A summer camp may have only one position for a lead kayaker and want someone with incredible stamina. The camp director might then ask two applicants for the position to kayak as far as they can over the next ten hours. The stronger of the two kayakers could choose a distance that the weaker kayaker cannot achieve, guaranteeing that the test separates. Both models give us conditions for when signals will separate and when they will not. They therefore give us deeper insights than anecdotal accounts of people, animals, politicians, and governments signaling, by providing explicit characterizations of when they signal and how costly those signals are. For example, the models will make clear why students work so hard to signal their worthiness for college and medical school. In the conclusion, we discuss the contributions of signaling models as well as some other implications. We also discuss how signaling occurs in ecology, anthropology, and business.

Discrete Signals

We begin with a *discrete signaling model,* in which a person decides whether to take an action or not. You can buy an expensive watch to prove wealth. You can major in physics to prove your intelligence. You can swim the English Channel to prove your physical health. You cannot go halfway: you either send or do not send the signal. The model assumes two types of people, denoted as *strong* and *weak*. The types could correspond to physically fit or unfit marine recruits, or monolingual and bilingual employees.

The cost of sending the signal, which might be engaging in a fitness regimen for a month for a potential marine or the aforementioned presentation in Spanish for the job applicants, depends on a person's type. Strong potential marines find it less costly to complete the fitness regimen. In the model, we assume that everyone who sends the signal shares equally in the total benefit. This assumption can be interpreted in either of two ways. In some cases a resource might be split among those who send the signal. For example, everyone who donates $1,000 to a school (a signal of generosity) may get their name on the wall. In other cases, such as for the marines and job applicants, the winner(s) of some prize may be randomly chosen from the set of people who send the signal.

The model supports three types of outcomes: *pooling*, where everyone sends the same signal; *separating*, where each type sends a unique signal; and *partial pooling*, where some types separate and others do not.

Discrete Signals Model

A population of size N consists of S **strong** types and W **weak** types whose costs of sending a signal are c and C respectively, with $c < C$. Those members of the population who send the signal equally divide a benefit of $B > 0$. The model has three possible outcomes:

Pooling $(C < \frac{B}{N})$: Both types signal.

Separating $(c < \frac{B}{S}$ and $\frac{B}{S+1} < C)$: Strong types signal.

Partial pooling $(c < \frac{B}{N} < C < \frac{B}{S+1})$: Strong types and a fraction of the weak types signal.

In the model, we assume that individuals make optimal choices given the actions of others. That is, we treat this as a game and solve for an equilibrium. In the pooling equilibrium, everyone sends the signal. This equilibrium exists when the benefit is high and the cost for the weak type to send the signal is low. The exact condition is that the benefit divided by the number of people must exceed the cost for the weak type. Suppose that a benefactor donates $1 million for student scholarships to be split among all graduates of a high school with 100 students. Suppose that 50 of the students are weak and 50 are strong and that a strong student can graduate from high school by studying two hours per week, while a weak student must put in ten hours. We can approximate the costs of studying to be $2,000 for the strong students and $5,000 for the weak students. If all 100 students graduate, each receives a $10,000 scholarship. Given these assumptions, it makes sense for both types of students to study.

Suppose, though, that we reduce the size of the scholarship pool to $200,000. Now, if all of the students graduated, each would receive only $2,000. Studying is no longer in the weak students' self-interest. It still makes sense for the strong students, who now each receive $4,000. That amount is still not sufficient to induce even a single weak student to graduate (the second condition in the box above). In this case, the scholarship offer produces separation.

Last, suppose that the scholarship pool contains $400,000. Once again, if everyone graduated, the weak students would receive less than their $5,000 cost of studying. It follows that they would not all choose to study. However, if none of them studied, the strong students would each receive $10,000. That amount would be enticing to the weak students. In equilibrium exactly 30 of the weak students graduate along with all 50 strong students. In total, 80 students graduate, and each receives $5,000, the costs of studying for a weak student. We refer to this outcome as partial pooling, as only a portion of the weak students pool with the strong students.

The partial pooling equilibrium is more complicated than the others as it requires coordination among the weak students. We could assume that there exists some process by which weaker students communicate to others that they plan to take actions that ensure graduation. Or, we could assume that the weaker students put in just enough effort so that whether they graduate is a random event such that, on average, 30 of them graduate. This second scenario seems less plausible. In general, we should interpret partial pooling equilibria as benchmarks, as what would happen if people optimized. Whether a partial pooling equilibrium is attained likely depends on the situation, in particular, on whether people can communicate intended actions.

Continuous Signals

In the partial pooling equilibrium of the discrete signals model, the strong types may feel frustration. If they could send a strong signal, they could separate fully from the weak types and earn a higher payoff. To include that possibility in the model, we can alter our assumptions and allow the strong types to choose the magnitude of the signal they send. This requires only slight modification in our model. We reinterpret the costs of sending the discrete signal as a cost per unit of the continuous signal. We assume that for any fixed amount of the signal, the strong types have a lower per-unit cost of signaling.

To create separation in this new model, the strong types must be willing to choose a magnitude that is prohibitively costly for the weak types, yet still worth sending given the benefits and the costs. Reasoning through the model, we find that at least some of the strong types, though not necessarily all, can separate themselves from the weaker types.

Surprisingly, the magnitude of the signal the strong types must send becomes smaller as the size of the strong group increases. This occurs because the benefits to a weak type of sending the signal decrease in the size of the strong group. Being part of a larger group provides fewer benefits. The condition for full separation implies that it will be more likely when there are few strong types or when the strong types have much lower costs of signaling.

Separation with Continuous Signals

A population of size N consists of S **strong** types and W **weak** types with per-unit costs of signaling c and $C > c$, respectively. The individuals who send the largest signal split a benefit B. Any signal of magnitude $M \geq \frac{B}{SC}$ **separates** the strong types. If $CW \geq cN$, then all of the strong types separate. If not, a partial pooling equilibrium exists in which a portion of the strong types signal.[2]

This model can explain why expensive watches and jewelry function as signals of wealth. A person's house or car also signals her wealth, but people cannot carry their houses and cars around with them at all times. Clothes can signal wealth as well, but may not be able to create separation. For a few hundred dollars, a person could dress like someone of significant wealth. Watches and jewelry, because of their high costs, are more effective as signals. A poor or middle-class person could not afford a $10,000 watch. By wearing one a person proves she has wealth. The benefit she derives may be that people treat her with more respect—assuming that people see wealth as correlating in some way with importance (an inference we might question).

The Uses and Value of Signals

We signal in an attempt to reveal hidden attributes to others. We take actions to signal our fitness, wealth, intelligence, and generosity. Sometimes our actions produce signals as by-products. A person who competes in marathons for the pure joy of running signals fitness and dedication even if that was not her intent. The signaling model provides an alternative lens for interpreting a large number of actions. Did a person choose to attend an event, master a skill, or purchase an item out of personal interest, or to signal a type? We might not be able to distinguish between the two cases.

Signaling models also provide alternative explanations for empirical regularities—even for the value of earning a college diploma. Data on incomes show that college graduates earn significantly higher salaries. We could infer that the higher wages result from skills and knowledge acquired at college. That data also show that math and science majors earn even higher salaries. From that we could infer that the skills learned in those majors have greater economic value. However, if we look at the tasks that math majors perform, we may find that few use calculus. Moreover, almost no one interviewing for a job is ever asked the derivative of the cosine function or to explain Boyle's law. In light of that, we might infer that college degrees in general and science and math degrees in particular represent signals of a person's ability to acquire knowledge. The higher pay that graduates receive stems entirely from the signaling value of degrees rather than what the graduates have learned.[3]

Consider the signaling required to become a doctor. Students must pass classes in physics, organic chemistry, and calculus. But do doctors use any calculus? Has your doctor ever looked in your ears and nose and then scratched a calculus equation on her notepad? Of course not. Knowledge of calculus, for the most part, may not be relevant to being a doctor, but it may be a good signal of the ability to master a corpus of knowledge. If so, passing calculus becomes a useful signal even though the subject material has little direct relevance.

When constructing institutions or rules in which people send signals, we would prefer the actions that produce signals build useful skills. To be a successful doctor, a person must be able to memorize facts. To signal the ability to memorize, one could require that applicants memorize the capitals and currencies of every country. Success would signal a candidate's ability to memorize but would be functionally useless. When rushed to the emergency room with a strange feeling in your stomach, you would not care if your doctor knew that Bratislava is the capital of Slovakia, but you would want your doctor to know the various parts of the digestive system. For this reason, medical boards require that doctors pass exams on anatomy. Passing anatomy signals the ability to memorize. More important, knowing the parts of the body is useful for those who do pass. Thus, passing the anatomy exam provides a *functional signal*.

Summary

The signaling model can be applied widely. As already noted, the peacock's feathers signal robust health. Otherwise, the decorative fans of feathers have little functional value. They may, in fact, be worse then useless. Far better that the peacocks had chosen to develop stronger talons. Strong talons would be more difficult for peahens to notice from afar, so tail feathers won out evolutionarily.[4] The colorful rear ends of male fruit flies serve a similar function as the peacock's feathers, as does the chirping of grasshoppers and birds. Chirping takes energy. Only a well-fed grasshopper can devote time to chirping rather than chasing down food. Hence, chirping functions as a signal.

Human societies engage in a variety of actions to signal their fitness. Anthropologists differentiate between three forms of costly signals: *unconditional generosity, wasteful subsistence behavior,* and *craft traditions*.[5] The potlatch, a ritual carried out by the indigenous people of the Pacific Northwest, may be the most salient example of signaling generosity. To celebrate an event, say a birth or death, a chieftain would give away (or destroy) great amounts of wealth and would challenge other chiefs to match those amounts. Failure to match was a loss of prestige. Giving away wealth can be seen as socially good, but burning it is wasteful.

Wasteful subsistence behavior occurs whenever people, typically men, choose to engage in hunting behaviors with lower expected payoffs than harvesting seeds or berries. The men do so because they earn extra respect. The successful hunter signals his strength and bravery, which could be useful in other settings. The successful berry picker signals good eyesight and patience—useful traits, to be sure, but not as predictive of certain types of fitness as turtle-hunting skill. A study of the Meriam, who live on a group of islands of northern Australia, found that male turtle hunters had more than twice as many surviving offspring at age fifty as non-turtle-hunters.[6]

Relatedly, many craft traditions involve commitments that require sufficient time and resources. These activities can result in useful objects—blankets, for example. Craft traditions also involve the making of ceremonial objects that have little practical value. Some anthropologists interpret their creation as signaling. The meaning that accrues to such objects—they can be endowed with substantial cultural significance—need not depend on their functionality.

Some advertising also can be interpreted as costly signaling. By purchasing an expensive Super Bowl commercial, a company signals legitimacy. It implies that it believes consumers will like its product enough for the company to earn back the cost of the advertisement in profits. For example, imagine that two manufacturers each introduce a new coffeemaker. One of the manufacturers knows it has a superior product. The other knows that despite the best efforts of its engineers, its product is likely to break down and leave customers less than satisfied. This second manufacturer expects to have 20% of its products returned.

Millions of people may buy a coffeemaker during a given year. Without advertising, the two firms might well split the market. Suppose that the manufacturer of the better product spends $2 million on an advertisement to signal the quality of its product. The manufacturer anticipates that early purchasers will buy the product and, in the long run, that this will lead to more sales. The manufacturer may well have a version of the Polya Process model in her head. In contrast, the manufacturer of a lousy product would not take out such an advertisement because its product has little chance of large sales. Spending money to signal product quality is sometimes referred to as *burning money*. Burning money attracts buyers much as the peacock's feathers attract mates.

In all of these cases, signaling involves a cost. Those who send the signal find that the benefits of identifying their greater wealth, ability, or even generosity are reduced by the cost of the signal. In addition, the time and effort spent signaling can be thought of as opportunity costs: resources could have been used in some other way, perhaps creating greater social surplus. For example, a teenager may spend hours deciding which clothes to wear so as to signal his social awareness or devote time to nonproductive activities that he believes will improve his chances of acceptance at an elite college.

To make signaling less socially costly, we try to make signals as functional as possible. Better to have young people signal their physical fitness and bravery by playing team sports, where they learn rules of sportsmanship and collective interests, than to have them risk death jumping motorcycles. As previously noted, better to require doctors to memorize human anatomy than a random collection of J. R. R. Tolkien's Elvish languages. Nevertheless, try as we might, wasteful signaling will persist. Our challenge is to use models—in particular, the tools of mechanism design—to construct institutions and protocols so that the signals people send are as functional as possible.

26. *Learning Models*

The most important attitude that can be formed is that of desire to go on learning.

—John Dewey

In this chapter, we study models of individual and social learning. We apply each in two contexts. The first setting involves learning the best choice in a set of alternatives. In that setting, both types of learning, individual and social, converge on the optimal choice. The choice of learning rule only affects the rate of convergence. We then apply the learning rules to actions in games. In a game, an action's payoff depends on the action of the other player or players. In that setting, both learning rules favor risk-averse equilibrium outcomes over efficient ones. We also find that individual and social learning need not produce the same result and that neither performs better in all environments.

These findings bolster our many-model approach to representing behavior. Learning models lie in between the rational-choice models, which assume that people think through the logic of situations and games and take optimal actions, and rule-based models that assign behaviors. Learning models do assume that people follow rules, but those rules enable behavior to change. In some cases, the behavior converges to optimal behavior. In those cases, learning models can be used to justify the assumption that people optimize. However, learning models need not also converge to equilibria; they might produce cycles or complex dynamics. If the models do converge, they may select some equilibria more than others.

The chapter begins by describing a reinforcement learning model and applies it to the problem of choosing the best alternative. The model reinforces actions with higher rewards. Over time, the learner takes only the best action. This is a baseline model that proves ideal for learning about learning. It also fits quite well with experimental data, and not just for humans. Sea slugs, pigeons, and mice all reinforce successful actions. It may be a better model of sea slugs, which possess fewer than 20,000 neurons, than of humans, who have more than 85 billion.

That extra capacity allows humans to consider counterfactuals when learning, a phenomenon left out of the reinforcement learning model.

We then introduce social learning models, where individuals learn from their own choices and the choices of others. Individuals copy the actions or strategies that are most prevalent or that are performing above average. Social learning requires observation or communication. Some species create social learning through *stigmergy*: a process in which successful actions leave a trace or residue that others can follow, such as when goats who roam a mountain range leave trampled grass, reinforcing routes to water or food.

In the third section, we apply both types of learning models to games. As already noted, games present a more complicated learning environment. The same action might produce a high payoff in one period and a low payoff the next. As might be expected, we find that both social and individual learning models can fail to converge to efficient equilibria. They can also produce different outcomes. We conclude with a discussion of more sophisticated learning rules.[1]

Individual Learning: Reinforcement

In *reinforcement learning,* an individual chooses actions based on the weights of those actions. Actions with a lot of weight are chosen more often than actions with little weight. The weight assigned to an action depends on the reward (payoff) that a person has received from taking that action in the past. This reinforcement of high-reward payoffs leads to better actions being taken. The question we explore is whether reinforcement learning converges to only choosing the alternative with the highest reward.

At first, it may seem that to choose the most rewarding alternative is a trivial task. If the rewards are expressed in numerical form, such as money or time, we would expect people to choose the best. In Chapter 4, we invoked that line of thinking to argue that a person choosing a route to work in Los Angeles would settle on the shortest one.

If rewards do not take numerical form, which is generally the case, people must rely on memory. We grab lunch at a Korean restaurant. We find the kimchi delicious, so we are more likely to eat there again. On Monday, we eat an oatmeal cookie an hour before running and find we can sustain a strong pace for ten kilometers. If prior to Wednesday's run we again grab an oatmeal cookie and perform well, we add weight to that action. We learn that cookies improve our performance.

Other species do the same. Edward Thorndike, an early psychologist who studied learning, conducted an experiment in which cats who pulled a lever to escape a box were rewarded with fish. When returned to the box, the cats pulled the lever within seconds. Thorndike's data revealed a process of continued experimentation. He found that cats (and people) learned faster when he increased the reward. He called this the *law of effect*.[2] This finding has a neurological explanation. Repetition of an activity builds neurological pathways that induce that same behavior in the future. Thorndike also found that more surprising rewards, rewards that far exceeded past or expected outcomes, produced faster learning in people, a phenomenon known as the *surprise principle*.[3]

In our reinforcement learning model, the weight assigned to a chosen alternative is adjusted based on how much the reward from that alternative exceeds our expectations (our *aspiration level*). This construction embeds both the law of effect (we take actions that produce higher rewards more often) and the surprise principle (the amount of weight we add to a choice depends on how much its reward exceeds the aspiration level).[4]

A Reinforcement Learning Model

A collection of **alternatives** $\{A, B, C, D, \ldots, N\}$ have associated **rewards** $\{\pi(A), \pi(B), \pi(C), \pi(D), \ldots, \pi(N)\}$ and a set of strictly positive **weights** $\{w(A), w(B), w(C), w(D), \ldots, w(N)\}$. The probability of choosing K is as follows:

$$P(K) = \frac{w(K)}{w(A) + w(B) + w(C) + w(D) + \cdots + w(N)}$$

After choosing K, $w(K)$ increases by $\gamma \cdot P(K) \cdot (\pi(K) - A)$, where $\gamma > 0$ equals the **rate of adjustment** and $A < \max_K \pi(K)$ equals the **aspiration level**.[5]

Notice that the aspiration level must be set below the reward for at least one alternative. Otherwise, any alternative chosen becomes less likely to be chosen in the future and all of the weights converge to zero. It can be shown that if the aspiration level is below the reward for at least one alternative, eventually almost all of the weight will be placed on the best alternative. This occurs because each time the best alternative is selected, its weight increases by the most, creating stronger reinforcement of that alternative. This occurs even if we set the aspiration level below the reward from each alternative. In that case, every alternative increases in weight when selected. Thus, the model can capture *habituation,* where we do more of something just because we have done it in the past. Even with a low aspiration level, the alternatives with the highest rewards increase in weight the fastest, so the best alternative wins out in the long run. However, the time required for convergence on the best alternative may be long. It will also be true that as we add more alternatives, time to convergence also increases.

To avoid these complications, we can build in *endogenous aspirations*. We emend the model so that the aspiration level adjusts over time by setting it equal to the average reward. Imagine a parent learning whether a child prefers apple pancakes or banana pancakes. Assign a reward of 20 to apple pancakes and 10 to banana pancakes. Set the initial weights on both alternatives to 50, the rate of adjustment to 1, and the aspiration level to 5. Assume the parent makes banana pancakes the first day. The weight on banana pancakes will increase to 55. Suppose that parent makes banana pancakes the next day as well. The reward of 10 equals the new aspiration level, so the weight on banana pancakes does not change.

Suppose that on the third day the parent makes apple pancakes. These produce a reward of 20, 10 above the aspiration level. This increases the weight on apple pancakes to 60, making them the more likely choice. The high reward also increases the average payoff, and therefore the aspiration level, above 10. Thus, if the parent makes banana pancakes again, the weight on banana pancakes decreases because the reward from banana pancakes lies below the new aspiration level. Reinforcement learning therefore converges to only apple pancakes being selected.

It can be proven that reinforcement learning will converge toward selecting the best alternative with probability 1. That means that the weight on the best alternative will become arbitrarily large compared to the weights on all other alternatives.

Reinforcement Learning Works

In the **learning-the-best-alternative framework,** reinforcement learning with the aspiration level set equal to the average earned reward (eventually) almost always selects the best alternative.

Social Learning: Replicator Dynamics

Reinforcement learning assumes an individual acting in isolation. People also learn from watching others. Social learning models assume that individuals see the actions and rewards of others. This can speed the rate of learning. The most widely studied model of social learning, *replicator dynamics,* assumes that the probability of taking an action depends on the product of its reward and its popularity. We can think of the former as a *reward effect* and the latter as a *conformity effect*.[6] Most often replicator dynamics models assume an infinite population. We can then characterize the actions taken as a probability distribution across the various alternatives. In the standard construction, time advances in discrete steps so that we can capture learning by changes in the probability distribution.

Replicator Dynamics

A collection of **alternatives** $\{A, B, C, D, \ldots, N\}$ have associated **rewards** $\{\pi(A), \pi(B), \pi(C), \pi(D), \ldots, \pi(N)\}$. The actions of a population at time t can be written as a probability distribution across the N alternatives: $(P_t(A), P_t(B), \ldots, P_t(N))$. The probability distribution changes according to the **replicator equation**:

$$P_{t+1}(K) = P_t(K) \cdot \left(\frac{\pi(K)}{\bar{\pi}_t} \right)$$

where $\bar{\pi}_t$ equals the average reward in period t.

Consider a community in which parents choose between apple, banana, and chocolate chip pancakes. Assume that that all of their children have identical preferences and that the three types of pancakes produce rewards of 20, 10, and 5. If initially 10% of parents make apple, 70% make banana, and 20% make chocolate chip, the average reward equals 10. Applying replicator dynamics, the probabilities of choosing each of the three alternatives in period two are as shown in the table below:

The Replicator Equation

Alternative	π	P_1	$\pi / \bar{\pi}_t$	P_2
Apple	20	0.1	$\frac{20}{10}$	0.2
Banana	10	0.7	$\frac{10}{10}$	0.7
Choc. Chip	5	0.2	$\frac{5}{10}$	0.1

Applying the replicator equation, in the next period twice as many parents make apple pancakes. This occurs because the reward for apple pancakes equals double the average reward. Half as many parents make chocolate chip pancakes because that reward equals half of the average reward. Finally, the proportion of parents making banana pancakes, which produce exactly the average reward, does not change. Combining all of these changes, we can show that the average reward increases to 11.5.

As noted above, replicator dynamics includes a conformity effect (more popular alternatives are more likely to be copied) as well as a reward effect. In the long run, the reward effect dominates, because high-reward alternatives always grow in proportion to lower-reward alternatives. In replicator dynamics, the average reward performs a function similar to that of the aspiration level in reinforcement learning when the aspiration level adjusts to equal the average reward. The only difference is that in replicator dynamics, we calculate the average reward for a population. In reinforcement learning, the aspiration level equals an individual's average reward. That distinction matters insofar as a population provides a larger sample. Thus, replicator dynamics produce less path dependence than reinforcement learning.

In our construction of replicator dynamics, we assume that every alternative exists in the initial population. Given that the highest-reward alternative always has a higher-than-average reward and its proportion increases in every period, (eventually) replicator dynamics converge to the entire population choosing the best alternative.[7] Thus, in a setting of learning the best alternative, both individual and social learning converge to the alternative with the highest reward. That will not be true in games.

> # Replicator Dynamics Learns the Best
>
> In learning the best from a finite set of alternatives, replicator dynamics with an infinite population converges to the entire population choosing the best alternative.

Learning in Games

We now apply our two learning models to games.[8] Recall that in a game, a player's payoff depends both on her own action and on the actions of the other players. The payoff from a given action, such as cooperating in the Prisoners' Dilemma, could be high in one period and low in the next depending on the action of the other player. We begin with the *Guzzler Game,* a two-person game in which each player must choose whether to drive an economy car or a gas guzzler. Choosing the gas guzzler always produces a payoff of 2. Choosing an economy car when the other player also chooses an economy car produces a payoff of 3—both drivers have good lines of sight, require less fuel, and have no fear of being crushed by an enormous gas guzzler. If the other player chooses a gas guzzler, a player driving the economy car must be cognizant of the other driver. To capture that effect, we assume that her payoff falls to zero. We represent these payoffs in figure 26.1.

	Guzzler	Economy
Guzzler	2, 2	2, 0
Economy	0, 2	3, 3

Figure 26.1: The Guzzler Game

The Guzzler Game has two pure strategy equilibria: both players can choose economy cars or both players can choose gas guzzlers.[9] The equilibrium in which both choose the economy car produces the higher payoff. It is the efficient equilibrium.

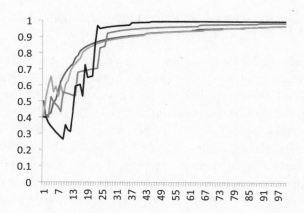

Figure 26.2: Reinforcement Learning ($\gamma = \frac{1}{3}$) Probability of Choosing Guzzler

We first assume that both players use reinforcement learning. Figure 26.2 shows results from four numerical experiments with the initial weights on each action set equal to 5, an aspiration level of zero, and a learning rate (γ) of $\frac{1}{3}$. In all four

311

experiments, both players learn to select the gas guzzler, the inefficient pure strategy equilibrium. To see why this occurs, we need only look at the payoffs. The gas guzzler always returns a payoff of 2. The economy car sometimes returns a payoff of 3 and sometimes returns a payoff of zero. By assumption, both actions will be equally represented in the initial population. Therefore, the economy car produces an average payoff of only 1.5 to the gas guzzler's payoff of 2. More players choose the gas guzzler, and the payoff from selecting the economy car decreases further.

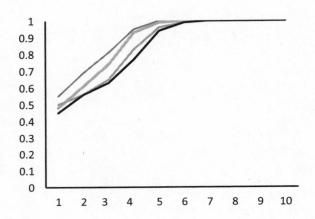

Figure 26.3: Replicator Dynamics (100 Players): Probability of Choosing Guzzler

Next, we apply replicator dynamics to the same game. Again we assume an initial population consisting of equal proportions of people choosing gas guzzlers and economy cars. We further assume that each player plays the game against every other player in the population. People who choose the gas guzzler receive higher payoffs, and because initially equal numbers choose each action, in the second period more people will choose gas guzzlers.[10] Applying the replicator equation a second time, shows that the number of players choosing gas guzzlers would again increase. Continued application of the replicator equation results in the entire population choosing guzzlers. Figure 26.3 shows results from four runs of discrete replicator dynamics with 100 players. By assuming a finite population, we introduce a small amount of randomness. The proportions adopting each action may not be exactly equal to those stated in the replicator equation. In each of the four runs, all of the players choose the gas guzzler after only seven periods. Convergence occurs quickly because both the conformity effect and the reward effect push people to choose the gas guzzler after the first period. For example, when 90% of the population chooses gas guzzlers, the payoff from choosing an economy car will be less than one-sixth of that from choosing the gas guzzler. The conformity effect amplifies the reward effect, making social learning much faster than individual learning, which took, on average, more than 100 periods to reach 99% guzzlers.

In this game, both learning rules converge to choosing the gas guzzler because it has the higher payoff when both actions are equally likely. Such actions are called *risk dominant*. Both learning rules favored the risk-dominant equilibrium over the efficient equilibrium. We next construct a game in which our two learning rules converge to different equilibria.

The Generous/Spiteful Game

Our next game, *The Generous/Spiteful Game,* builds on a much-analyzed question about human behavior: Do we care more about our absolute or relative payoffs? A person who would prefer a $10,000 bonus when all of his colleagues receive $15,000 bonuses over an $8,000 bonus when all of his colleagues receive only $5,000 cares more about his absolute payoff. A person who would accept less money to have the largest bonus cares more about his relative payoff. An extreme preference for relative payoffs is captured in the story of the spiteful man and the magic lamp.

> # The Spiteful Man and the Magic Lamp
>
> *A spiteful man finds a bronze lamp while on an archeological expedition. He rubs the lamp and a genie appears. The genie proclaims, "I will grant you one wish for anything that you desire, and because I am a benevolent genie, I will give everyone you know double what I give you." The man ponders the proposition, grabs a stick, and says, "Poke out one of my eyes."*

The spiteful man takes an action that gives him a low absolute payoff and a high relative payoff.[11] A similar tension exists in foreign affairs. Neoliberals believe that countries want to maximize absolute payoffs measured by military power, economic prosperity, and domestic stability. Another camp, known as neorealists, believes that countries value relative payoffs. A country would rather have a lower absolute payoff but be stronger than its enemies. Kenneth Waltz, a neorealist, wrote at the height of the Cold War, "The first concern of states is not to maximize power but to maintain their positions in the system."[12] Neorealists would claim that during the height of the Cold War, had either Russia or the United States rubbed the magic lamp, each would have handed the genie a stick.

313

We can embed the conflict between absolute and relative gains in an N-person game with a generous action that increases absolute payoffs for everyone along with a spiteful action that only increases one's own payoff. This game differs from a collective action game, where generosity comes at a cost.[13] The formal game with payoffs is shown in the box. The generous action is a dominant strategy. Regardless of the actions of the other players, a player choosing generous receives a higher payoff. However, on average the players choosing spiteful earn higher payoffs.

These may at first appear to be contradictory statements. They are not. By being generous a player raises his absolute payoff by 3 but also raises the payoffs of all other players by 2. A player who chooses to be spiteful raises his payoff by only 2 but does not raise the payoffs of the other players. Each player improves his payoff by choosing to be generous. When a player chooses to be spiteful, he reduces his payoff, but (and here's the key assumption) he reduces the payoff to everyone else by an even larger amount.

The Generous/Spiteful Game

Each of N players chooses to be *generous* G or *spiteful* S.

$$\text{Payoff}(G, N_G) = 1 + 2 \cdot N_G$$
$$\text{Payoff}(S, N_G) = 2 + 2 \cdot N_G$$

If we apply reinforcement learning in the Generous/Spiteful Game, the players learn to be generous. To see why, suppose that the players have almost converged to an equilibrium, with N_G of the players choosing to be generous. A spiteful player earns a payoff of $2 + 2 \cdot N_G$. This will be his aspiration level. If he chooses G, which occurs with small probability, he earns a payoff of $1 + 2 \cdot (N_G + 1) = 3 + 2 \cdot N_G$, which is above his aspiration level. He will become more likely to be generous. By continuing to apply this logic, we see that all players will learn to be generous.

If we apply replicator dynamics, the population learns to be spiteful. This can be seen by referring to the replicator equation. In every period, players who choose to be spiteful earn higher payoffs than players who choose to be generous. Therefore, the proportion of players choosing to be spiteful increases in each period.

These findings highlight a key difference between individual and social learning. Individual learning leads people to choose the better action, so people learn a dominant action if one exists. Social learning leads people to choose actions that perform well relative to other actions. In most cases, those actions would also produce higher payoffs. That is not the case in the Generous/Spiteful Game, where the spiteful action has a higher average payoff, while the generous action is dominant. Notice that our analysis arrives at the rather paradoxical finding that if people learn individually, they learn to act more generously than if they learn socially. That occurs because in social learning the players copy the actions of players who perform relatively well.

We might now take a moment to consider an earlier comment: that we can think of replicator dynamics as an adaptive rule or as the selection of fixed rules. If we assume the latter, then our model says that selection could favor spiteful types. Selection need not produce cooperation. This result runs counter to what we found when studying the repeated Prisoners' Dilemma, where repetition led to cooperation. In that case, we considered repeated games and allowed for more sophisticated strategies.

Combining Learning Models

We have seen how individual and social learning both find the best solution among a fixed set of alternatives, but that when applied to games, they can produce different outcomes. This lack of agreement is a strength. Imagine a giant set consisting of all possible games. Imagine a second set consisting of all learning models. We could apply every learning model in the second set to every game in the first set and evaluate their performance. We can then partition the set of all games into two sets: those in which the learning rule produces the efficient outcome and those in which it does not. We could also look to experimental data and evaluate each learning rule as a predictor of actual behavior. That exercise would undoubtedly reveal contingencies. Each learning rule would result in efficient outcomes for some games but not for others. Each learning rule would also vary in the contexts in which it accurately describes behavior. Hence, we advocate many models.

In this chapter, we covered two canonical models. Each includes only a few moving parts. Our goal was to provide a gentle introduction to a large and exciting literature. By adding more details to either learning model, we would better fit experimental and empirical data. Recall that in the reinforcement learning model, individuals add or subtract weight to an alternative or action depending on whether its reward (payoff) exceeds the aspiration level. Individuals do not add weight to actions not taken: we do not increase the probability of taking some action that would have given a high payoff had we taken it.

That assumption may not make sense in all cases. Consider the case of an employee who decides not to take his cell phone on vacation. While he is away his boss calls with an important question. The employee misses the call and is passed up for a promotion. In the reinforcement learning model, the employee would not attach more weight to bringing his phone on vacation in the future. The *Roth-Erev learning model* amends the standard model so that alternatives that are not chosen also receive weight based on their hypothetical payoffs. In the example, the employee would attach more weight to bringing his phone.

This modification creates a *belief-based learning* rule. The amount of the increase in weight for the alternatives not chosen is determined by an experimentation parameter. The higher the experimentation parameter, the more individuals take into account the effect of others' actions and the more they increase the weights on those actions. Roth and Erev also discount the past to take into account that other players are learning as well and their strategies likely change.[14]

These additional assumptions make intuitive sense and have empirical support, but they do not fit all cases. If we go back to our example of the parent making pancakes, the first assumption implies that after the parent makes banana pancakes, additional weight is added to the alternative of making apple pancakes and that weight is proportional to the payoff from apple pancakes. Such an assumption makes sense only if the parent knows the payoff from apple pancakes. That would be true only if people can see or intuit the payoffs of unchosen actions.

A model by Camerer and Ho creates a functional form that admits both reinforcement learning and belief-based learning as special cases. A parameter that can be fitted to data allows a determination of the relative strength of each type of learning rule.[15] The ability to combine models was one motivation for mastering many models. That said, combining models necessarily leads to a better fit because of the increase in parameters. Even taking into account the parameter increase, Camerer and Ho's model produces better predictions and deeper explanations.

Modeling learning creates several challenges. Learning rules that work well in one setting may not capture other situations as well. Furthermore, what people learn to do can depend on their initial beliefs, so two people may learn differently in the same setting and the same person may learn differently in different settings. Even if we could construct an accurate learning model, we again confront the *exploitability principle*: if a model explained how people learned, then others could apply that model to anticipate (and in some cases exploit) that knowledge. It is then likely that people would learn not to be exploited, and our original learning model would no longer be accurate. We encountered this phenomenon earlier when discussing the Lucas critique and in our analysis of the efficient market hypothesis. We cannot necessarily conclude that because people learn that they optimize. We can assume learning will winnow out poor actions in favor of better ones.

Does Culture Trump Strategy?

We now apply contagion models and learning models to address the long-standing claim from organizational theory that culture trumps strategy.[16] In brief, the claim states that strategic incentives to change behaviors fail. The pull of culture, the existing set of repertoires and beliefs, proves too powerful. Economists argue the opposite: that incentives drive behavior.

To turn these opposite proverbs into conditional logic, we first apply a version of the network contagion model. In this model, the manager, or possibly the CEO, announces a new strategy and produces evidence for the benefits of the change. The CEO may even redefine the organization's core principles to reflect this new behavior. Individuals in the organization then choose whether or not to adopt the behavior based on how compellingly the manager makes her case. Some initial proportion of people buy into the initiative. When they make contact with others in their work network, they spread their enthusiasm. There also exists a pull against the new strategy, causing people to no longer adopt the new strategy. The three features that determine if the new strategy spreads—the contact rate, the spreading rate, and the rate of abandonment—map naturally into the parameters in the basic reproduction number, R_0:

$$R_0 = \frac{P_{spread} \cdot P_{contact}}{P_{recover}}$$

If we add in the possibility of superspreaders, then we might conclude that culture trumps strategy provided any of three conditions hold: if people do not believe in the new strategy, if they are quick to abandon it, or if the strategy's advocates are not well connected. Otherwise, strategy may well trump culture.

Our second model applies replicator dynamics to a *Culture/Strategy Game* that models interactions between pairs of employees. We can represent these choices in game form as a *cultural action* (doing what they currently do) and as an *innovative strategic action*. We assume that the manager constructs payoffs so that both players earn higher payoffs if both choose to be innovative. However, a single innovative player earns less.

	Culture	Innovate
Culture	200, 200	220, 180
Innovate	180, 220	300, 300

The Culture/Strategy Game

The game has two strict pure-strategy Nash equilibria: one in which both innovate (strategy trumps culture) and one in which neither innovates (culture trumps strategy). The manager appears to have constructed incentives so that the employees will take the innovative new action, as it has the higher payoff. If we write down a learning model, we see that the manager needs sufficient initial buy-in for the innovation to take hold. In the game above, it can be shown that if the initial buy-in, that the proportion that adopts the innovation in the first period, does not exceed 20%, then culture trumps strategy.[17] If we were to increase the payoff from the innovative strategic action, then initial buy-in could be even lower yet still result in the efficient outcome.

These two models show that the opposing proverbs "Culture trumps strategy" and "People respond to incentives" can both be correct conditionally. According to the first model, charismatic CEOs who can convince well-connected employees can introduce new strategies that trump culture. According to the second model, culture trumps weak incentives but not strong ones.

27. *Multi-Armed Bandit Problems*

There's one thing I'm really good at, and that's hitting the ball over a net, in a box. I'm excellent.

—Serena Williams

In this chapter, we add uncertainty to the problem of learning the best alternative to create a class of models known as *multi-armed bandit problems*. In bandit problems, rewards from alternatives are distributions rather than fixed amounts. Bandit problems apply to a wide variety of real-world situations. Any choice among actions that has an uncertain payoff—pharmaceutical drug trials, choice of where to place advertisements, choice among technologies, decisions as to whether to allow laptops in the classroom—can be modeled as bandit problems; so too can the problem of choosing a profession at which we can excel.[1]

A person facing a bandit problem must experiment with alternatives to learn the payoff distributions. This feature of bandit problems creates a trade-off between exploration (searching for the best alternative) and exploitation (choosing the alternative that has performed best so far). Finding an optimal balance in the explore-exploit trade-off requires sophisticated rules and behaviors.[2]

The chapter consists of two parts followed by a discussion on the value of applying models. In the first part of the chapter, we describe a special class of Bernoulli bandit problems, in which each alternative is a Bernoulli urn with unknown proportions of gray and white balls. We describe and compare heuristic solutions, and then show how these solutions can improve comparison tests of drug treatments, advertising plans, and teaching strategies. In the second part, we describe a more general model, where the reward distributions can take any form and the decision-maker has a prior distribution over their types. In that part, we also show how to solve for the Gittins index, which determines the optimal choice.

Bernoulli Bandit Problems

We begin with a subclass of bandit problems in which each alternative has a fixed probability of producing a successful outcome. This first class of bandit problems is equivalent to deciding among a set of Bernoulli urns, each containing different proportions of gray and white balls. Therefore, we refer to these as *Bernoulli bandit problems*. These are also called *frequentist problems* because the decision-maker knows nothing about the distributions. As the decision-maker tries alternatives (explores), she learns about those distributions.

Bernoulli Bandit Problems

Each of a collection of **alternatives** $\{A, B, C, D, \ldots, N\}$ has an unknown probability of producing a successful outcome, $\{p_A, p_B, p_C, p_D, \ldots, p_N\}$. In each period, the decision-maker chooses an alternative, K, and receives a successful outcome with probability p_K.

For example, suppose a chimney cleaning company has a list of phone numbers of recent home buyers. The company tests three sales pitches: the scheduled appointment approach ("Hello, I'm calling to arrange a time for your annual chimney cleaning"), the concerned questioner approach ("Hello, did you know a dirty chimney can be a fire hazard?"), and the personal touch approach ("Hello, my name is Hildy, and I started this chimney sweeping company with my father fourteen years ago").

Each sales pitch has an unknown probability of success. Suppose that the company first tries the scheduled appointment approach, and it fails. It then tries the concerned questioner approach and gains a client. That approach also works on the next call but then fails on the next three calls. After that, the company tries the third approach, which works on the first call but fails on the next four. After ten calls, the second approach has the highest success rate, but the first approach was only tried one time. The decision-maker faces a choice between exploiting (choosing the alternative that has worked best) or exploring (returning to the other two alternatives to get more information). This same problem is faced by a hospital selecting among surgical procedures and a pharmaceutical company testing various drug protocols. Each protocol has an unknown probability of success.

To gain insight into the explore-exploit trade-off, we compare two heuristics. The first, *sample-then-greedy,* tries each alternative a fixed number of times, M, and thereafter chooses the alternative with the highest average payoff. To determine the size of M, we can refer back to the Bernoulli urn model and the square root rules. The standard deviation of the mean proportion is bounded above by $\frac{1}{2\sqrt{M}}$. If each alternative is tested 100 times, the standard deviation of the mean proportion will equal 5%. If we apply a two-standard-deviation rule to identify a significant difference, we can confidently distinguish between proportions that differ by 10%. If one alternative produced successful outcomes 70% of the time and another produced successes 55% of the time, we could place more than 95% confidence on the first being better.

The second heuristic, an *adaptive exploration rate heuristic*, allocates ten initial trials to each alternative. The next twenty trials are allocated in proportions corresponding to the success rates. If in the first ten trials one alternative produced six successes and the other produced only two, then the first alternative would receive three-fourths of the next twenty trials. The second set of twenty trials could also be allocated according to the ratio of the squared success probabilities. If successes continued in the same proportions, the better alternative would then receive $\frac{(0.6)^2}{[(0.6)^2+(0.2)^2]}$, or 90%, of the third set of twenty trials. For each successive set of twenty trials the exponent of the probabilities could be increased at some rate. By increasing the rate of exploitation over time, the second algorithm improves on the first. If one alternative had a much higher probability of success than another, say 80% to 10%, the algorithm would not waste a hundred trials on the second alternative. On the other hand, if the two probabilities of success were close, the algorithm would continue to experiment.[3]

Adherence to the *sample-then-greedy* heuristic not only is inefficient, it can even be unethical. When Robert Bartlett tested an artificial lung, its success rate far surpassed those of the other alternatives. Continuing to test the other alternatives when the artificial lung performed best would have resulted in unnecessary deaths. Bartlett stopped experimenting with the other alternatives. Everyone was given the artificial lung. In fact, that can be shown to be an optimal rule: if an alternative is always successful, keep choosing that alternative. Experimentation can have no value because no other alternative could perform better.

Bayesian Multi-Armed Bandit Problems

In a *Bayesian bandit problem,* the decision-maker has prior beliefs over the reward distributions of the alternatives. Given these prior beliefs, a decision maker can quantify the trade-off between exploration and exploitation and (in theory) make optimal decisions in each period. However, except for the simplest of bandit problems, determining the optimal action requires rather tedious calculations. In real-world applications, these exact calculations may be impractical, obliging decision makers to rely on approximations.

Bayesian Multi-Armed Bandit Problems

A collection of **alternatives** $\{A, B, C, D, \ldots, N\}$ have associated **reward distributions** $\{f(A), f(B), f(C), f(D), \ldots, f(N)\}$. The decision-maker has prior beliefs over each distribution. In each period, the decision-maker chooses an alternative, receives a reward, and calculates new beliefs based on the reward.

Determining the optimal action relies on a four-step process. First, we calculate the expected immediate reward from each alternative. Second, for each alternative, we update our beliefs about the reward distribution. Third, based on our new beliefs about reward distributions, we determine the best possible actions in all subsequent periods based on what we know. Last, we add the expected reward from the action in the next period to the expected rewards from the optimal future actions. That sum is known as the *Gittins index*. In each period, the optimal action has the largest Gittins index.

Notice that the calculation of the index quantifies the value of exploration. If we try an alternative, the Gittins index does *not* equal the expected reward. It equals the sum of all future rewards assuming we take optimal actions given what we have learned. Computing a Gittins index is difficult. For a (relatively) simple example, suppose there exists a safe alternative that is certain to pay $500 and a risky alternative that with a 10% probability will always pay $1,000. The remaining 90% of the time, the risky alternative pays nothing.

To calculate the Gittins index for the risky alternative, we first ask what could happen: either it always pays $1,000 or it always pays nothing. We then think through how each outcome would influence our beliefs. If we knew the risky alternative paid $1,000, we would always choose it. If we knew that the risky arm paid nothing, we would always choose the safe arm in the future.

It follows that the Gittins index for the risky arm corresponds to a 10% probability of a reward of $1,000 in each period and a 90% probability of a reward of $500 in every period but the first. For a situation in which we get to choose an alternative many times, this averages out to approximately $550 per period. The risky alternative is therefore the better choice.[4]

The Gittins Index: Example

To show how to compute Gittins indices, we consider the following example with two alternatives. Alternative A produces a certain reward in $\{0, 80\}$ with 0 and 80 equally likely. Alternative B produces a certain reward in $\{0, 60, 120\}$ with each equally likely. We assume that the decision-maker wants to maximize reward over ten periods.

Alternative A: With probability $\frac{1}{2}$ the reward equals 0, so alternative B, which has an expected reward of 60, will be chosen in all remaining periods. This produces an expected reward of 540 (9 times 60). With probability $\frac{1}{2}$ the reward equals 80. The optimal choice in the second period even with this outcome is to choose alternative B in the second period. With probability $\frac{1}{3}$, B produces a reward of 120, so the total payoff equals 1,160 (80 plus 9 times 120). With probability $\frac{1}{3}$, B produces a reward of 60. In that case, alternative A is optimal choice in all subsequent periods generating a total payoff equal to 780 (60 plus 9 times 80). Finally, with probability $\frac{1}{3}$, B produces a reward of 0. In this case as well, alternative A is optimal choice in all subsequent periods. The total payoff will 720 (9 times 80).

Combing all three possibilities, it follows that the Gittins index in period one for alternative A equals the following:

$$\text{Gittins index}_1(A) = \frac{1}{2}540 + \frac{1}{2}\left(\frac{1}{3}1160 + \frac{1}{3}780 + \frac{1}{3}720\right) = \frac{2140}{3}$$

Alternative B: With probability $\frac{1}{3}$ the reward equals 120. If that occurs, the optimal choice in all future periods will also be B. Over ten periods the total reward will equal 1,200. If the reward equals zero, then the optimal choice in all future periods will be alternative A, which has an expected reward of 40. The expected total reward will equal 360 (9 times 40). If the reward equals 60, then the decision-maker could choose alternative B in all future periods, for a total return of 600. However, if she chooses A in the second period, half of the time it will always produce a reward of 80, for a total return of 780 (60 plus 9 times 80). The other half of the time it produces a reward of zero, and the optimal choice in all subsequent periods will be B, which produces a reward of 60, resulting in a total reward of 540 (9 times

60). It follows that the expected reward from making optimal choices after choosing A in the second period equals 660 ($\frac{1}{2} \times 780 + \frac{1}{2} \times 540$).

Combing all the possibilities, it follows that the Gittins index in period one for alternative B equals the following:

$$\text{Gittins index}_1(B) = \frac{1}{3}1200 + \frac{1}{3}660 + \frac{1}{3}360 = \frac{2220}{3}$$

Given these calculations, alternative B is the optimal first period choice. The optimal long run choice depends upon what is learned in the first period. If alternative B produces an outcome of 120, we stick with B forever.

The analysis shows that when taking an action, we care more about the probability that an alternative will be the best than about its expected reward. Moreover, if an alternative produces a very high reward, we should be more likely to select it in the future. In contrast, if it produces an average reward, even a reward above the expected reward of another alternative, we may be less likely to stick with the alternative. That is particularly true in early periods, where we want to look for high-reward alternatives. These insights hold across the many applications discussed. Provided there are not risks or high costs associated with actions, the model tells us to explore potentially high-reward actions even if they are low probability.

Summary

A key takeaway from this book is that by learning models a person can make better decisions. We can see that in stark relief by comparing what people should do in the bandit problem with what people actually do. Most people do not try to estimate a Gittins index when confronted with a bandit problem. They fail to do so, in part, because they do not keep data. Only recently, for example, have doctors begun to keep data on the efficacy of the many procedures—the efficacy of the various types of artificial joints or, say, the advantages of stents. Without that data, a doctor cannot determine which action has the highest expected payoff.

Doctors, and everyone else, need data to apply the lessons produced by the model. So if you wanted to learn whether taking a walk before dinner or after dinner resulted in better sleep patterns, you would need to keep track of how well you had slept, and by using a sophisticated heuristic, you could learn which probably works best. That may seem like a lot of effort. And it is, but less so now. New technology enables us to gather data on sleep patterns, pulse rate, weight, and even mood.

Most of us will not gather the necessary data and compute Gittins indices for life choices like when to exercise. The point is only that we could, and if we did, we would see improvement in life choices—in our sleep patterns and general health. Psychologist Seth Roberts performed self-study for twelve years and found that standing at least eight hours a day improved his sleep (though he slept less) and that standing along with getting morning light reduced his upper respiratory infections.[5] We may lack his dedication to self-experimentation. By not keeping data and comparing outcomes, we may go through life skipping breakfast when we would have been better off having grapefruit.

On high-stakes business, policy, and medical decisions where data are easier to gather, applying bandit models is common practice. Businesses, policy makers, and nonprofits experiment with alternatives and then exploit those that perform best. In practice, the alternatives may not remain fixed. A government mailer to increase participation in a farm subsidy program may be altered from year to year— say, swapping out a picture of a man with a picture of a woman.[6] This type of continued experimentation can be captured by the models we take up in the next chapter: rugged-landscape models.

Presidential Elections

We now apply three models to analyze outcomes in presidential elections: the spatial model, the category model, and the multi-armed bandit model.

Spatial model: To attract voters, candidates compete in an ideological issue space. Thus, we should expect candidates to tend toward moderate positions, elections to be close, and the winning sequence of parties to be random. Presidential elections are, with a few exceptions, close. To test if the sequence of winners is random, we first construct a time series of the thirty-eight winning parties from 1868 through 2016.

RRRRDRDRRRRDDRRRDDDDDRRDDRRDRRRDDRRDDR

We can then measure the *block entropy* of subsequences of various lengths. Subsequences of length 1 have entropy 0.98. Subsequences of length 4 have entropy of 3.61. Statistical tests show that we cannot reject that the sequence is random. For comparison, a random sequence of length 38 had block 1 entropy of 1.0 and block 4 entropy of 3.58.

Category model: If we think of each state as a category and assume heterogeneity across states, the spatial model implies that once the candidates choose initial positions, some states will not be competitive. The model predicts fierce competition in a handful of moderate states. In 2012, Obama and Romney spent over 96% of their television advertising budgets in ten states. Each spent nearly half of their advertising budgets in three moderate states: Florida, Virginia, and Ohio. In 2016, Clinton and Trump also spent more than half of their television dollars in three moderate states: Florida, Ohio, and North Carolina.[7]

Multi-armed bandit model (retrospective voting): Voters will be more likely to reelect a party that produces good outcomes. Voting for effective parties corresponds to pulls of a lever that generate a high payoff. A strong economy should benefit the incumbent party. Evidence shows that voters are more likely to reelect the party in power when the economy performs well. The effect is larger for the incumbent candidate than for a nonincumbent candidate from the party in power.[8]

28. *Rugged-Landscape Models*

Amazing the things you find when you bother to search for them.

—attributed to Sacagawea

In this chapter, we study the rugged-landscape model. Like spatial and hedonic models, the rugged-landscape model defines an entity as a collection of attributes. Each set of attributes maps to a value. The goal is to modify attributes to construct an entity of highest value. This model originated in ecology to study evolution. It is now also used to study problem solving, competition among firms, and innovation. That will be our focus here. We apply the model to reveal how interdependence in the effects of attributes makes innovation difficult and leads to path dependence in the solutions found and also leads to a greater variety of solutions. We also see how more difficult problems benefit from a greater diversity of problem solving approaches.

The chapter consists of three parts followed by a discussion of how to extend the model to capture competition. In the first part, we describe the ecological model of a fitness landscape and show how we can reinterpret it as a model of problem solving and innovation. In the second part, we discuss the implications of ruggedness within a one-dimensional model. In the third part, we present the NK model of rugged landscapes, which extends the one-dimensional model to an arbitrary number of binary dimensions.

The Fitness Landscape

The *fitness landscape model* assumes species have features or traits that contribute to their fitness, loosely defined as their reproductive potential, and that individual members of a population differ in how much they have of a particular trait. If we plot the amount of the trait on the horizontal axis and the fitness of species on the

327

vertical axis, we produce a graph known as a fitness landscape in which points of high elevation correspond to high fitness.

To draw a landscape in which the trait corresponds to the length of a coyote's tail, we hold all other attributes of a coyote the same, vary the length of the tail, and measure the effect on fitness. To draw the graph, we need to know how the tail contributes to fitness. Suppose that a coyote's tail helps to balance the coyote when it jumps, and that it signals happiness, fear, or aggression. We begin at the left of the horizontal axis with a tail of length zero. It cannot carry out either function, so it has a fitness of zero. As the tail becomes longer both balance and signaling improve. Thus, fitness increases with tail length. At some point, say eighteen inches, the tail may be an ideal length for contributing to balance. If the tail becomes longer, the coyote will be less agile. Longer tails may continue to improve signaling value, so perhaps a tail of length twenty inches produces the most overall fitness. Once the tail becomes longer than twenty inches, fitness falls. The resulting graph, shown in figure 28.1, has a single peak.

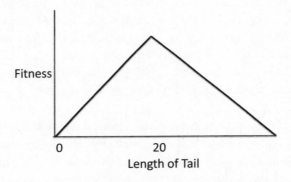

Figure 28.1: A Mount Fuji Landscape

This landscape is known as a *Mount Fuji*. Such landscapes often occur in the real world. Mount Fuji problems are considered easy. We expect that evolution or learning will always find the peak when encountering one. Imagine a population of coyotes with different lengths of tails. Selective pressure would result in coyotes with tails of approximately twenty inches. Coyotes with tails of that length optimally blend balance and signaling. They have the highest fitness and produce the most offspring, resulting in more coyotes with twenty-inch tails. If we think of this as an optimization problem, we see that any hill-climbing algorithm would locate the peak.

We can apply one-to-many thinking and reinterpret this as a problem of product design—specifically, the problem of designing a coal shovel. Suppose that we have already decided on the length of the handle and the shape of the pan. The remaining design decision is how large to make the pan. Pan area will correspond to the trait on the horizontal axis. On the vertical axis, we plot how much coal a worker could shovel in an hour given that pan size.

As before, we start at the far left, which corresponds to a shovel pan with zero surface area. The technical term for a zero-surface-area shovel is "stick." A stick cannot shovel coal and has value zero. As we increase the pan—say, to the size of a teaspoon, then a tablespoon, then a toy shovel—we make the shovel more and more effective. The graph of shovel fitness slopes up. At some point, the pan area becomes too large. Lifting the shovel becomes a chore, and the amount of coal that a person can shovel in an hour decreases with further increases in pan size. When the pan area becomes sufficiently large, no one would be able to lift the shovel and fitness will be zero. Once again, we have a Mount Fuji landscape. And once again, we should expect to be able to find the peak, the ideal pan size for our shovel.

The idea of plotting the efficiency of shovels as a function of pan size to determine the optimal shovel was developed by Frederick Taylor. In the 1890s, Taylor and others ushered in an era of scientific management in which manufacturing decisions—how fast to move the assembly line, how strong to make the weld, how many breaks to give workers—were modeled as rugged-landscape problems. Many of the great industrialists of the twentieth century including Henry Ford, John D. Rockefeller, and Andrew Carnegie contributed to this movement toward efficiency, or what now is commonly called Taylorism.

The move away from artisans making individual and distinct products to large-scale manufacturing, in which processes were broken into parts and each part was optimized and then routinized, led to increases in efficiency but also, in the eyes of many, the dehumanization of labor. Herein lies a welcome reminder about the need for multiple models. Any single model simplifies the world and highlights only some dimensions. Scientific management models focused on process efficiency. This led to criticism. Making decisions based on efficiency of output caused other objectives, such as the happiness and well-being of workers, to fall by the wayside.

The landscape model may seem to be a relatively obvious idea: plot the fitness, efficiency, or value of a characteristic as a function of a trait or attribute and then climb up the hill to find the optimal amount of the trait. Thinking of solving a problem as climbing up a hill may also seem little more than a metaphor. The validity of these critiques is not in question. However, by constructing a formal landscape model we will produce nontrivial insights.

Rugged Landscapes

When we allow for multiple attributes and for the contribution of one attribute to interact with those of others, we produce a rugged landscape—that is, a landscape with multiple peaks. Consider designing a couch in which we must choose the thickness of the cushions and the width of the arms. Let the value of a design equal its expected sales in the market, which correlates with aesthetic quality. If the couch has thick cushions, then wide arms may create a more appealing aesthetic. If the cushions are thin, then the ideal couch may have thin arms. A two-dimensional plot of expected sales as a function of arm length and cushion thickness will have two peaks. One peak corresponds to a couch with thin arms and thin cushions. The other will have thick arms and thick cushions.

Interdependent effects between variables create ruggedness on the landscape. Ruggedness has several implications. First, different approaches to finding the highest point on a rugged landscape may locate different peaks. So too may different starting points. Thus, ruggedness creates sensitivity to initial conditions and the possibility of path dependence. Each of these implies that landscape ruggedness contributes to outcome diversity. Ruggedness also implies the possibility of suboptimal outcomes. These are represented as *local peaks* on the landscape.

Figure 28.2 shows a rugged landscape with five peaks. Four of these peaks are local peaks, points whose neighboring points all have lower values, and one is the *global peak,* the point with the highest value. To see how search could land on a local peak that depends on the initial point, imagine beginning from a point and then climbing uphill. This is known as a *gradient heuristic* or a *hill-climbing algorithm*. On a rugged landscape, gradient heuristics get stuck on a local peak.

If the starting point is at the far left, the gradient heuristic would locate local peak 1, which is not optimal. If the gradient heuristic starts in the region denoted by Basin 2 in figure 28.2, then it locates local peak 2. Each of the other peaks, including the global peak, has a region such that if the gradient heuristic begins in that region, it will locate that local peak. These regions are called *basins of attraction* and are identified in figure 28.2. The global peak has the smallest basin of attraction. If we were to choose a random starting point and apply the gradient heuristic, the global peak is the least likely to be found.

The basins of attraction depend on the heuristic. A different heuristic may produce different basins. For example, consider the heuristic *go to the right* which moves to the right until finding a local peak. This heuristic produces identical local peaks as the gradient heuristic, but those peaks have different basins of attraction, as can be seen by comparing figure 28.3 with figure 28.2.

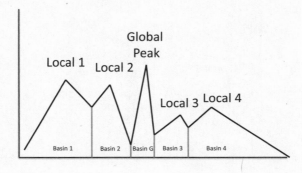

Figure 28.2: A Rugged Landscape with Five Peaks

To find an optimal or near-optimal peak on a rugged landscape requires either diversity or sophistication. The value of diversity should be self-evident. If distinct heuristics locate different peaks, then applying multiple, diverse heuristics to a problem will produce multiple, diverse local peaks, and one can choose the best from among these.[1] The same result will occur if one applies the same heuristic from different starting points: distinct local optima will be found and the best among them can be chosen.

Note also that the ruggedness of the landscape, as measured by the number of peaks, correlates with problem difficulty. However, a problem can be difficult to solve yet not have a rugged landscape. The problem of finding a gold coin in a cornfield would be represented by a flat landscape with a single peak at the coin's location. The landscape would not be rugged, but the coin would be very difficult to find.

The NK Model

We now describe the *NK model*, which allows us to formalize the connection between interactions and ruggedness.[2] The model represents objects, or what we might call alternative solutions, as binary strings of length N. The value of an object equals the sum of the contributions of each bit on the string. The K term in the model refers to the number of other bits that interact with each bit to determine its value. If K equals zero, the value function is linear. If K equals $N-1$, then every

331

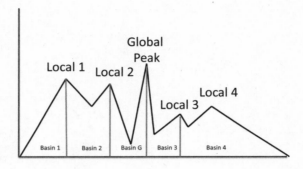

Figure 28.3: Basins of Attraction Produced by the "Go to the Right" Heuristic

bit interacts with every other and the value of each string is random. Thus, we can think of increasing K as tuning the ruggedness of the landscape to somewhere between Mount Fuji and random.

NK Model

An **object** consists of N bits, $s \in \{0, 1\}^N$.

The **value** of an object is $V(s) = V_{k1}(s_1, \{s_{1k}\}) + V_{k2}(s_2, \{s_2 k\}) + \cdots + V_{k2}(s_2, \{s_{2k}\})$ where $\{s_{ik}\}$ equals a randomly selected set of k bits other than i, and $V_{k1}(s_1, \{s_{1k}\})$ is a random number drawn from the interval $[0, 1]$.

$K = 0$: Results in a linear function of the bits.

$K = N - 1$: Any bit change produces a new random contribution from each bit.

The NK model framework provides a wonderful space to explore ideas and ask questions. The first question we ask is how the number of local optima depends on the number of interaction terms. We then ask how the height of the global optimum depends on the number of interaction terms. At the moment, both of those questions are ill-posed because we have yet to define how we are searching the space of possibilities, that is what heuristic we are using. Recall that what the set of peaks depend on our choice of heuristic.

In what follows, we rely on the *single-flip algorithm*. This algorithm chooses each attribute in sequence and switches the attribute's state. If changing this attribute results in a higher value, the switch is adopted. Otherwise the attribute is returned to its original state. The choice of this algorithm can be motivated in two ways. It can be interpreted as a crude model of genetic mutation, where good variants take over in the population and bad ones die. It is also the most natural way to represent a hill-climbing algorithm in this space.

We first evaluate the NK model with $N = 20$ and $K = 0$. When $K = 0$, each attribute's contribution to the total value is independent of the other attributes. The single-flip algorithm can identify the better state for each attribute and the global optimum. Thus $K = 0$, no interactions, corresponds to a Mount Fuji landscape. Each state's value is uniformly distributed in the interval $[0, 1]$. It can be shown that the higher of two random draws from a uniform distribution has an expected value of $\frac{2}{3}$. If we average the contributions across twenty attributes, the global optimum will also have an expected value of $\frac{2}{3}$.

At the other extreme, $(N = K - 1)$, each attribute is connected to every other attribute. When the state of an attribute is switched, the contribution of every attribute will change. It will be a new random number drawn uniformly from the interval $[0, 1]$. The value of the object will be the sum of these twenty new random numbers (one for each attribute), meaning that each flip of an attribute results in a value for the entire object that is uncorrelated with the earlier value. Thus, the landscape will be incredibly rugged—just as likely to go up at any point as it is to go down.

By applying that insight, we can derive the expected number of local peaks. If we start from any alternative, the single-flip algorithm compares that alternative to each of N alternatives. For example, starting from the alternative with all bits taking value zero, the algorithm will evaluate the N alternatives in which exactly one bit takes value one.

Initial alternative	00000000000000000000
Switch attribute 1	10000000000000000000
Switch attribute 2	01000000000000000000
...	
Switch attribute 20	00000000000000000001

A local peak must have a higher value than each of these N alternatives. The probability that the original alternative has the highest value equals $\frac{1}{N}$. Therefore, the number of local peaks approximately equals the number of possible alternatives,

2^N, divided by N. For $N = 20$, that calculation yields fifty thousand local peaks. With so many local optima, the single-flip algorithm rarely locates the global peak.

The relevant question is not the number of these local optima, but their values. It remains only to compare the expected average value of these local optima with the expected value of the global optimum. That comparison will determine how well the single-flip algorithm performs. To calculate theses values, we can use the central limit theorem. It is not difficult to show that the expected value of a local optimum equals approximately 0.6 while the expected value of the global optimum equals a little more than 0.75.[3] Comparing these to the global optimum for the case $K = 0$, which equals $\frac{2}{3}$, reveals that the local peaks on the rugged landscape have lower values than on Mount Fuji, but the global peak has a higher value.

This begs the question of what happens in between these two extremes, as we increase the number of attribute interactions, K, from zero to $N-1$. The answer is that we see both effects. The increased number of interactions produces a higher global peak, but more, and therefore lower-value, local peaks. Assuming that the search uses the single-flip algorithm, then computational investigations of this model show that for small K, the benefit of the interactions—a higher global peak—outweighs the increase in local peaks. So initially, the expected value of a local peak increases in K. The growing number of local peaks means that the average value will decrease. So if you were stuck using the single-flip algorithm, you would prefer a relatively small K value, around, say, 3 or 4. But why should we be constrained to use this simple heuristic of switching a single attribute? Evolution by mutation may be constrained to this heuristic, but we are not. We could switch the state of two attributes or even three. A more sophisticated algorithm will reduce the number of local optima.

Ruggedness and Dancing Landscapes

The NK model implies that we want a moderate degree of interdependence as that creates higher peaks. Many-model thinking demands that we step out from the particular assumptions of the model and consider the logic that drives that result. We find that the logic consists of two parts. The first rests on combinatorics: the number of pairs of combinations increases with the square of the number of pairs and the cube of the number of triples. Thus, interdependent effects create more possibilities of beneficial interactions. The second part rests on the fact that we need only keep the better combinations. Imagine grabbing any four food items to make a snack. Four items implies six possible combinations of two items. Suppose that we grab the following set of four: {*pickles, bananas, chicken, caramel*}. Of the resulting six pairs—bananas and pickles, pickles and chicken, caramel and pickles, bananas and chicken, caramel and bananas, and caramel and chicken—only one sounds remotely appealing. We only need choose that option. We enjoy the caramel bananas. We ignore the rest.[4]

A similar logic applies in evolutionary systems. Phenotypic combinations that produce positive interactions—a hard shell and short sturdy legs—remain in the population, while survival of the fittest works against combinations that produce negative interactions. We do not encounter many slow-footed, tasty animals with vibrant colors. If they ever existed, they have been caught and eaten.

We encountered a similar intuition in our model of search. When we have an abundance of possibilities, we prefer variation. The same logic applies here: combining pairs (and triples) produces abundant possibilities. And we would have preferred that these many possibilities had high variation in value. We then have a greater likelihood that one of them has a very high value. Given that interdependent effects increase variation, on the whole they are advantageous, but only up to a point. As we have just seen, too many make the landscape random. Ideally, then, we have a moderate number of interactions. Some argue that if the number and size of interactions can evolve or adapt, then systems should naturally evolve to rugged landscapes with high peaks. This would suggest that systems tend toward complexity and not equilibrium or randomness.[5] When and whether that is true is exactly the sort of question that is fun to explore with models.

One final point: We have taken the landscape as fixed. In ecological and social systems, the landscape that a species or firm confronts depends on the actions and attributes of others. An adaptation by a competing species, or a change in strategy by another firm, will shift and rearrange the fitness landscapes of competitors. We can now reinterpret our earlier models of spatial and hedonic competition as movements on dancing landscapes. Those movements could lead to an equilibrium, where each player stands on a local or global peak, or, competition on dancing landscapes may lead to complex patterns of actions and outcomes. Even a cursory glance at ecosystems, politics, and economics suggests that we see more of the latter. One reason that we see so much complexity may well be that much of our world consists of adaptive and purposive actors maneuvering on dancing landscapes. To make sense of that complexity, we need many models.

Do We Patent Knowledge?

Our well-being rests atop a centuries-long accumulation of knowledge that includes the laws of physics, the combustion engine, double-entry accounting, the germ theory of disease, X-rays, and HTML. Knowledge is often a public good. Knowledge is always non-rival. It may or may not be excludable. Exclusion requires verification, which is easier when a physical artifact embeds the knowledge. Verifying that someone used an algorithm or technique for solving a problem may be impossible. Verifying that someone embedded that algorithm in a software program is not.

When knowledge can be excluded, we confront a choice. We could treat knowledge like roads and national defense, and tax people to fund its production. Governments pay people to think through grants and indirectly by supporting universities. Governments also allow people to patent knowledge. A patent creates an incentive to produce knowledge by creating a period of exclusive rights to use the knowledge and to charge others for its use. In the United States and Europe, patents last twenty years from the date of filing.[6] Advocates of patents argue that private entities would have little incentive to spend years developing a better mousetrap, computer algorithm, or sound system if their discoveries could be used by anyone for free. They argue that patents overcome the incentive problems inherent in knowledge production.

Boldrin and Levine construct an argument that borrows ideas from several models to argue against patents.[7] In models in which ideas can be combined, patents can hinder innovation by limiting recombinations. One firm's patent on touch-screen technology may reduce the number of other firms that design new products that incorporate that technology. Absent patent protection, the technology could be incorporated into more products. Innovation would increase.

Proponents of patents push back by noting that while slowing innovation may be bad, without patents the reduction in investment would be much larger. To counter that claim, Boldrin and Levine use a logic partly based on our diffusion model. A useful product based on new knowledge will spread quickly through the population of buyers. That was true of the radio, television, Google's search engine, and Facebook. This creates a *first-mover advantage*. The innovator can still benefit, but only by producing something with the idea. With a patent, an inventor can wait for others to implement the idea and profit.

Boldrin and Levine also question how much credit the inventor deserves anyway. If breakthroughs were the result of a solitary genius, and most ideas would never have been produced without incentives, then the case for patents is stronger. The rugged landscape model suggests that difficult problems may have multiple workable solutions. New inventions, particularly those that combine existing ideas and technology such as the car, the telephone, and online auctions, may be natural occurrences not acts of genius. Any number of people might have made these innovations given the ideas swirling around in the community of thinkers. The simultaneity of major discoveries—calculus (Isaac Newton and Gottfried Leibniz), the telephone (Alexander Graham Bell and Elisha Gray), and the natural selection theory of evolution (Charles Darwin and Alfred Russel Wallace)—supports that inference. In sum, many-model thinking shows advantages and disadvantages to patents. The deeper, more nuanced understanding the models provide argues for a more flexible patent policy. Perhaps some ideas—those that many people might have discovered and those that could recombine with other ideas—should have different lengths or types of patents, or even not be patentable at all.

29. *Opioids, Inequality, and Humility*

Everything is complicated; if that were not so, life and poetry and everything else would be a bore.

—Wallace Stevens

In this final chapter, we apply many-model thinking to two salient policy issues: the opioid epidemic and economic inequality. We show how by engaging multiple models, we can better reason through these issues and better communicate why both have proven so difficult to solve. We can also see how, particularly in the case of opioids, experts might have used multiple models to anticipate the crisis before it occurred. That said, we do not want to oversell the potential for models to avoid disaster. Our treatment of the opioid epidemic is superficial, meant only as a template for how to apply many models when reasoning through a proposed policy or action. We do not gather data or calibrate any models. Rather, we apply the models qualitatively to gain insights.

Our analysis of income inequality, on the other hand, includes more detail and connects more tightly to the academic literature. It represents the other extreme of many-model thinking, where we deeply engage a variety of models. In both cases, thinking with many models makes us more knowledgeable and wiser. The chapter concludes with a brief comment on the need for humility. Models can make us wiser, but complex systems by definition are difficult to predict and understand. We will make mistakes. And we can learn from those mistakes to become even wiser.

Many Models and the Opioid Epidemic

To give some sense of the scale of the opioid epidemic in 2015, in the state of Massachusetts over 4% of the population above age 11 had an opioid use disorder according to one estimate. Nationwide, in 2016, doctors wrote more than 200 million prescriptions for opioids, between 10 and 12 million people misused opioids, over 2 million people were classified as having an opioid use disorder, and more than 30,000 people died from opioid-related causes.

The primary reason that so many opioids were prescribed was that they work: they reduce pain. Given the 100 million Americans with chronic pain, opioids had an enormous potential market. The danger with opioids was, of course, the potential for people to become addicted. To make sense of how opioids received approval and how the epidemic arose, we apply four models to generate some core intuitions as to how the crisis came to be.

The first model, the multi-armed bandit model, explains why opioids were approved for use. When seeking drug approval, a pharmaceutical company runs clinical trials to demonstrate drug efficacy and a lack of deleterious side effects. We can model a clinical trial as a multi-armed bandit problem where one arm corresponds to prescribing the new drug and the other arm corresponds to a placebo or the existing drug.

A Model of Opioid Approval
Multi-Armed Bandit Model

To demonstrate their efficacy, opioids were tested against placebos. In clinical trials, patients were randomly assigned to take either opioids or a placebo. The assignment of the opioid can be modeled as one arm of a two-armed bandit and the placebo as the other arm. At the end of treatment, each trial is classified as a success or a failure. Clinical tests found that patients who received opioids experienced (statistically) significantly less pain. Tests on patients who had hip replacements, dental surgery, and cancer treatments all found that opioids outperform placebos.

With any drug, the potential for addiction is a concern. Tests showed that a small percentage of patients, fewer than 1%, became addicted, allowing the drug to be approved. those tests did not take into account the possibility that doctors would write longer prescriptions, in some cases a month's supply. The longer an individual takes opioids, the more likely that person becomes addicted. Empirical addiction rates exceeded 2.5% for patients with longer prescriptions. The Markov model shown in the box below shows how an increase in those rates from 1% to 2.5% can increase the equilibrium number of addicts 5-fold.

The model's transition probabilities are only loosely calibrated to data. We are using the model to build intuition for how a relatively small rate of addiction can lead to a large number of addicts. By experimenting with the model, we find that if we lower the probability of leaving the addict state and increase the probability of moving from the no-pain state to the opioid state, then the proportion of addicts can increase dramatically. If, for example, we lower the transition from addict to no pain to 1% in the second model, the proportion of addicts increases to 35%. One implication of this type of model thinking has been that some health care providers, such as Blue Cross, now limit the number of pills a doctor can prescribe. In addition, some states, including the State of Michigan, have passed laws restricting the number of pills in a single prescription.

Transition-to-Addiction Model
Markov Model

A three-state Markov model reveals a nonlinear relationship between transitions to addiction and overall addiction rates. The model's states represent people not in pain, people using opioids, and addicts. We estimate the transition probabilities between those states, which we represent as arrows. The model on the left assumes that 1% of of people who use opioids become addicted and that 10% of addicts revert to the no-pain state. It also assumes that 20% of the people in the no-pain state become opioid users. In equilibrium, only 2.2% of the population are addicts. To account for longer prescriptions, the model on the left assumes that 2.5% of people who use opioids become addicted and that 5% of addicts revert to the no-pain state. It also assumes that 20% of the people in the no-pain state become opioid users. Now, in equilibrium, 10% of the population are addicts.[1]

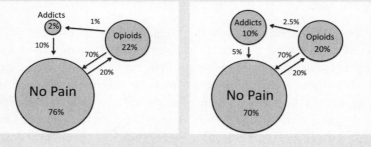

Our third model relies on systems dynamics. This model, like the Markov model, assumes that there are people in pain, people who use opioids, and people no longer in pain. Rather than write transition probabilities between these states, however, it imagines a flow from people in pain to opioid use to people not in pain. More elaborate systems dynamics models can include sources for other drug providers, and allow for movements between opioid and heroin users. In addition, a richer model could include heterogenous types of potential users. The fact that people who suffer from anxiety and depression are more likely to become addicted could therefore be included in the model.[2]

Paths to Heroin Addiction
Systems Dynamics Model

A population of people in pain produces opioid users and addicts. People on opioids flow into the no-pain state and also flow into the addict state. Addicts, in turn, can become heroin users. One reason that people use heroin is that they can no longer get opioids. Thus, as the flow of opioids increases, so does the number of heroin users.

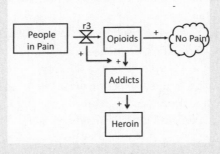

A possible final model, which we do not write down formally, relies on social networks to explain why maps of per-capita opioid use show clustering in rural counties. From our analysis of the square root rules, we know that smaller populations should have higher variation. (Recall the example of the best and worst performing schools being small.) Higher use in rural areas could also be due to doctors writing longer prescriptions for rural patients who live farther from pharmacies. Those explanations aside, the clustering exceeds what would occur randomly.

Clustering could arise if people provide or sell opioids to neighbors. Unlike used furniture, which people can sell by placing ads, opioids most often sell through personal connections. A model of opioid selling would assume a social network of family and friends within which people distribute opioids. Such a model would produce the local clusters of opioid abusers seen in data.[3]

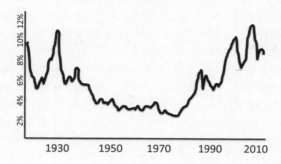

Figure 29.1: Income Shares of the Top 0.1% 1916–2010. Source: Piketty 2011.

Many Models of Inequality

Our final many-model exercise delves relatively deep and wide into the causes of economic inequality. We undertake this effort for three reasons. First, inequality is one of the most important policy issues of our time. Income and wealth correlate with human flourishing. Higher-income individuals enjoy better health, longer life expectancy, and higher life satisfaction and happiness. Those at the bottom of the income distribution experience higher rates of homicide, divorce, mental illness, and anxiety.[4] We must be careful not to confuse correlation with causation: a substantial part of this correlation can be explained by the fact that healthier, happier people earn more money. Nevertheless, almost all studies show a connection between income and flourishing. No one prefers to be poorer. Second, we have a plethora of models of inequality written by a diversely tooled collection of economists, sociologists, political scientists, and even physicists and biologists. Third, we have abundant data on income and wealth within and across countries. We have both current data and time series stretching back hundreds of years.

We start by summarizing some empirical regularities. First, in all countries at all times, the distribution of income has an elongated tail, with many low-income people and a small percentage of people who earn large incomes. Historically, income distributions were calibrated to lognormal distribution or Pareto distributions. Recently, more granular data reveals the tail to be longer than lognormal, though not quite that described by a power law. Wealth distribution is similarly skewed.

Second, within most developed countries, income and wealth inequality, however measured, have been rising in recent decades. Current levels of income and wealth inequality in the United States approach those of the Gilded Age. Shifts in entire distributions can be hard to discern, so, following convention, we describe those shifts with respect to the share of income that goes to the upper tail of the distribution. Figure 29.1 shows how the top 0.1% has increased its share of income. The share of income to the top 0.1% of families fell steadily through 1950, and remained stable at less than 4% until around 1980, when it began to climb. In 2018, the proportion of total wealth own by these super rich was around 10%.

Third, globally, the number of people living in poverty has dropped precipitously. We should see no logical contradiction in these opposing trends. Fast-rising incomes in poor countries reduce cross-country differences and more than offset within-country increases in inequality. Our model of group selection produced similar effects. The growth in the number of altruistic communities outpaced the trend toward selfishness within each community.

Inequality has multiple, interwoven causes. Economic forces, sociological trends, exercises of political power, and the weight of history all contribute to disparities. Thus, as Steven Durlauf points out, we should not try to explain the levels or trends in disparities with a single equation. Nor should we base policy on one.[5] We must be nuanced in our thinking. The processes concentrating wealth and income in the top 1% or top 0.1% may be unrelated to the forces trapping the bottom 20% in a cycle of poverty. To understand the disparate causes, we need many models.

We start by describing models that explain the changing distribution of income. Income has four sources: wages and salaries, business income, capital income, and capital gains. The relative sizes of those shares varies with income level. Low-income people earn few capital gains or capital income. Many of the highest earners receive substantial income from every category. They earn income from wages, businesses, and capital.

Our first model extends the Cobb-Douglas production model to include two types of labor: educated and uneducated. The wage paid to a type of labor depends on the relative supply of that type and on technology.[6] This model explains the recent rise in inequality based on supply and demand. During the 1950s, the rise in manufacturing increased demand for uneducated workers. At the same time, increased college enrollments due in part to the GI Bill increased the supply of educated workers. In the 1980s, decreased incentives to attend college slowed the growth in the number of college graduates, and a subsequent inflow of immigrants with low education levels increased the supply of low-skilled workers. At the same time, technological changes—the rise of automated manufacturing and the transition to a more digital economy—increased the relative value of educated workers, and their rising wages reflected this value.

Time series data on average incomes by education level fit this model reasonably well. For this reason, many economists rely on the model to guide policy. The model advocates increasing access to education, as that will depress the wages of educated workers and reduce inequality. This model explains broad trends well, but it cannot explain the increase in variation within each income class.

Technology and Human Capital Model
Growth Model

Output depends on *physical capital* (K), *educated labor* (S), and *uneducated labor* (U) as follows:

$$Output = AK^\alpha S^\beta U^\gamma$$

The parameters A, α, β, and γ capture the technology and the relative value of the three types of labor. The relative market wage for high- and low-skilled workers is:[7]

$$Wage_S = \left[\frac{U}{S}\right] \cdot \left[\frac{\beta}{\gamma}\right] \cdot Wage_U$$

Cause of inequality: Technological changes that favor educated workers increase β and decrease γ. These changes, along with increases in the supply of low-skilled workers, increase inequality.

The next model, the *positive feedback model*, can explain the increased variation within professions. It focuses on the tail of the distribution and, in particular, on entrepreneurs. In 2011, entrepreneurs made up 70% of the 400 wealthiest individuals in the United States.[8] The model assumes that technologies—the internet and smartphones in particular—have made us more connected and more influenced by the choices of others.[9] A person buying wireless stereo speakers can read reviews online and select "the best" from among a dozen choices. In the past, that person might have had a single option at her local stereo store. Now, a person who twists her knee can search the web and learn the identity of her favorite athlete's doctor. That behavior creates a positive feedback and more inequality. We model socially influenced economic choices by reframing the preferential attachment model as a model that links positive feedbacks to talent.

Though the positive feedbacks model cannot be fitted to time series data with the same fidelity as the previous technology model, we can look to experiments to see how feedbacks contribute to inequality. Recall the music lab experiments described in Chapter 6. College students sampled and downloaded music under two treatments. In the first treatment, subjects could not see what music others had downloaded. This treatment captures the pre-internet world. In the second treatment, subjects could see the download numbers for each song. In the treatment without social information, no song receives more than two hundred downloads and only one song receives fewer than thirty. When people can see downloads, one song receives more than three hundred downloads and over half receive fewer than thirty. Information and social influence amplify the Matthew effect. The rich get even richer, and the poor become relatively poorer.

Positive Feedbacks to Talent
Preferential Attachment Model

There exist N producers, and each begins with zero sales. The first consumer buys from a random producer with zero sales, giving that producer positive sales. Each subsequent consumer with probability p buys from a producer with zero sales and with probability $(1-p)$ buys from a producer with positive sales. When buying from a producer with positive sales, a customer selects randomly, with the probability of choosing a particular producer that is proportional to that producer's sales.

Cause of inequality: Increased connectedness increases social influences, creating a positive feedback.

We can apply that same logic to the economy writ large.[10] The potential for positive feedbacks through social networks to contribute to inequality depends in part on the nature of what people buy. Weightless goods such as movie and music downloads, web applications, and some technologies can be scaled quickly, if not immediately. Tractors, cars, and washing machines cannot be duplicated by clicking on an icon. So while a new smartphone application can scale up with little to no capital outlay, a best-selling car cannot. As a benchmark, in May 2015, Volvo announced that it would build its S60 sedan in South Carolina. The company broke ground on the plant in September 2015 with the expectation that the first cars would roll off the line in late 2018.

Our next model applies the spatial voting model to explain the rise in CEO pay, which is not determined by social forces. In 2012, the average income of a CEO at a Fortune 500 company exceeded $10 million, or roughly 300 times the average pay of a worker. By comparison, in 1966, the CEO made only about 25 times the average worker's salary. CEOs in other countries earn much less. In Japan, CEOs earn about 10 times what the average worker does. In Canada and throughout Europe, CEOs earn approximately 20 times the pay of the average worker.

At most companies, the CEO's pay is set by a compensation committee consisting of members of the board of directors. That pay includes salary, bonuses, and stock options. The people who determine the pay of CEOs are often other CEOs. They have an incentive for the pay of other CEOs to be high in order to drive up their own pay. We can use the spatial model to represent the preferences of the compensation committee. According to the spatial model, the salary will be set at the median voter's preferences. The difference in CEO pay by country can be explained by the composition of boards and the compensation committees. In Germany, boards of directors include workers, who prefer that the CEO be paid less.

CEO Political Capture
Spatial Voting Model

CEO pay is determined by a vote of a compensation committee. In the United States, compensation committees include many current and former CEOs, who prefer higher pay, as well as compensation experts (X). Other countries include workers (W) on compensation committees, resulting in a median voter who prefers much lower pay.

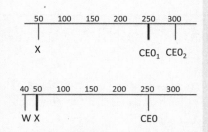

Cause of inequality: CEOs determine their own pay through capture. Increases in the pay of any one CEO shifts preferences toward higher pay for all CEOs.

The model explains the rise in CEO pay based on the preferences of board members about what appropriate CEO pay should be. Here we can refer back to multiple models of value. It could be that the preferences of compensation committee members rely on model based thinking informed by data. However, those preferences might also be socially constructed or part of an elaborate log roll, in which CEOs vote to raise one another's pay.

Our next model of income inequality comes from Thomas Piketty's best-selling book *Capital in the Twenty-First Century*. This is less a model than an observation that the rate of return on capital exceeds the growth rate of capital. When that holds, the portion of income that high-income individuals receive from returns on capital will increase over time. By constructing more elaborate versions of the growth models from Chapter 8, it can be shown that the return to capital should always exceed the rate of growth in the broader economy. Over the long haul, an economy might grow at less than 2% or 3%, but returns to capital may be more than double that.

It follows that in an economy that consists of workers who earn wages and capitalists who earn income from rents, the share of income going to capitalists will increase. To be a bit more formal, the rate at which capital increases will depend on three rates: the consumption rate, the tax rate, and the return on capital. Consumption depends on the level of capital. A person with little capital will consume a large percentage of her income. A person who owns a substantial amount of capital will consume a small percentage of her income. As shown formally in the box below, if we make the consumption level constant, the consumption rate will equal that amount divided by the level of capital. Thus, wealthier people will consume at a lower rate making it more likely that their net capital increases.

Rent-from-Capital Model (Piketty)
Rule of 72

The economy consists of **workers** and **capitalists**. The wages of workers increase at a rate g, the growth rate in the economy. The capitalists have wealth W_t at time t and earn return r (net of taxes) and consume a constant amount A. The income of capitalists will increase faster than that of workers if and only if

$$r - \frac{A}{W_t} > g$$

Cause of inequality: In a market economy, the rate of return on capital exceeds the overall rate of growth ($r > g$). Capitalists with large accumulations of wealth spend a small proportion of their income from capital on consumption, so their share of total income increases over time.

To see how the difference in rates produces inequality, we can apply the rule of 72. If initially the incomes of workers equal those of capitalists and wages grow at 2% while capital grows at 6%, then in thirty-six years wages double but income from capital increases 8-fold. Within seventy-two years, capitalists earn sixty-four times the income of workers.

Piketty applies this model to explain long-term trends in inequality of both income and wealth. The model calibrates remarkably well with three centuries of data from France and England. The model also sheds light on patterns of inequality over the past century in the United States and Europe, in which the two world wars destroyed capital stocks in Europe, evening out the income and capital distributions there. One reason the model fits the data as well as it does is that it omits two effects that cancel out. By excluding entrepreneurs, the model understates inequality. In assuming that all succeeding generations of capitalists invest wisely—not all do—the model overstates capital accumulations contribution to inequality. The creation of a new class of rich individuals and the loss of an old class of rich individuals need not balance out. A more granular model would include movement in and out of the wealthy class.

That caveat aside, the model's implication is that so long as capital increases, capitalists earn an increasing portion of the economic pie. If we keep applying the rule of 72, we find that the income of the capitalists eventually dwarfs that of the workers. The problem of capital accumulation has a straightforward solution: impose a wealth tax. That may not be politically possible. As an alternative, we might wait for a war or revolution to redistribute wealth by force or for some technological breakthroughs that produces a new set of wealthy capitalists.

Our next two models give priority to sociological forces. Both also have strong empirical support. The first explains rising inequality based on *assortative mating*. A family's income depends on the incomes of both partners. If a low-income person marries a high-income person, then that marriage will contribute toward equalizing income distributions. If high-income people marry other high earners, then income disparities will increase. Most people marry at an age when a potential partner's lifetime income cannot be known with certainty. People do know the education level and general health of potential partners and get signals of their ambitions. Evidence shows that as men and women become more educated and earn higher incomes—refer back to the technology and human capital model—they choose life partners who also have higher education levels.

The increase in inequality results from the following factors. First, women increasingly earn college degrees. Second, relative income increases with education level. Third, educated men and women prefer educated partners. It follows that families with two educated people will be more likely to have two high incomes contributing to household-level income inequality. The logic is airtight. The only question concerns the size of the effect.[11]

Sociologists calibrate the model by categorizing people into five education levels: dropout, high school, some college, college degree, and postgraduate. They then calculate the average income for each education level and fit the data for the number of marriages between each pair of education levels, resulting in a crude approximation of the impact of assortative mating.

Assortative Mating
Sorting Model and Categories

Each individual has an education level: $\{1, 2, 3, 4, 5\}$ where $1 =$ dropout, $2 =$ high school diploma, $3 =$ some college, $4 =$ college degree, and $5 =$ postgraduate.

Let $P(m, j)$ and $P(w, j)$ denote the probability that a man and woman have education level j. Income(g, ℓ) equals the (estimated) income of a person of gender g and income level ℓ. Household income for a couple consisting of a man with education level ℓ_M and a woman with education level ℓ_W earns the following estimated household income:[12]

$$\text{Income}(M, \ell_M) \; + \; \text{Income}(W, \ell_W)$$

Cause of inequality: Increases in the number of educated women, increased pay for workers with higher levels of education, and assortative mating (the tendency for people to marry others of the same income level) result in an increase in household-level income inequality.

Had marriages been random rather than assortative, income inequality would be much less. One study finds that inequality as measured by the Gini coefficient, a common measure of inequality, would have decreased by 25%.[13]

Our next model analyzes movements between income categories using a Markov model. It categorizes people (or households) by income level: high, upper middle, lower middle, and low. Each category contains one-fourth of the distribution. Given a time period—it could be a year, a decade, or a generation— we can then estimate the transition probabilities between income categories to capture mobility.

Intergeneration Income (Wealth) Dynamics
Markov Model

The population can be divided into four income (or wealth) categories with equal numbers of people. We can estimate the **transition probability** that an individual (or family) in one category moves to another category within a generation, as shown in the figure below. More equal transition probabilities correspond to greater **social mobility**.

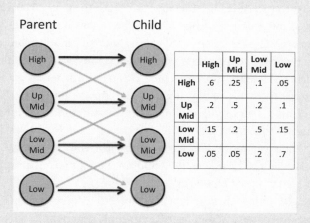

	High	Up Mid	Low Mid	Low
High	.6	.25	.1	.05
Up Mid	.2	.5	.2	.1
Low Mid	.15	.2	.5	.15
Low	.05	.05	.2	.7

Transition Probabilities Between Income Levels

Cause of inequality: Social skills, tacit knowledge, attitudes toward risk and education, and bequests reduce mobility between income classes.

If there were no stickiness across generations, then the income of the child of a high-income parent would be equally likely to belong to any of the four income classes—all of the transition probabilities equal $\frac{1}{4}$. In the most extreme case of no mobility, transition probabilities would consist of only 1s along the diagonal. Empirical estimates suggest that the reality lies between these extremes.

We can run experiments by taking 100 randomly selected low- or high-income families and computing the probability distribution of incomes in subsequent generations. Using the probabilities shown in the box, the children of high-income parents have a 60% chance of being high-income and only a 5% chance of being low-income. The grandchildren of the high-income parent have less than a 43% chance of being high-income and more than a 10% chance of being low-income.[14]

The income dynamics model also serves as a baseline from which to evaluate the causes of income mobility. We might use a linear model to estimate a child's income as a function of parental wealth, parental income, and parental ability levels (assuming we had data). The Piketty model would imply a positive coefficient on parental wealth. The ability-based model would imply a positive coefficient on parental ability given that there exists some correlation between parental ability and ability of offspring.

Note that determining the coefficient on parental income requires data on the income of each child and each parent. Scholars have individual-level income data only for the past few decades. In *The Son Also Rises,* Gregory Clark (2014) found a novel solution to the problem of lack of data: he relies on surnames. He calculates the average income of everyone named, say, Thatcher, in 1888 and compares this to the average income of everyone named Thatcher in 1917. The thirty-year increment represents the length of a work life. He finds substantial correlation across surnames' average incomes, suggesting a lack of income mobility.

This type of model allows us to identify racial differences in intergenerational transfers. African Americans exhibit less persistence of wealth at the top of the income distribution and more persistence at the low end. A wealthy African American will be less likely to have wealthy children, and a poor African American will be more likely to have poor children.[15]

Our last model based on neighborhood effects, Durlauf's *persistent inequality model,* leverages the empirical regularity that people segregate by income category— that is, high-income people live in communities with other high-income people and low-income people live near low-income people. Segregation by income produces economic, sociological, and psychological externalities that reduce mobility. In the model, an individual's income depends on ability, educational spending, and spillovers.

The educational attribute captures public spending on education, which empirically correlates with average income: high-income locations spend more on education than low-income locations, resulting in better educational outcomes and higher incomes for children in high-income neighborhoods.

The spillover term can be interpreted as socially transmitted knowledge of appropriate tools to acquire. Here we can link Durlauf's model to how people who live in high-income communities gain awareness of appropriate tools. We can also link the model to our network model and the *strength of weak ties* phenomenon: people who live in high-income communities will be connected indirectly to more people with access to economically valuable information. This will produce a positive feedback on income.

We can also interpret the spillover as socially transmitted behaviors, such as the number of hours spent studying or working. If income includes a random component, then a person in a low-income community will see (correctly) a low return to time spent on self-improvement. Relatedly, the spillover could include psychological attributes—a positive or negative outlook on life, a feeling of safety, or a belief in oneself.

Persistent Inequality (Durlauf)
Schelling Segregation Model + Local Majority Model

Individuals belong to income classes and segregate residentially by income. Individuals allocate a portion of their income to education, resulting in positive spillovers that increase with community income level. The future income of a child living in community C depends on her innate ability, spending on education, and spillovers. The contributions of education and spillovers depend on the level of income within the community, I_C.

$$\text{Income}_C = \text{F}(\text{ability}, \text{education}(I_C), \text{spillover}(I_C))$$

Cause of inequality: Children who grow up in low-income neighborhoods receive fewer educational opportunities and economic spillovers.

In the complete model, Durlauf solves for equilibrium levels of educational spending and derives conditions in which persistent inequality arises. That inequality results from what he calls *poverty traps*. Individuals living in low-income communities lack the educational resources and levels of spillovers necessary to earn high incomes regardless of their ability levels. Durlauf's model can help to explain the enormous racial gaps in income levels. African Americans disproportionately live in poor neighborhoods, and as a result, they may become trapped in low income trajectories because of a lack of resources.

These various models describe distinct causes of income inequality. In a sense, each is correct, but, as we know, each model is also wrong. Empirically, the models differ in how much and what part of the variation in income they can explain. For the upper end of the income distribution, the empirical evidence most strongly supports the models that rely on technological change.[16] For over twenty years, the IRS has tracked the highest 400 incomes. Those at the top of the distribution come from technology, mass retail, and finance, three industries that can scale quickly. That high growth rate could stem from winner-take-all markets for search engines (Google) or social networking sites (Facebook). These models tell us little about the lower end of the income distribution. Nor do they say much about income mobility, or explain why CEO pay in the United States far exceeds that in other countries.

To explain these other features of the data, we need the other models, such as the income mobility model, Durlauf's persistent inequality model, and the spatial voting model. By constructing a dialogue between multiple models and data, we come away with a deep, multifaceted understanding of the causes of inequality. We identify multiple processes that produce and maintain inequality and see how they overlap and intersect. Our understanding of the complexity of inequality and the self-reinforcing causal forces that sustain it should make us dubious of quick fixes. Reducing inequality will require concentrated efforts on multiple fronts.

Into the World

We have just learned how by applying many models as an ensemble we can explicate the multiple causes of the opioid epidemic and income equality and reveal the limits of any one framing. Were we policymakers, we could fit some of these models to data to gauge effect sizes. We could then run natural experiments to help us guide policy choices based on what we have learned.

We could also take the many model approach to any number of social challenge including reversing trends in obesity, improving school performance, mitigating climate change, managing water resources, and improving international relations. In each case, even adding a single new model could have enormous consequences. Take for example the problem of predicting financial collapses. The United States Federal Reserve relies on traditional economic models using national accounting data on inflation, unemployment, and inventories. Those data suffer from lags. They are released weekly, quarterly, or annually. Those data also come from surveys, that is, samples of the entire economy.

Complexity scholar, J. Doyne Farmer argues for creating a second class of models based on real time data scraped from the web. These new models would rely on more granular, instantaneous data, and therefore differ from traditional models. Farmer argues that such models could prove much better than existing models. He may be right. Yet, these new models need not be more accurate to be of use in predicting and preventing financial disasters. Given the new models would use different data and rely on different assumptions, they would make different predictions. As we know from the diversity prediction theorem so long as the new models are not far less accurate, when combined with existing models, these new models would improve predictions. Policy makers, to use Farmer's turn of phrase, would be more collectively aware.[17]

An executive might engage in a similar exercise when making a business decision. She could apply multiple models informed by data to decide on product attributes, time product launches, design compensation plans, construct supply chains, and forecast sales. Because each of these actions occurs within a complex system, any one model would be wrong. Many models would lead to better actions.

To sum up, when confronted with a choice, when asked to make a prediction, or when faced with a design challenge, we should take a many-model approach. Many-model thinking produces better performance than taking actions based on hunches and gut instincts. That said, we have no guarantee of success. Even with many models, we may not identify the most relevant logical chain. The domain of interest might be so complex that even ensembles of models can only explain a small portion of the variation.

The same holds when applying models to aid in design, we may find ourselves unable to construct useful abstractions. The simplicity of models may, in those cases, be their undoing. In the face of complexity, it is possible that we find models not up to the tasks of communicating ideas, making accurate predictions, or pointing us toward the best actions. Even our explorations with models may reveal little of value. In those instances, the REDCAPEs that the book has promised will not provide much lift. Nevertheless, even in those cases, we benefit from contemplating and applying many models. In doing so, we uncover interdependencies. We learn why a complex process can frustrate our attempts to understand, explain, or communicate.

Thus, we must maintain a degree of humility. Even aided by many models, our abilities to reason have limits. For that reason, we must remain curious. We must continue to build new models and to improve upon existing ones. If a model leaves out key features of the world—such as social influences, positive feedbacks, or cognitive biases—then we should build other models that include those features. By doing so, we can begin to discern when those attributes matter and how much. The fact that all models are wrong should not take the wind out of our sails but be seen as motivation for building a crowd of models capable of producing wisdom.

Last, we should seek joy in those efforts. Though the book has emphasized pragmatic aims—to become better thinkers, to be more effective at work, and to operate as more informed citizens of the world—it has also had an implicit goal of revealing the beauty of models and the fun of modeling. The practice of modeling can be a beautiful game. We make the assumptions, write the rules, and then play within those rules bound also by the laws of logic. Through our logical explorations, we improve ourselves and become wise. May we take that wisdom out into the world and help to change it in positive ways.

Notes

I have too many colleagues and friends to thank properly. This book was improved by conversations about modeling with Eric Ball, Andrea Jones-Rooy, Michael Mauboussin, Carl Simon, John Miller, Lu Hong, Helene Landemore, Jim Johnson, Skip Lupia, Josh Berke, Patrick Grim, Bob Axelrod, PJ Lamberson, Jessica Steinberg, Jessica Flack, Charlie Doering, Michael Ryall, Robert Deegan, Jay Grusin, Sarah Silvestri, Zev Berger, Ken Kollman, Jean Clipperton, Michael Barr, Benjamin Bly, Elizabeth Bruch, Abbie Jacobs, Mark Newman, Cosma Shalizi, Kent Myers, and Josh Cooper Ramo. I would also like to thank the Guggenheim Foundation for funding a sabbatical as well as the faculty, staff, and students at INSEAD and the people of Fontainebleau, France. In Ann Arbor, the staffs at Lab Cafe, Mighty Good Coffee, and all four Sweetwaters Coffee and Tea locations provided coffee and quiet support. This book has also benefited from online conversations with thousands of people who gave feedback through Model Thinking. My agent, Max Brockman; editor, TJ Kelleher; and Melissa Veronesi kept me focused over the final year. Mita Gibson and Linda Wood provided unending support and encouragement, covering for my many lapses in scheduling as well as demonstrating facility with the printer, and, in the final stages, Lucy Fleming and John Burt solved vexing typesetting issues.

Chapter 1: The Many-Model Thinker

[1] See, for example, O'Neil (2016) for an account of how simple models based on data can ignore some segments of the population and can ignore the adaptive feedbacks we discuss Chapter 4.

[2] See Paarsch and Shearer 1999, which analyzes the timber industry. Raw data on tree planting shows that the piece rate is negatively correlated with the number of trees planted; the more someone is paid to plant a tree, the fewer trees that person plants. The inference runs counter to standard economic logic. If you pay planters more per tree, they should work harder. In Paarsch and Shearer's model, timber companies pay workers a per-tree piece rate so that the hourly market wage equals $20 an hour. That assumption results in the following equation for dollars per tree:

$$\$20 = \text{Number of Trees Planted per Hour} \times \text{Dollars per Tree}.$$

If a person can plant ten trees in an hour, then the pay per tree will equal $2. If a person can plant twenty trees in an hour, then the pay per tree will be $1. Thus, the model predicts that the piece rate will be *negatively correlated* with the number of trees planted. It also predicts that the piece rate times the number of trees will equal a constant.

[3] For evidence on models outperforming people, see Dawes 1979, Tetlock 2005, Silver 2012, and Cohen 2013. See Kahneman 2011 on biases.

[4] See Slaughter 2017 and Ramo 2016.

[5] Studies show that the most impactful research and patents disproportionately draw ideas from multiple disciplines. An analysis of 35 million papers shows that in the long run, interdisciplinary papers have greater impact (Van Noorden 2015). A combination of ideas is not necessarily a combination of models, but in many cases it is; see Jones, Uzzi, and Wuchty 2008 and Wuchty, Jones, and Uzzi 2007. Freeman and Huang 2015 show that ethnic diversity also correlates with citations. If we interpret patents as evidence of innovation, then two separate strands of research link diversity of thinking to success. Shi, Adamic, Tseng, and Clarkson 2009 show that patents that cross categories produce more citations. Youn, Strumsky, Bettencourt, and Lobo 2015 show that a majority of patents include multiple subcategories. Interdisciplinary research has increased steadily to the point where, on average, social scientists cite more papers from other disciplines than from their own.

[6] See Box and Draper 1987.

[7] See Page 2010a.

[8] I am not equating knowledge with models. I am claiming that models can represent knowledge and provide a clear way to communicate those understandings. The term "knowledge" encompasses rather broad terrain and includes physically embedded and tacit skills such as how to play tennis, speak French, or negotiate a contract. I apply a narrower definition. For a broader conception, see Adler 1970.

[9] You can arrive at this approximation by noting that falling skydivers reach terminal velocities of 200 mph. Terminal velocities scale with the inverse of mass. Assume that a skydiver has a mass 400 times larger than the stuffed cheetah. The square root of 400 equals 20. Therefore, the terminal velocity of the stuffed cheetah will equal 200 mph divided by 20, or 10 mph.

[10] He was correct. For the record, Fresno is 30% larger than Iceland. Ball and LuPima 2012 describe how one can take lessons from the academy to the business world.

[11] See Lo 2012. For a general argument see Myerson 1992.

[12] Versions of this story can be found in the writings of William James, Stephen Hawking, and Antonin Scalia.

Chapter 2: Why Model?

[1] See Epstein 2008 for a finer categorization of reasons to model. Lave and March (1975) describe three categories of use: to explain empirical phenomena, to predict other, new phenomena, and to build and design systems. Implicitly, they also advocate using models to explore.

[2] See Harte 1988. This categorization borrows from Johnson's 2014 treatise on the uses of models in the social sciences. These two approaches are also known as the *Galilean* and *minimalist idealizations*. See Weisberg 2007. For more on analogies, see Pollack 2014 and Hofstadter and Sander 2013; the latter refers to analogies as the "fuel and the fire" of thinking. See also Schelling 1978, 87, for elaboration on classes of models. Daniel Little's blog, *Understanding Society*, provides an entree into the topic of the ontology of social science.

[3] See Arrow 1963. A collective ranking is possible if we limit the possible individual rankings. If, say, everyone had the same ranking, then the collective would as well. In general, we have no way of mapping individual rankings to a coherent collective ranking.

[4] The best minds of my generation surely noticed that I borrowed "this actually happened" from *Howl*. See Bickel, Hammel, and O'Connell 1975.

The figure below shows one of many possible examples of adding a node and reducing total edge length. The network on the left shows four points as corners of a square. The one on the right includes a fifth point in the center. If we set the length of a side of the square to 1, the edges in the graph on the left have a total length of 3, and those on the right have a total length equal to $4 \cdot 0.71$, which is less than 3.

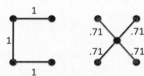

Simpson's paradox arises when there are more men than women who apply to departments with higher admission rates. For example, consider a university with a medical school and a veterinary school. Suppose that 900 men apply to medical school and 480 (or 53%) are accepted, that 300 women apply and 180 (or 60%) accepted, that 100 men apply to veterinary school and 20 (or 20%) are accepted, and that 300 women apply and 90 (or 30%) are accepted. In each school, a larger percentage of women are accepted, but overall, 50% of men (500 out of 1,000) while only 45% of women (270 out of 600) are accepted.

For an example of Parrondo's paradox, suppose that the first bet always loses $1 and the second bet loses $2 in any period whose number is not divisible by three but wins $3 in periods 3, 6, 9, 12, etc. Each bet produced an expected loss, but if you make the second bet only in the periods when it wins and make the first bet in the other periods, you will win $1 every three periods.

[5] See Kooti, Hodas, and Lerman 2014.

[6] Suppose that each person makes the same income I and pays a constant tax rate of t. Let c denote the percentage reduction and r denote the increase in income. Current government revenue equals $I \cdot t$. After the tax cut, revenue equals $I(1+r) \cdot t(1-c)$. Revenue will increase if and only if $I \cdot t < I(1+r)t(1-c)$. Rearranging terms gives $r > c(1+r)$.

[7] See Ledyard, Porter, and Wessen 2000 for a market-based mechanism that produces a better solution to multidimensional payload problems.

[8] I borrow the adjective "unreasonable" from physicist Eugene Wigner (1960), who described the mathematical models used in the physical sciences as *unreasonably effective*.

[9] See Ziliak and McCloskey 2008 for a discussion of the ability of social science models to explain variation.

[10] See Porter and Smith 2007 for a history on the spectrum auction.

[11] See Squicciarini and Voigtländer 2015. See Mokyr 2002 for a full historical account of the importance of knowledge transfer.

[12] See www.treasury.gov/initiatives/financial-stability/Pages/default.aspx.

[13] For example, during the mid-1990s, about 60% of the restaurants that opened in Columbus, Ohio, failed. None received a government bailout, nor should they have. A healthy market economy includes failures. See Parsa et al. 2005.

[14] Taken from the IMF's 2009 *Global Financial Stability Report*. The strength of a connection was based on the correlation in portfolio values. That correlation was based on extreme events; days on which the institutions performed particularly well or particularly poorly. This measure was thought to capture the likelihood of one failure spreading to another. In point of fact, the correlations in performance could result from similarities in investment portfolios as well as one bank holding assets in another.

[15] See Geithner 2014.

[16] See Weisberg 2012 for a description of the San Francisco Bay model and its usefulness in policy.

[17] See Stone et al. 2014 for a full account.

[18] I thank Josh Epstein for the first example.

[19] See Dunne 1999 and Raby 2001.

Chapter 3: The Science of Many Models

[1] Levins 1966.

[2] See Page 2007, 2017 for more detailed descriptions and derivation.

[3] See Suroweicki 2006 on the wisdom of crowds; Tetlock 2005 on how foxes outperform hedgehogs; Kalyvas 1999 on the failure of political science to predict the fall of the Soviet Union; and Patel et al. 2011 on ensemble methods in computer science.

[4] Hong and Page 2009 show that independent models require a unique set of categorizations. That is, there exists only one way to create a set of independent predictions in a binary category model.

[5] See three of my earlier books—*The Difference* (Page 2008), *Diversity and Complexity* (Page 2010), and *The Diversity Bonus* (Page 2017)—for elaboration on the diversity prediction theorem. For data on economic predictions, see Mannes, Soll, and Larrick 2014.

[6] Consider the four bungalows A, B, C, and D in the figure below, along with their market values. Create two categories based on whether or not the bungalow contains a recording studio denoted by a circle above the door). Bungalows A and B do not contain recording studios, so they belong to one category, while bungalows C and D do contain studios, so they belong to the second category.

Four Bungalows and Their Market Values

We first compute the *total variation* in the bungalows' prices. This equals the sum of the squared differences from each value to the mean. We make all calculations in units of $1000. The mean value of the four bungalows equals 400, so total variation equals 100,000: Total Variation $= (200-400)^2 + (300-400)^2 + (500-400)^2 + (600-400)^2 = 100,000$.

To calculate the *categorization loss*, we assume that we know the true mean within each category: $250,000 for the first category and $550,000 for the second category (the bungalows that have been turned into recording studios). The categorization lumps together houses of different values. That remaining variation equals the **categorization loss**: Categorization Loss A & B $= (200-250)^2 + (300-250)^2 = 5,000$ and Categorization Loss C & D $= (500-550)^2 + (600-550)^2 = 5,000$. The total categorization loss equals the sum of these two numbers, or 10,000.

To compute *valuation error*, assume that the model predicts prices of $300,000 for bungalows A and B and $600,000 for bungalows C and D. Valuation error equals the squared differences between the predictions for each category and the true mean. Valuation error A & B $= (300-250)^2 + (300-250)^2 = 5,000$ and Valuation Error C & D $= (600-550)^2 + (600-550)^2 = 5,000$. The total valuation error equals 10,000.

Total *model error* equals the squared differences between the predictions and the actual values: Model Error $= (200-300)^2 + (300-300)^2 + (500-600)^2 + (600-600)^2 = 20,000$. Notice that the model error equals the sum of categorization loss and valuation error.

[7] See Brock and Durlauf 2001 for a spin-glass model of social interactions in two dimensions. Glaeser, Sacerdote, and Schenkman 1996 have a one-dimensional model used to examine crime. Fudenberg and Levine 2006 construct an economic model of the brain.

[8] Niarchos was not the first shipper to attempt to exploit scale. In 1858, Isambard Kingdom Brunel, the legendary British engineer who built the Great Western Railway, launched a nearly 700-foot-long ship, the SS *Great Eastern*. That ship proved a failure. A lack of hydrodynamic models resulted in a poor overall design. The boat proved seaworthy at only the slowest of speeds. It finally found use laying transatlantic cables. See West 2017 for how multiple models aid in designing ships.

[9] BMI can also be written as 703 times weight (in pounds) divided by height (in inches) squared.

[10] LeBron James stands 6 feet 8 inches and weighs around 250 pounds, giving him a BMI of 27.5. Kevin Durant, who stands 6 feet 9 inches and weighs 235 pounds, has a BMI of 25.2. Aston Eaton, the 2012 and 2016 Olympic decathlon gold medalist, measures 6 feet 1 inch and weighs 185 pounds, resulting in a BMI of 24.4, on the cusp of being overweight. His predecessor, 2008 Olympic decathlon gold medalist Brian Clay, had a BMI of 25.8.

[11] See Flegal et al. 2012.

[12] We assume a mouse 3 inches long, 1 inch high, and 1 inch wide, and an elephant 10 feet tall, 10 feet long, and 5 feet wide. The elephant has a surface area of 400 square feet, or 57,600 square inches. The elephant has a volume of 500 cubic feet, which equals 864,000 cubic inches.

[13] Geoffrey West and colleagues have constructed more elaborate and accurate models that predict that metabolism should scale with mass raised to the three-quarters power. See West 2017.

[14] Controlled experiments that send identical resumes but change the name demonstrate that women receive lower salary offers and lower evaluations than men (Moss-Racusina et al 2012).

[15] The probability of a man becoming CEO equals the probability he receives fifteen promotions in a row, or P_M^{15}. The probability that a man becomes a CEO relative to a woman, the *likelihood ratio*, equals $\left(\frac{P_M}{P_W}\right)^{15}$. Given our assumptions of 50% and 40%, the likelihood ratio equals $(1.25)^{15} = 28.4$.

[16] See Dyson 2004.

[17] See Breiman 1996.

Chapter 4: Modeling Human Actors

[1] See Haidt 2006.

[2] Assume the individual has a budget of M, that the price of one unit of the consumption good, C, is \$1, and that the price of one unit of housing equals P_H. We can write her budget constraint as $M = C + P_H \cdot H$. This implies $C = M - P_H \cdot H$, so that we can then write her utility as a function of H.

$$U(H) = (M - P_H \cdot H)^{\frac{2}{3}} H^{\frac{1}{3}}$$

To find the H that maximizes utility, we take the derivative with respect to H and set it equal to zero. This requires applying the chain rule for derivatives.

$$-\frac{2}{3} P_H \left(\frac{H}{M - P_H \cdot H}\right)^{\frac{1}{3}} + \frac{1}{3} \left(\frac{M - P_H \cdot H}{H}\right)^{\frac{2}{3}} = 0$$

Moving the first term to the other side of the equation and then cross multiplying gives the following:

$$2P_H \cdot H = (M - P_H \cdot H)$$

Substituting $2P_H \cdot H$ for $(M - P_H \cdot H)$ in the budget constraint gives $M = 2P_H \cdot H + P_H \cdot H$, or that $M = 3P_H \cdot H$. It follows that the individual spends $\frac{1}{3}$ of her income on housing.

[3] In the United States, this is a close approximation. Source: US Bureau of Labor Statistics 2013.

[4] The formal theorem is written as follows: Let $X = \{A, B, C, ..., N\}$ denote a finite set of **outcomes**, and let a **lottery** be a probability distribution over outcomes: $L = (p_A, p_B, ..., p_N)$. If preferences ($\succ$) over lotteries satisfy **completeness**: any two lotteries, L and M, can be compared; **transitivity**: if $L \succ M$, and $M \succ N$, then $L \succ N$; **independence**: if $L \succ M$, then given any lottery N and any probability $p > 0$, the lottery $pL + (1-p)N \succ pM + (1-p)N$; and **continuity**: if $L \succ M$ and $M \succ N$, then there exists a probability p such that $pL + (1-p)N \sim M$, then preferences can be represented by a continuous utility function that assigns a real number, that is a utility, to each lottery. A sketch of the proof goes as follows: Assume that there exists a best outcome, B, and a worst outcome, W. Set the utility of B equal to 1 and the utility of W to zero. Given any other outcome A, by the *continuity axiom*, there exists a probability p that makes a person indifferent between receiving A for sure and receiving B with probability p and W with probability $(1-p)$. We write this as $A \sim pB + (1-p)W$. We then assign a utility of A equal to p. A little introspection (and a bit of math) shows that the more a person likes an outcome (or lottery), the larger p will be. And almost by magic, we have turned rankings into numbers. For a complete proof, see Von Neumann and Morgenstern 1953.

[5] See Rust 1987.

[6] See Camerer 2003.

[7] See Harstad and Selten 2013.

[8] See Myerson1999 on the use of rational choice as a benchmark.

[9] See Camerer, Loewenstein, and Prelec 2005 for an early survey.

[10] See Kahneman 2011 for an overview of this research.

[11] The original paper by the Open Science Collaboration (2015) has led to more attempts at replication showing similar percentages.

[12] See Medin, Bennis, and Chandler 2010 for the need for greater diversity in subject pools.

[13] See Berg and Gigerenzer 2010. They give full voice to this line of criticism, arguing that it undermines mathematically based psychological models.

[14] Kahneman and Tversky 1979.

[15] *Gain frame:* Treatment A will save 40% of the patients for certain. Treatment B has a 50% chance of saving everyone. *Loss frame:* Treatment A′ will cause 60% of the patients to die for sure. Under treatment B′, there is a 50% chance that no one dies and a 50% chance that everyone dies. According to prospect theory, most doctors choose treatment A in the gain frame and B′ in the loss frame.

[16] See Thaler 1981 and Laibson 1997 for early papers on the implications of hyperbolic discounting.

[17] The formula for hyperbolic discounting can be written in the more general form

$$H(r,t) = \frac{1}{(1+rt)^{\frac{\beta}{\alpha}}}$$

[18] See Gigerenzer and Selten 2002.

[19] See Gode and Sunder 1993.

[20] See Gigerenzer and Selten 2002.

[21] In his Nobel Prize lecture, Vernon Smith notes that "Ecological rationality uses reason—rational reconstruction—to examine the behavior of individuals based on their experience and folk knowledge. People follow rules without being able to articulate them, but they can be discovered." See Smith 2002.

[22] See Arthur 1994.

[23] See Lucas 1976 and Campbell 1976.

[24] See de Marchi 2005 for a discussion of how models illuminate what could happen. See Gilboa and Schmiedler 1995 and Bednar and Page 2007, 2018 for equation- and rule-based models in which actors apply behaviors in similar games.

Chapter 5: Normal Distributions: The Bell Curve

[1] Given a set of data $\{x_i, ..., x_N\}$, the variance equals the average squared distance to the mean, μ, which is written as:

$$VAR = \frac{\sum_{i=1}^{N}(x_i - \mu)^2}{N}$$

The standard deviation equals the square root of the average squared distance to the mean, μ:

$$s.d. = \sqrt{\frac{\sum_{i=1}^{N}(x_i - \mu)^2}{N}}$$

[2] Any of several conditions are sufficient. One of the more common, the *Lindeberg condition*, requires that the proportion of the total variation that comes from any one variable will converge to zero as the number of variables grows large.

[3] See Lango et al. 2010.

[4] For the general case, assuming independent random variables, we have the following expressions:

$$\bar{\mu} = \frac{\sum_{i=1}^{N}\mu_i}{N}$$

$$\bar{\sigma} = \sqrt{\frac{\sum_{i=1}^{N}\sigma_i^2}{N^2}}$$

Setting $\sigma_i = \sigma$ for all i gives $\bar{\sigma} = \sqrt{\frac{N\sigma^2}{N^2}} = \frac{\sigma}{\sqrt{N}}$.

[5] Wainer 2009 provides a more thorough analysis of the policy choices involved.

[6] The threshold of two standard deviations (5% significance) is a convention open to criticism, but it is what social scientists generally use. A large coefficient significant at the 6% level is likely more worth noting than a small coefficient at 4.9% significance. See Ziliak and McCloskey 2008.

[7] See Gawande 2009.

[8] Distributions of products of random variables are called lognormal because the logarithm of the distribution will be normally distributed. A sketch of the logic as to why goes as follows: First, we write a product of numbers, $y = x_1 \cdot x_2 \cdot x_3 \cdots x_n$, as a product of terms written as powers of 10:

$$10^{log_{10}(y)} = 10^{log_{10}(x_1)} \cdot 10^{log_{10}(x_2)} \cdot 10^{log_{10}(x_3)} \cdots 10^{log_{10}(x_n)} = 10^{log_{10}(x_1)+log_{10}(x_2)+log_{10}(x_3)+\cdots+log_{10}(x_n)}$$

We then take the logarithm base 10 of both sides to obtain the following:

$$log_{10}(y) = log_{10}(x_1) + log_{10}(x_2) + log_{10}(x_3) + \cdots + log_{10}(x_n)$$

Thus, the logarithm of the variable y can be written as the sum of the logarithms of random variables. The logarithms of those random variables are also random variables, and so long as their variances satisfy the conditions of the central limit theorem, their sum, which is $log_{10}(y)$, will be normally distributed.

[9] See Limpert, Stahel, and Abbt 2001.

[10] This idea can be traced back to Gibrat 1931.

Chapter 6: Power-Law Distributions: Long Tails

[1] See Parrish 2017 for an account of the impact and cultural meanings of this flood.

[2] This numerical example assumes an exponent of 2 and is borrowed from Clauset, Young, and Gleditsch 2007.

[3] For a technical description of the models presented in this chapter as well as references to many examples of power laws, see Newman 2005.

[4] See Newman 2005 and Piantadosi 2014 for surveys.

[5] The constant C makes the total probability over all outcomes equal to 1. Given this definition, power-law distributions satisfy *scale invariance*. If we change the units with which we measure outcomes, the shape of the distribution does not change.

[6] These probabilities can be calculated by first solving for the probability of the event not happening within a year. If an event has probability $\frac{1}{1000}$, the probability of that event not happening within a year equals $(0.999)^{365} = 0.69$. So the probability of that event happening equals 31%. The probability of a one-in-a-million event not happening in a century is calculated similarly.

[7] See Cederman 2003; Clauset, Young, and Gleditsch 2007; Roberts and Turcotte 1998. The probability of a terrorist act with x deaths can be written as a constant term of approximately .06 divided by x squared. For a discrete distribution where x only takes on integer values, the power-law distribution can be written $p(x) = 0.608x^{-2}$. The coefficient 0.608 is chosen so that the probabilities sum to 1: $\sum_{n=1}^{\infty} \frac{1}{n^2} = 1.644934$. The product of 0.608 and 1.644934 equals 1.

[8] For the power law, we take logs of both sides and transform $y = Cx^{-a}$ into $\log(y) = \log(C) - a\log(x)$, a linear equation for $\log(y)$ in terms of $\log(x)$. When we plot the values of $\log(y)$ and $\log(x)$ we obtain a straight line. For an exponential distribution, $y = C \cdot A^{-x}$. If we take logarithms of both sides, we obtain $\log(y) = \log(C) - x\log(A)$, which means that $\log(y)$ is linear in x. That means that $\log(y)$ will decrease rapidly in $\log(x)$, creating a concave graph.

[9] If we take the logarithm of a lognormal distribution, we obtain an equation of the following form: $\log(y) = C - b \cdot \log(x) - \frac{\log(x)^2}{\sigma^2}$, where σ is the natural logarithm of the standard deviation of the lognormal distribution, a proxy for the variance of the distribution. For large σ the contribution of $\log(x)^2$ will be small until $\log(x)$ becomes sufficiently large value to cause a downturn in the graph.

[10] To see how to formally distinguish between a lognormal and a power law see Broido and Clauset (2018), who show that many networks thought to have power-law distributions may not.

[11] See Piantadosi 2014 for a survey of Zipf's law in word frequencies and a range of candidate models. If the event size distribution satisfies a power law, then so too do the ranks. A general proof goes as follows: A power-law distribution with exponent a on the open interval $[1, \infty)$ has the form $p_a(x) = ax^{-a}$. Assume 100 events. Let S_R denote the expected size of the Rth largest event. The probability of an event larger than S_R must equal $\frac{R}{100}$. For example, if $R = 3$, then the probability of an event larger than S_3 must equal 3%. Therefore,

$$\int_{S_R}^{\infty} ax^{-a} = \frac{R}{100}$$

Solving gives $S_R^{-(a-1)} = \frac{R}{100}$, which can be rewritten as:

$$S_R = \left(\frac{R}{100}\right)^{\frac{-1}{(a-1)}}$$

In the special case where $a = 2$, this expression becomes $S_R = (\frac{R}{100})^{-1}$.

[12] See Bak 1996. How widely the model applies remains an open question. Scholars have used the concept to explain economic fluctuations, war deaths, terrorist acts, punctuated equilibria in evolution, and traffic jams. See, for example, Paczuski and Nagel 1996, Sneppen et al. 1995.

[13] See Salganik, Dodd, and Watts 2006 for the original study, and Ormerod 2012 for an alternative analysis. A power-law distribution also implies a multitude of small events that constitute a large portion of the probability distribution. These small events can combine to produce economic value of the same magnitude as the large events. See Anderson 2008b. The internet has made it possible for retailers to stock enormous catalogs of books, movies, and music even though some items appeal to small numbers of people. A publisher that sells 5 million copies of a best-seller earns the same revenue as a publisher that sells 500 copies each of 10,000 different books.

[14] For a specific model that shows how this could arise, see Denrell and Liu 2012.

[15] Geologists measure earthquake magnitudes using the Richter scale, which equals the logarithm of the size. An earthquake measuring 6 on the Richter scale is ten times as large as an earthquake measuring 5. See Merriam and Davis 2009 on using Zipf's law to predict earthquake sizes but not their timing.

[16] See Eliot, Golub, and Jackson 2014 for an explicit model of how increasing connectedness can lead to a decreased likelihood of failure.

[17] See May, Levin, and Sugihara 2008 for the full argument.

[18] See Stock and Watson 2003.

[19] The following explanation summarizes Carvalho and Gabaix 2003.

[20] See Clarida, Galí, and Gertler 2000.

[21] I thank Seth Lloyd for bringing this example to my attention.

[22] We let distribution of salaries equal $100,000 times a random variable x with $p(x) = 2x^{-3}$ from 1 to ∞. The variable x has a mean equal to 2, so the salary distribution has a mean value of $200,000.

[23] See Weitzman 1979 for a model showing this result in more generality.

[24] See Bell et al 2018.

Chapter 7: Linear Models

[1] Valuable vintages, such as Bordeaux wines, receive rankings by experts. These wines also get priced in the marketplace. Prices and rankings can function as proxies for quality. Ashenfelter has fit (log) linear models for Bordeaux quality based on the amount of the winter rainfall, harvest rainfall, and average temperature in September. See Ashenfelter 2010. A log linear model expresses the logarithm of the dependent variable as a linear sum of the logarithm of the independent variables:

$$\log(y) = b_0 + b_1\log(x_1) + b_2\log(x_2)$$

This expression implies that the dependent variable can be written as a product of the independent variables. We can see this by making each side of the equation the exponent of *e*, resulting in the following equation:

$$y = e^{b_0}(x_1)^{b_1}(x_2)^{b_2}$$

By taking logarithms, multiplication becomes addition and one can apply the tools of linear regression. Using the price of the vintage as the dependent variable, Ashenfelter's model has an *R-squared* (that is, the percentage of variation that is explained) of 83%. Evidence shows it predicts wine prices more accurately than the judgments of more qualitative wine experts. His model even predicts changes in experts' assessments. Robert Parker, a well-known wine evaluator, initially assigned scores of 95 (out of 100) to the Pomerol and St. Emilion 1975 vintages. Ashenfelter's model predicted lower quality rankings. In 1983, Parker lowered his rankings to below average, as Ashenfelter predicted. See Storchmann 2011.

[2] See Xie 2007.

[3] See Ryall and Bramson (2013) for an introduction to causal models.

[4] Mauboussin 2012 shows how the equation can guide sound managerial decision-making.

[5] See Bertrand and Mullainathan 2001.

[6] See Shapiro, Meschede, and Osoro 2013. They are not conflating correlation and causality here. If two variables are not correlated, then we should not expect a causal relationship.

[7] To find the best line to classify data, many analysts use *support vector machines (SVMs)*, an approach similar to regression. The key difference is that SVMs find the line that maximizes the distance to the closest points in each set that separates the data into positives or negatives. If no such line exists, as will often be the case, penalties are assigned to violations. Regression, by comparison, considers the distance to all of the data points and minimizes total distance.

Chapter 8: Concavity and Convexity

[1] See Arthur 1994.

[2] Thirty doublings equals 2^{30}, which exceeds 1 billion.

[3] See Karlsson (2016) for contextualization of Escobar's hippos.

[4] See Ebbinghaus 1885.

[5] Brain researchers have found that even with chocolate, there exists an amount of consumption at which people begin to form an aversion (Small et al. 2001).

[6] This, like many examples, I borrow from Lave and March 1975.

[7] Suppose that you invest $3,000 a year. If the stock were a constant $15, then you could buy 200 shares each year. If the price alternated between $20 and $10, then in the high-price years you could buy 150 shares, and in the low-price years you could buy 300 shares. On average, you could buy 225 shares, which is more than you could buy with constant prices.

[8] The key assumption is that the exponents in the Cobb-Douglas production function, a and $(1-a)$, sum to 1. This implies that if we double the number of workers and the amount of capital, total output also doubles.

$$\text{Output} = \text{Constant} \cdot (2 \cdot \text{Workers})^a (2 \cdot \text{Capital})^{(1-a)}$$

Expanding terms and some rearranging gives a doubling of output:

$$\text{Output} = 2 \cdot \text{Constant} \cdot \text{Workers}^a \text{Capital}^{(1-a)}$$

[9] Output per day in the second year equals $100\sqrt{2} = 141$. In the third year, it equals $100\sqrt{3} = 173$. The growth rate equals the percentage increase in output from year to year.

[10] The calculations are as follows: **Year 2:** machines: 290, output: 1,702 ($= 100\sqrt{290}$). Investment $= (0.2) \cdot 1702 = 340$, so consumption $= 1362$. Depreciation $= (0.1) \cdot 290 = 29$. **Year 3:** machines: 601, output: 2,453 ($= 100\sqrt{601}$). Investment $= (0.2) \cdot 2453 = 491$, so consumption $= 1962$. Depreciation $= (0.1) \cdot 601 = 60$.

[11] The long-run equilibrium can be solved for by solving for M^*, such that $0.2 \cdot 100\sqrt{M^*} = 0.1M^*$. This occurs at $M^* = 40,000$.

[12] The full Solow model replaces the square root functions with a parameter a, as in the Cobb-Douglas model, and also includes a labor market.

[13] To solve for the equilibrium, we set investment equal to depreciation: $s \cdot A\sqrt{L}\sqrt{K^*} = d \cdot K^*$. Therefore, the equilibrium number of machines K^* satisfies the equation $\sqrt{K^*} = A\frac{s}{d}\sqrt{L}$. Placing this back into the output function, output equals $A^2 L \frac{s}{d}$.

[14] Gordon (2016) takes the pessimistic view that new technologies on the horizon will produce a large increase in A. More advanced growth models make technology a function of other variables. A model by Paul Romer (1986) assumes that growth arises from an increased variety of goods: as the economy grows, so does variety. Weitzman (1998) explicitly models the generation and recombination of ideas.

[15] See Arthur 2011 on lags in technology.

[16] For example, countries that never connected rural villages with telephone lines were able to build radio towers and provide cell phone service. Gerschenkron (1952) refers to this as the *advantage of backwardness*.

[17] Easterly and Fischer 1995.

[18] Piketty 2014 shows that world GDP has grown only 1.6% on average from 1700 to 2012 and that half of that growth is due to population growth. Applying the rule of 72 to an 0.8% growth rate, we find that over a 300-year period average standards of living increase about 10-fold.

Chapter 9: Models of Value and Power

[1] This value is found by multiplying $\frac{1}{6}$ the probability of the rower adding value times the rower's expected added value, which equals $\frac{4}{5} \cdot 10$ plus $\frac{1}{5} \cdot 2$. Note that the Shapley values sum to 10, the total value of the game.

[2] Formal calculations are as follows: Of Arun's six ideas, two are unique, one is also proposed by Betty, and three are proposed by everyone. If he enters first, which occurs with probability $\frac{1}{3}$, he adds value 6. If he enters second, he adds two unique ideas, none of the ideas shared by all three, and has a one-in-two chance of arriving before Betty and adding one more idea. So, he adds 2.5 ideas. If he arrives third, he adds his two unique ideas. Thus, his Shapley value equals 3.5. Betty proposes four ideas shared with one other person and three proposed by everyone. Therefore, her Shapley value equals 3. Finally, Carlos proposes three ideas proposed by one other person along with the three proposed by everyone. His Shapley value equals 2.5. Note that the Shapley values sum to 9, the total number of ideas.

[3] An alternative measure of voting power, the *Banzhaf-Penrose index*, counts the total number of parties that can be pivotal given all possible winning coalitions and assigns a value to each party equal to the number of times it is pivotal divided by this number. See Banzhaf 1965.

[4] See Groseclose and Snyder 1996 for the formal analysis.

Chapter 10: Network Models

[1] See Newman 2010 for comprehensive treatment of networks; see Jackson 2008 and Tassier 2013 for network effects in social science.

[2] Given any node, its minimal paths have length 1 to 4 nodes, minimal paths of length 2 to 4 nodes, and minimal paths of length 3 to 4 nodes, producing a total of 12 nodes on the minimal paths from

each node. On average, the minimal paths from a node visit one other node. It follows that average betweenness for each node equals $\frac{1}{12}$. By symmetry, all nodes must have the same betweenness.

[3] See Newman 2010 for an overview of community detection algorithms. The partitions constructed by these algorithms likely differ because of the number of possibilities. A network with 100 nodes can be partitioned over 190 million ways. Due to randomness in the order in which edges are removed, the same algorithm will often produce different partitions. By applying multiple algorithms and applying each one multiple times, we increase the robustness of our inferences.

[4] From the central limit theorem, we know that the degrees will be normally distributed and that the mean will be $\frac{2E}{N}$, because each edge connects two nodes.

[5] Watts and Strogatz 1998.

[6] See Newman 2010 for formal analysis of network formation models.

[7] Ugander et al. 2011.

[8] Given a network with N people, let d_i equal the number of neighbors of node i, that is, the degree. The average degree, $\bar{d}$, can be written as follows:

$$\bar{d} = \frac{\sum_{i=1}^{N} d_i}{N}$$

The average degree, $\bar{d}$, equals the expected number of neighbors of a node. When counting the average number of neighbors of neighbors, a node with degree d_i will be counted d_i times, once for each neighbor. Therefore, the total number of neighbors of neighbors N_2 of a node can be expressed as follows:

$$N_2 = \sum_{i=1}^{N} d_i^2$$

To obtain the average degree of neighbors of a node we have to divide this by the total number of neighbors, which equals $N\bar{d}$. Therefore, it suffices to show

$$\frac{\sum_{i=1}^{N} d_i^2}{N\bar{d}} \geq \bar{d}$$

This can be rewritten as

$$\frac{\sum_{i=1}^{N} d_i^2}{N} - \bar{d}^2 \geq 0$$

The term on the left equals the variance in the degree distribution. If any two nodes have different degree, the degree distribution has positive variance, therefore, the average degree of neighbors of a node exceeds the average degree of a node.

[9] See Eom and Jo 2014 for the formal model.

[10] Dodds, Muhamad, and Watts 2003.

[11] Newman 2010, Jackson 2008.

[12] See Granovetter 1973.

[13] The number of friends of degree four is calculated by summing the following eight sets of nodes: $C \cdot R \cdot C \cdot R = 4,000,000$, $C \cdot R \cdot R \cdot C = 4,000,000$, $R \cdot C \cdot R \cdot C = 4,000,000$, $C \cdot R \cdot R \cdot R = 800,000$, $R \cdot C \cdot R \cdot R = 800,000$, $R \cdot C \cdot R \cdot R = 800,000$, $R \cdot R \cdot R \cdot C = 800,000$, and $R \cdot R \cdot R \cdot R = 160,000$.

[14] Albert, Albert, and Nakarado 2004.

[15] See Groysberg 2012. Our model of streaks explains their lack of success as regression to the mean.

[16] See Burt 1995 on the value of filling structural holes.

[17] See Frank et al. 2018 for a survey on the implications of teacher networks.

[18] The coalitions that have nonzero value are $\{A, B\}$, $\{B, C\}$, and $\{A, B, C\}$. The value of each of the first two coalitions equals ten. The added value of the third coalition equals minus six. The coalition's

stand-alone value equals fourteen, but the coalition consists of both other coalitions, each of which has a value of ten. Therefore, the value of the coalition equals fourteen minus twenty: $-6 = (14-10-10)$. We can then assign the following Shapley values for each player in each coalition: Coalition $\{1,2\}$: Player 1: 5, Player 2: 5, Coalition $\{2,3\}$: Player 2: 5, Player 3: 5, Coalition $\{1,2,3\}$: Player 1: -2, Player 2: -2, Player 3: -2. Summing these values produces Myerson values.

Chapter 11: Broadcast, Diffusion, and Contagion

[1] The models assume discrete time steps, like days or weeks, and use difference equations that describe the number of infected (or informed) people tomorrow as a function of the number infected (or informed) today. Continuous time models require differential equations and calculus. None of the results of our models would change qualitatively if we switch to continuous time.

[2] Plugging the first equation into the second gives the following expression: $36,000 = 20,000 + 20,000 - P_{broad} \cdot 20,000$, which reduces to $4,000 = P_{broad} \cdot 20,000$, so $P_{broad} = 0.2$ and $N_{POP} = 100,000$.

[3] See Griliches 1988.

[4] Initial sales, I, equal 100 for each app. First, set $P_{diffuse} = 0.4$ and POP = 1000. New sales in period three equal $0.4 \cdot \frac{100*900}{1000} = 36$. Similar calculations yield sales data for future periods. For the second set of data, let $P_{diffuse} = 0.3$ and POP = $1,000,000$. New sales in the second period equal $0.3 \cdot \frac{100*999,900}{1,000,000} = 30$. Subsequent sales are solved for similarly.

[5] Bass (1969) refers to the people who adopt the technology or buy the product as *innovators* and the people who copy them as *imitators*.

[6] The formal derivation of R_0 begins with the observation that for small numbers of infected people, the number of susceptible people is approximately equal to the size of the relevant population. To reduce the number of variables, we can substitute the number of susceptible people with the relevant population. We can then write the change in the number of infected people as a linear function of the number of initially infected people (see box). R_0 can be derived formally as follows: when a new disease appears, it infects a small number of people. Denote this small amount by I_0. Plugging into the SIR model, the number of infected people in period one equals

$$I_1 = I_0 + P_{contact} \cdot P_{spread} \cdot \frac{I_0}{N_{POP}} \cdot S_0 - P_{recover} I_t$$

Approximate S_0 as N_{POP} and the expression becomes $I_1 = I_0 + P_{contact} \cdot P_{spread} \cdot I_0 - P_{recover} I_0$. It follows that the number of infected increases if and only if $P_{contact} \cdot P_{spread} > P_{recover}$, which is equivalent to

$$\frac{P_{contact} \cdot P_{spread}}{P_{recover}} \geq 1$$

[7] Quarantine cuts the probability of contact to near zero, lowering the basic reproduction number, but has high cost. In the early 1900s, tuberculosis ($R_0 \approx 3$) caused over one hundred thousand deaths per year in the United States. States raised property taxes to build sanitariums to house patients because surgical techniques such as removing a lung and plumage, the collapsing of the lungs and subsequent refilling with ping-pong balls, proved ineffective. See Dubos 1987.

[8] To solve for the vaccination threshold, we must take into account those vaccinated. For the disease to spread, there must be contact (probability $P_{contact}$) with an unvaccinated person (probability $(1-V)$) and the disease must spread (P_{spread}), giving the following period one difference equation:

$$I_1 = I_0 + P_{contact} \cdot P_{spread} \cdot \frac{I_0}{N_{POP}} \cdot (1-V) \cdot S_0 - P_{recover} I_0$$

Using the approximation $S_0 = N_{POP}$, as in the derivation of R_0, this becomes

$$I_1 = I_0 + P_{contact} \cdot P_{spread} \cdot I_0 \cdot (1-V) - P_{recover} I_0$$

The number of infected people increases if and only if $P_{contact} \cdot P_{spread} \cdot (1-V) > P_{recover}$. This can be rewritten as $R_0(1-V) \leq 1$. Expanding and rearranging terms gives $R_0 - 1 < V \cdot R_0$. Dividing both sides by R_0 gives the result.

[9] For an analysis of the SIR model and herd immunity see Tassier 2013.

[10] Stein 2011.

[11] Updike 1960.

[12] Tweedle and Smith 2012.

[13] See Lamberson and Page 2012b.

[14] Data source: wikinoticia.com.

[15] See Christakis and Fowler 2009.

[16] Centola and Macy (2007) refer to diffusion that requires multiple exposures as *complex contagion.*

Chapter 12: Entropy: Modeling Uncertainty

[1] See Smaldino 2013 for further elaboration.

[2] The logarithm base 2 of a number x equals the power to which 2 must be raised to produce x, so $\log_2(4) = 2$ and $\log_2(2^N) = N$. In the general case, $\log_a(x)$ equals the power that a must be raised to in order to equal x. Thus, if $a^y = x$, then $\log_a(x) = \log_a(a^y) = y$.

[3] We can write the information entropy in long form as follows:

$$H = -\frac{1}{2} \cdot \log_2\left(\frac{1}{2}\right) - \frac{1}{8} \cdot \log_2\left(\frac{1}{8}\right) - \frac{1}{8} \cdot \log_2\left(\frac{1}{8}\right) - \frac{1}{8} \cdot \log_2\left(\frac{1}{8}\right) - \frac{1}{8} \cdot \log_2\left(\frac{1}{8}\right)$$

This simplifies to

$$\frac{1}{2}(1) + \frac{1}{8}(3) + \frac{1}{8}(3) + \frac{1}{8}(3) + \frac{1}{8}(3) = \frac{1}{2}(1+3) = 2$$

[4] The *diversity index,* the inverse of the sum of the squares of the probabilities, satisfies the first two axioms plus the multiplication axiom. Thus, the diversity index of a known outcome equals 1 and not 0 (Page 2007, 2010a).

[5] See Wolfram 2001 or Page 2010a for a more detailed description.

[6] Alexander lists fifteen such properties in all. His ideas are presented along with beautiful photographs in four self-published books: *The Nature of Order, Book 1: The Phenomenon of Life* (2002), *The Nature of Order, Book 2: The Process of Creating Life* (2002), *The Nature of Order, Book 3: A Vision of a Living World* (2005), and *The Nature of Order, Book 4: The Luminous Ground* (2004). The second in this sequence is the most germane to the current discussion.

Chapter 13: Random Walks

[1] See Mlodinow 2009 for an engaging tour of random walks.

[2] See Taleb 2001.

[3] See Turchin 1998 and Suki and Frey 2017.

[4] Note that the law of large numbers says that the mean proportion converges, whereas the central limit theorem tells us that the distribution over the proportion of white balls will be normal.

[5] A player who makes 46% of his three-pointers has about a $\frac{1}{1000}$ probability of making nine in a row (0.46^9). If that player keeps taking three-pointers, then in a ten-year NBA career (about 800 games), the odds of *not* making nine in a row at least once (0.999^{800}) are about 47%.

For over three decades, statisticians have pondered the question of whether basketball players and other professional athletes exhibit "hot hands," that is, whether the probability of making any one shot or free throw is *not* independent of the success of the previous attempt. See, for example, Chance 2009 for an analysis of Joe DiMaggio's fifty-six game hitting streak. In considering the evidence for hot hands, we must take behavior into account. If a player believes he has a hot hand, he may attempt more difficult shots. Further, if the defense thinks the player has a hot hand, they can tighten their defense. These behavioral responses can be accounted for by coding shots by difficulty. Gilovich, Tversky, and Vallone (1985) found no evidence of hot hands. Miller and Sanjurjo (2015) discovered an inference error in previous calculations of conditional probabilities revealing that previous studies that purported to not show a hot hand actually support the hot hand hypothesis. The error in previous analysis stemmed from the sampling technique. Those studies gathered sequences of made and missed shots for multiple players. They then calculated the probability of a basket following a randomly selected sequence of baskets. This sampling procedure introduces a subtle statistical bias that can be seen by examining a situation in which many players each take exactly four shots and each shot has an equal likelihood of being made or missed. There exist sixteen possible sequences of makes and misses. Let B represent a basket and M a missed attempt. Of those sixteen sequences, six include two consecutive makes followed by another shot: $BBBB$, $BBBM$, $MBBB$, $BBMB$, $BBMM$, and $MBBM$. These comprise the sample of cases for two consecutive baskets followed by a third shot. If $BBBB$ is drawn, then regardless of which sequence of two B's is chosen, the probability of a basket equals 100%. If $MBBB$ is chosen, then the probability of a B following a BB also equals 100%. If $BBBM$ is chosen, the probability of an M after BB equals 50%. Last, if $BBMB$, $BBMM$, or $MBBM$ occurs, then an M necessarily follows a B. Averaging across the six cases gives the conditional probability of an M following a B as

$$p(B \mid BB) = \frac{1}{6}\left[1 + 1 + \frac{1}{2} + 0 + 0 + 0\right] = \frac{5}{12}$$

The bias arises because there are two BB's that could be chosen in the sequence $BBBB$ but only one in the other sequences such $BBMB$. The sampling procedure makes each of the two sequences in $BBBB$ half as likely to be chosen as the single sequence in $BBMB$. The implication of this bias is that if there were no hot hands, the sampling procedure would have shown misses more likely following made baskets. The fact that it did not means that made baskets were in fact more likely following made baskets.

[6] Madoff announced 1.5% positive monthly returns to his clients for decades. He claimed that his investments rose each month regardless of changes in the broader market. Posting positive returns during a downturn is more difficult than outperforming the market. When the market falls, a person could beat the market yet still post negative returns. Madoff posted positive returns when the broader market fell in more than eighty months. If we make the heroic assumption that Madoff was somehow able to post positive returns even when the broader market fell three-fourths of the time, his odds of succeeding eighty periods in a row (0.75^{80}) are about 1 in 10 billion.

[7] We can solve for the standard deviation by recognizing that the value of a random walk is the sum of identical, independent random variables. Each of those random variables has a mean of zero and takes value either +1 or -1. Therefore, each has a standard deviation equal to 1. Setting $\sigma = 1$ and applying the sigma square root formula for a sum gives the result.

[8] See Newman 2005 for a formal proof.

[9] See Levinthal 1991 and Axtell 2001.

[10] See Newman 2005 and Sneppen et al. 1995.

[11] Lakes are measured by surface area. Our model produces a power-law diameter. Surface area equals a constant times the diameter squared, so area would also be power-law distributed. See Downing et al. 2006 for data.

[12] On a balanced roulette wheel, all pockets have equal likelihood. If the table has any slant, then the ball will more likely fall off the outer edge as it heads uphill. For an account of how J. Doyne Farmer, Norman Packard, and friends constructed a wearable computer to exploit this phenomenon and beat roulette, see http://en.wikipedia.org/wiki/Eudaemons.

[13] After N bets, the expected value of this random walk equals $N \cdot \left(\frac{18}{38} - \frac{20}{38}\right) = \frac{-N}{19}$. Given that the probability of winning is approximately $\frac{1}{2}$, we can write the standard deviation of the value as $\sqrt{N}$. The

exact value equals $2\sqrt{\frac{18}{38} \cdot \frac{20}{38}N} = 0.9986\sqrt{N}$.

[14] Peel and Clauset (2015) model each game as a single sequence and find that the sequences of scores exhibit *anti-persistence*: the team that scored last is less likely to score next. Given that the teams alternate possessions, this should be expected.

[15] Baxter 2009.

[16] A proof of this result relies on computing the probability of returning to the starting point in N steps and summing this up over all possible N. For a complete proof see:

http://www.math.cornell.edu/~mec/Winter2009/Thompson/randomwalks.html.

[17] See Samuelson 1965.

[18] See Grossman and Stiglitz 1980.

[19] See Lo and MacKinlay 2007. Shiller 2005 shows that stocks with a relatively low price-to-earnings ratio outperform the market.

[20] See Mauboussin 2012.

[21] This window from 1967 to 2017 happens to begin at a peak in the S&P 500. Over most windows, the stock prices grow faster than the economy.

Chapter 14: Path Dependence

[1] Hathaway 2001.

[2] Pierson 2004.

[3] Bednar and Page 2018.

[4] See Page 2006.

[5] Here is a sketch from Page 2006. It suffices to show that the probability of drawing K white balls in the first N periods equals $\frac{1}{(N+1)}$. There exist $(N+1)$ possibilities because K could equal zero up to N. The probability of a given sequence of K white balls in N periods can be written as the product of N fractions. The denominators of those fractions are the numbers 2 through $N+1$. The numerators are the numbers 1 through K (white balls) and 1 through $(N-K)$ (gray balls). The product of the numerators equals $K!$ times $(N-K)!$ and the product of the denominators equals $(N+1)!$. That calculation gives the probability of a particular sequence of K white balls. The number of possible ways to arrange K balls across the N periods equals $\binom{N}{K} = \frac{N!}{K!(N-K)!}$. Therefore, the total probability of exactly K white balls equals

$$\frac{K!(N-K)!}{(N+1)!} \cdot \frac{N!}{K!(N-K)!} = \frac{1}{(N+1)}$$

[6] The proof is by contradiction. Suppose the result does not hold and that in the long run, 60% of the outcomes are white. It follows that 60% of the balls in the urn will be gray. But this would imply that 60% of the outcomes will be gray, a contradiction.

[7] See Lamberson and Page 2012b for a more elaborate model that relies on entropy as a measure of uncertainty.

[8] See Lamberson and Page 2012a.

[9] See Page 1997 for how to make decisions on public projects with positive externalities.

[10] VaR can also be calculated as the probability of losing more than $10,000 at any point during the year.

[11] These calculations follow from the fact that the standard deviation of the value of a random walk of length N equals $\sqrt{N}$; 2.5% corresponds to two standard deviations.

Chapter 15: Local Interaction Models

[1] In physics, the local majority model is known as the *Ising model*. The local majority model is a variant of the *voter model*, which assumes randomly selected neighbors of various sizes. See Castellano, Fortunato, and Loreto 2009.

[2] For the cells along the four edges, we connect the top edge to the bottom edge, creating a cylinder, and then connect the left edge to the right edge to create a torus (a donut).

[3] In alternative versions of the local majority model, cells can be activated simultaneously or according to the incentive to update, with the cells with the largest majority of neighbors in the opposite state moving first. If the cells update simultaneously, then the local majority model can produce cycles.

[4] See Miller and Page 2004 for a model of standing ovations.

[5] See Bednar et al. 2010 for a model of culture that includes consistency of actions across domains.

[6] Here too, attaching the top to the bottom to first create a cylinder and then attaching the ends to create a skinny donut.

[7] A class of models called *diffusion reaction models* also produce stripes on narrow shapes and splotches on wider shapes. Using these models, scientists can predict which animals will have stripes, which will have splotches, and which will be a solid color. The answer depends on the size of the mammal's embryo during the developmental stage when the patterns form and not on the animal's adult size. Otherwise, elephants would have splotches. See Murray 1988.

[8] I thank Bernardo Huberman for this imagery.

[9] Proving this involves writing a computer program that generates a random sequence of numbers or a random pattern and showing that the Game of Life mimics the computer program. The formal proof requires showing that the Game of Life is *universal* in the set of cellular automaton. See Berlekamp, Conway, and Guy 1982.

[10] See Dennett 1991 and Hawking and Mlodinow 2011.

Chapter 16: Lyapunov Functions and Equilibria

[1] See Nagel 1995 for the original experiments.

[2] I thank Jenna Bednar for the example of races to the bottom within federal systems as well as several others scattered throughout the text.

[3] See Page 2001 for a formal proof.

[4] The presence of externalities, even negative ones, need not preclude the construction of a Lyapunov function. Both the local majority model and the route selection model contain negative externalities. In the local majority model, when a cell changes its state it imposes a negative externality on its neighbors who are in the opposite state. However, the positive externality that it creates with the neighbors it now matches is larger.

[5] Guy 1983. If you are daring, start at 27.

Chapter 17: Markov Models

[1] A research article would use statistical techniques to estimate transition probabilities more exactly and calculate error ranges. It would also test to see if the transition rates remain fixed over the time period. They would not if transition probabilities depend on per capita income. See Przeworski et al. 2000.

[2] Flores and Nooruddin 2016.

[3] See Tilly 1998.

[4] The equilibrium percentage of people who have tile floors equals the probability of moving from linoleum to tile divided by the sum of the probabilities of moving from linoleum to tile or tile to linoleum. This equals $\frac{\frac{1}{4} \cdot \frac{1}{10}}{\frac{1}{4} \cdot \frac{1}{10} + \frac{3}{4} \cdot \frac{1}{60}} = \frac{\frac{1}{40}}{\frac{3}{80}} = \frac{2}{3}$. A general model can be written as follows: Let D denote the percentage of people who own a costly durable good and let C denote those people who own a cheaper good. Let BUY(C) $> \frac{1}{2}$ denote the probability that someone buys the cheaper good. Let REPLACE(C) and REPLACE(D) denote the probabilities of replacing the two types of products. If the following inequality holds:

$$\text{REPLACE}(C) \cdot (1 - \text{BUY}(C)) > \text{REPLACE}(D) \cdot \text{BUY}(C)$$

then more people buy the cheaper good but more people own the durable good. The first part of the claim holds by assumption. To prove the second part, we first solve for the transition probabilities. The probability that someone moves from D to C is $P(D, C) = \text{REPLACE}(D) \cdot \text{BUY}(C)$. The probability that someone moves from C to D is $P(C, D) = \text{REPLACE}(C) \cdot (1 - \text{BUY}(C))$. In equilibrium, $D - D \cdot P(D,C) + CP(C,D) = D$. Setting $C = (1 - D)$, we have $D = \frac{P(C,D)}{P(C,D)+P(D,C)}$, which is greater than 0.5 if $P(C, D) > P(D, C)$. That inequality is equivalent to $\text{REPLACE}(C) \cdot (1 - \text{BUY}(C)) > \text{REPLACE}(D) \cdot \text{BUY}(C)$.

[5] See McPhee 1963 and Ehrenberg 1969 for empirical evidence of double jeopardy. To prove the result, we need only show that consumers are equally likely to buy either product when switching brands.

[6] See Briggs and Sculpher 1998 for an overview.

[7] See Schrodt 1998.

[8] Khmelev and Tweedie 2001.

[9] Khmelev and Tweedie 2001.

[10] Reynolds and Saxonhouse 1995.

[11] See http://www.ams.org/samplings/feature-column/fcarc-pagerank.

[12] Skillful construction of such models lies in defining useful states and assigning accurate transition probabilities. See Langville and Meyer 2012.

[13] In a food web, a species links to those species it consumes. See Allesina and Pascual 2009.

[14] See Russakoff 2015.

Chapter 18: Systems Dynamics Models

[1] See Sterman 2000 for a more general introduction.

[2] See Wellman 1990 for an analysis of the value of qualitative systems dynamics models.

[3] The model assumes that all hares die by being eaten by foxes. Adding variables for hare death would complicate the model without changing the results as it would just lower the growth range of hares. The expression $\dot{H}$ denotes the rate of change in H per unit of time, or $\frac{\partial H}{\partial t}$. The equilibrium occurs when the rates of changes in the number of both hares and foxes equal zero, $\dot{H} = \dot{F} = 0$. To solve for the equilibrium take the equation $gH - aFH = 0$ and divide by H. This gives $g - aF = 0$. Solving for F gives the result. Next, take the equation $bFH - dF = 0$, and divide by F. This gives $bH - d - 0$. Solving for H gives the result.

[4] I thank Michael Ryall of the University of Toronto for this example.

[5] The model's findings are summarized in Meadows et al. 1972.

[6] The suite of models are often referred to as the *Club of Rome models*, as the Club of Rome, a group founded by David Rockefeller in 1968, funded reports on the models and promoted the models' findings.

[7] By manipulating more variables with small ranges, he can drive the model's prediction to nearly 30 billion. See Miller 1998.

[8] On the first point, see Hecht 2008. On the second point, see MacKenzie 2012.

[9] See Sterman 2006.

[10] Glantz 2008.

Chapter 19: Threshold Models with Feedbacks

[1] See Granovetter 1978.

[2] Airbnb's founders paid the costs of going door-to-door by selling Obama O's and Cap'n McCain's cereal boxes during the 2008 presidential election.

[3] See Jacobs 1989 for the revolving-door model. Empirical studies find that in jobs that require little formal education, bartending and gardening being examples, men leave (or choose not to enter) professions that include as little as 15% women (Pan 2015).

[4] See Syverson 2007.

[5] See Gammill and Marsh 1988 for a more detailed account.

[6] See Easley et al. 2012.

Chapter 20: Spatial and Hedonic Choice Models

[1] See Clark, Golder, and Golder 2008.

[2] See Martin and Quinn 2002 for analysis of judicial positions.

[3] Hotelling (1929) studied geographic location, Lancaster (1966) expanded Hotelling's model to study *hedonic competition*, and Downs (1957) applied the model to politics.

[4] The *nominate model* offers a more sophisticated approach to assigning ideologies based on this same idea (Poole and Rosenthal 1985).

[5] The constant term (C in the expression) is chosen so as to make all of the payoffs positive. To accomplish this it can be set to the maximal possible distance between any ideal point and an alternative.

[6] In constructing the cut line in this way, we assume that consumers equally weight the two attributes. We could include difference in weights. If people value sweetness more than the amount of cocoa, we would slant the cut line counterclockwise. At one extreme, if people only value sweetness, the cut line would be horizontal and divide A and B evenly on the vertical axis. The spatial model provides a clean example of how we convert an intuition—we prefer things closer to our ideal—into a formal model. Once we plot the alternatives (the chocolate bars) and the consumer's ideal point in space and define a preference ordering over alternatives (a ranking of the alternatives from best to worst) based on their distances from the ideal point, we have, in fact, written down a utility function based on the alternative. That utility of a product equals the inverse of the distance from the ideal point.

[7] Havel 1978.

[8] See Martin and Quinn 2002.

[9] See McCarty 2011. These percentages could change with shifts in levels of party loyalty or if the types of bills put to a vote change.

[10] Formally, if we assume an odd number of voters and the existence of a single voter ideal point at the two-dimensional median, the condition requires that any line through the two-dimensional median divides the remaining voter ideal points in two equal-sized sets. See Plott 1967.

[11] McKelvey (1979) showed that a sequence of elections could result in any policy in two or more dimensions, a finding that some referred to as a "chaos result." McKelvey was careful to state that his result was a statement about sequences of outcomes that were possible given preferences and not a prediction of what would happen in a series of elections. See Kollman, Miller, and Page 1997 for a computational version of the multidimensional spatial model in which candidates move to the center under a variety of behavioral assumptions.

[12] The proposer may have to offer 41.

[13] See McCarty and Meirowitz 2014 for a deeper analysis of this model and other models from game theory applied to politics.

[14] See Tsebelis 2002 on the effect of veto players more generally.

[15] A study of housing prices in Los Angeles estimated the cost of commuting at around \$28 per hour (Bajari and Kahn 2008).

[16] This can be calculated as follows: Initial revenue equals price times quantity, $p \cdot q$. In the first market, after the price drop, revenue falls by 10% and sales increase by 8% , so revenue equals

$$0.9p \cdot 1.08q = 0.972p \cdot q$$

In the second, more crowded market, quantity increases by 33%, so total revenue equals:

$$0.9p \cdot 1.33q = 1.197p \cdot q$$

Chapter 21: Game Theory Models Times Three

[1] The formal model and proof can be written as follows: Let E_i equal the effort level of player i. The payoff to player i equals $M - E_i$ if she wins and $-E_i$ otherwise. Assume that the probability that player i wins equals her proportion of total effort:

$$\text{Prob}(i \text{ wins}) = \frac{E_i}{\sum_{j=1}^{N} E_j}$$

To solve for the unique symmetric Nash equilibrium, we consider the effort of player i assuming that all of the other players choose a common effort level E^*. Player i's payoff equals

$$\frac{E_i \cdot M}{E_i + (N-1)E^*} - E_i$$

The first derivative of this payoff function equals

$$\frac{(N-1)E^* \cdot M}{E_i^2 + 2(N-1)E^*E_i + (N-1)^2(E^*)^2} - 1$$

To find a maximum, we set the first derivative equal to zero, giving $(N-1)E^* \cdot M = E_i^2 + 2(N-1)E^*E_i + (N-1)^2(E^*)^2$. In the symmetric equilibrium $E_i = E^*$. Substituting in gives the result. To prove that the first derivative gives a maximum, we need only check that the second derivative of the payoff function is negative.

[2] See Shalizi and Thomas 2011 on how to tease out network effects and on the difficulty of making claims about network effects with snapshot data. See Christakis and Fowler 2009 for many examples of clustered behaviors and attributes.

Chapter 22: Models of Cooperation

[1] See *Science* magazine's special 125th-anniversary edition, published in 2005.

[2] See Martin et al. 2008 and Biernaskie 2011.

[3] See Zaretsky 1998. In the bank example, ATMs would lower profits if the banks had been earning substantial geographic rents, that is, extra profits because they could exploit customers' costs of traveling across town to go to a competing bank. I thank Simon Wilkie for this example, among many others.

[4] Technically, we are showing that Grim Trigger is an equilibrium strategy in the probabilistically repeated Prisoners' Dilemma. Other strategies, such as Tit for Tat, can also be equilibrium strategies when paired with Grim Trigger.

[5] A player who defects against Grim Trigger earns a payoff of T in the first period and can earn at most zero in any future interactions. If the player uses Grim Trigger against Grim Trigger, she earns a payoff of R in each period in which the game is played. The probability of two periods of play is P, the odds of three periods are P^2, and the odds on N periods are P^{N-1}. Therefore, the expected payoff equals:

$$\sum_{t=1}^{\infty} P^{t-1} \cdot R = \frac{R}{1-P}$$

The proof that $(1 + P + P^2 + P^3 + \ldots) = \frac{1}{1-P}$ can be shown as follows: Assume the result is true and then multiply both sides of the equation by $(1-P)$. The right-hand side equals 1. The left-hand side equals $(1 + P + P^2 + P^3 + \ldots) - (P + P^2 + P^3 + \ldots)$, which also equals 1.

[6] We need only redo the calculations for maintaining cooperation, and having P fall to $\hat{P}$. If $\hat{P}$ does not support cooperation, then neither will several periods with a continuation probability P followed by a continuation probability $\hat{P}$.

[7] Nowak and Sigmund (1998) refer to this as *image scoring*. See also Bshary and Grutter 2006.

[8] Both forms of aggression can be mapped to the defect action if one assumes that they would increase the size of the warbler's domain. See Godard 1993.

[9] Four pairs of outcomes merit elaboration. When TFT plays GRIM, both cooperate forever, earning each an average payoff of 3. When TROLL plays itself, both defect in the first two periods and then cooperate forever, for an average payoff of a little less than 2. When GRIM plays TROLL, GRIM cooperates in the first round and TROLL defects; in the second round, both defect; in rounds three and four, GRIM defects and TROLL cooperates; thereafter, both defect. GRIM's sequence of payoffs can be written as 1, 2, 4, 4, and then a long sequence of 2s, for an average of 2^+. TROLL's sequence of payoffs can be written as 4, 3, 1, 1, and then a long sequence of 2s, for an average payoff of exactly 2. When TFT plays TROLL, TFT cooperates in round one, while TROLL defects. Both defect in round two. In round three, TROLL cooperates and TFT continues to defect. In round four, TROLL cooperates for the second time and TFT reverts to cooperation. Thereafter, both cooperate forever. Each receives one payoff of 1, one payoff of 4, one payoff of 2, and a long sequence of 3s, producing an average payoff of a little less than 3.

[10] In fact, both All C and GRIM are *dominated* by TFT. Against any strategy TFT does at least as well or better than either All C or GRIM. TFT dominates All C because TFT cannot be exploited by All D and TROLL. TFT dominates GRIM because it is able to cooperate with TROLL based on the fact that TFT forgives TROLL's defections whereas GRIM does not. In a famous experiment, Robert Axelrod asked scholars to submit strategies for a repeated Prisoners' Dilemma game with a continuation probability. Of the fourteen strategies submitted, TFT performed best. He then asked people to send in new strategies. This time sixty-two people submitted strategies. TFT won the second time as well. Axelrod chalks up the success of TFT to several properties of the strategy: it cooperates, it punishes, and it is also forgiving. GRIM is not forgiving, so GRIM is unable to reboot cooperation with TROLL. See Axelrod 1984. The table does not contain an average payoff against all of the strategies, as that would assume each strategy to be equally likely. One population may contain a majority of people who play TFT. Another may include a large proportion of TROLL. A third may include lots of All D and All C.

[11] Suppose, for example, that the temptation payoff equals four times the sucker payoff, $T = 4S$, and that 5% of the population cooperates. Then P must exceed $\frac{20T-R}{20T}$. Maintaining cooperation requires only that P exceeds $\frac{(T-R)}{T}$. In the case where $T = 4$ and $R = 3$, evolving coordination requires $P \geq \frac{77}{80}$. Maintaining cooperation requires $P \geq \frac{1}{4}$. For the general proof assume that a proportion θ of the population plays TFT (or GRIM) and a proportion $(1-\theta)$ plays All D. Assume that each person plays against the entire population. TFT (or GRIM) playing against TFT earns a payoff of R each period, producing an expected payoff of $\frac{R}{(1-P)}$. TFT playing against All D receives a payoff of $-S$. All D playing against All D receives a payoff of zero. And All D playing against TFT earns a payoff of T. Therefore, the average payoff for TFT equals $\theta \cdot \frac{R}{(1-P)} - (1-\theta) \cdot S$, and All D earns an average payoff of $\theta \cdot T$. TFT outperforms All D if and only if $\frac{R}{1-P} - S \cdot \frac{(1-\theta)}{\theta} \geq T$. Therefore, TFT earns a higher payoff if and only if the following holds:

$$P \geq \frac{T - R + S\frac{(1-\theta)}{\theta}}{T + S\frac{(1-\theta)}{\theta}}$$

If θ is small, then $\frac{(1-\theta)}{\theta}$ will be large and the condition will not be likely to hold. See Boyd 2006 for an analysis of the difficulty of evolving cooperation relative to maintaining it.

[12] We borrow this model and the analysis from Nowak (2006), who shows how repetition, reputation, and kin selection can also support cooperation.

[13] The open node copies the action of the highest-performing neighbor. By assumption, all neighboring defectors earn payoffs equal to zero. The cooperating neighbor earns a payoff of $K \cdot B - D \cdot C$. This exceeds zero if and only if $\frac{B}{C} \geq \frac{D}{K}$.

[14] See Wilson 1975 for the foundations of group selection theory. Wilson has several books that take up group selection in greater detail.

[15] Traulson and Nowak's model works as follows: Divide a population of N individuals into M distinct groups of equal size. Within each group, apply the *cooperative action model* and assign a performance to each individual. Let the probability of choosing individual i equal the performance of i divided by the sum of the performances of all N individuals. A clone of this individual is added to the same group. If the group's size now exceeds a threshold $\bar{S}$, then with probability $(1-q)$, a random individual from that same group is removed, and with probability q, the group splits into two groups with each member randomly placed in one of the groups. To keep the total number of groups constant, one of the existing groups is chosen at random and eliminated. For large M and rare splitting (small q) the number of cooperators increases if and only if the following holds: $\frac{B}{C} \geq 1 + \frac{\bar{S}}{M}$. See Nowak 2006.

[16] When Michelle Peluso, an advocate of agile management, became the chief marketing officer at IBM, she created competing teams whose performance was transparent to other teams. She then rewarded the best teams (Dan 2018). This type of agile management practice borrows ideas from Agile programming, which replaces the standard waterfall approach of sequential construction with simultaneous code writing, testing, and interactions with users.

[17] Another strategy, *Generous Tit for Tat,* cooperates initially and punishes defections only part of the time. In one set of experiments with errors, this strategy outperforms both Tit for Tat and Win Stay, Lose Shift. See Rand et al. 2009 and Wu and Axelrod 1995.

[18] See Axelrod, Axelrod, and Pienta 2006.

Chapter 23: Collective Action Problems

[1] See Hardin 1968 for an introduction to the tragedy of the commons.

[2] See Diamond 2005.

[3] See Ostrom 2005 and Ostrom, Janssen, and Anderies 2007.

[4] To solve for the social optimum, suppose that each person spends X on the public good. The total utility for the population equals

$$N\left[2\sqrt{NX} + \text{INCOME} - X\right]$$

Taking the derivative with respect to X and setting it equal to zero gives

$$N\left[\frac{\sqrt{N}}{\sqrt{X}} - 1\right] = 0$$

Solving gives $X = N$.

To solve for the symmetric Nash equilibrium, we assume that each other person contributes the same amount to the public good; call this amount A. Let Y denote the amount that the individual contributes. Her utility equals

$$2\sqrt{(N-1)A+Y} + \text{INCOME} - Y$$

Taking the derivative with respect to Y and setting it equal to zero gives $\frac{1}{\sqrt{(N-1)A+Y}} = 1$. Rearranging terms and squaring both sides gives $[Y + (N-1)A] = 1$. In the symmetric equilibrium, where everyone contributes the same amount ($Y = A$), then $Y = \frac{1}{N}$.

[5] Utilitarianism weights everyone's lot equally. Rawls (1971) proposed an alternative: the *maxmin principle,* in which the ideal social outcome maximizes the utility of the least well-off person. Rawls advocates evaluating outcomes from behind a *veil of ignorance* so that we do not know whether we will be rich, famous, and endowed with great capacities or hindered by circumstances.

[6] We can write individual j's utility as follows:

$$(1-\alpha) \cdot 2\sqrt{\text{PUBLIC}} + \text{PRIVATE}_j + \alpha \cdot \sum_{i=1}^{N} 2\sqrt{\text{PUBLIC}} + \text{PRIVATE}_i$$

To solve for the symmetric Nash equilibrium, we assume that every other person contributes an amount A to the public good. Let Y denote the amount that individual j contributes. Let I equal the common income level. Individual j's utility equals

$$(1-\alpha)\left(\sqrt{(N-1)A+Y} + I - Y\right) + \alpha\left(N\sqrt{(N-1)A+Y} + (N-1)(I-A) + (I-Y)\right)$$

Taking the derivative with respect to Y and setting it equal to zero gives

$$\frac{2(1-\alpha)}{\sqrt{(N-1)A+Y}} + \frac{2\alpha N}{\sqrt{(N-1)A+Y}} - 1 = 0$$

Rearranging terms gives: $(1-\alpha) + \alpha N = \sqrt{Y+(N-1)A}$. In the symmetric equilibrium, $Y = A$. It follows that $(1-\alpha) + \alpha N = \sqrt{NY}$. Squaring both sides gives $[(1-\alpha) + \alpha N]^2 = NY$, which implies $Y = \frac{[(1-\alpha)+\alpha N]^2}{N}$

[7] See Cornes and Sandler 1996 for a detailed analysis.

[8] A more realistic model would assume a nonlinear congestion cost, perhaps an S-shaped curve. That assumption would capture resources like roads, in which the first few other users have no effect on an individual's benefit and in which at some point the resource is so overcrowded as to be useless.

[9] The total utility from M people using the resource equals $(B - \theta \cdot M)$. Taking the derivative with respect to M and setting it equal to zero equals $(B - 2M\theta) = 0$. Solving gives $M = \frac{B}{2\theta}$. To solve for the Nash equilibrium, we set the value of abstaining equal to zero. People use the resource until the benefit equals the outside option: $M = \frac{B}{\theta}$.

[10] Note: we set the maximal benefit B equal to the population size N to reduce the number of variables. To solve for the socially optimal outcomes and Nash equilibrium, we first note that total utility equals $(N - M) \cdot M + 3(N - (N - M)) \cdot (N - M)$, which reduces to $4(N - M)M$. Taking the derivative with respect to M gives $4N - 8M = 0$. Solving gives $M = \frac{N}{2}$. Total utility equals $4(\frac{N}{2})^2 = N^2$. To solve for the equilibrium, we find an M such that the marginal utilities at the two parks are equal. This occurs when $(N - M) = 3N - 3(N - M)$, which can be rewritten as $N = 4M$. Total utility is calculated by plugging in the values for M and $N - M$ into the utility functions.

[11] To solve for equilibrium consumption, set $R^* = (1 - g)(R^* - C^*)$ and solve for R^*.

[12] See Kurlansky 1998.

[13] To see why the variation does not cancel out, we can consider a two-period model. A growth rate of 20% in the first year results in only 96 units of the resource ($80 \cdot 1.2 = 96$). A growth rate in the second year of 30% results in $98.8 = (96 - 20) \cdot (1.3)$ units of the resource. If we flip the growth rates, then after the first year, there exists 104 units of the resource, and after the second year there exists $100.8 = (104 - 20) \cdot (1.2)$ units of the resource.

[14] Ostrom, Janssen, and Anderies 2007.

[15] See Craine and Dybzinski 2013.

[16] For a short overview see Ostrom 2010; for a more complete account see Ostrom 2004.

Chapter 24: Mechanism Design

[1] See Ledyard, Porter, and Rangel 1997.

[2] For an example of Pareto efficiency, consider the following four payoff profiles for three people:

$$\{(3,3,4), (9,0,0), (0,8,1), (2,2,3)\}$$

All except $(2,2,3)$ are Pareto efficient. The allocation $(2,2,3)$ is dominated by $(3,3,4)$.

[3] See Hurwicz and Schmeidler 1978.

[4] The third bidder might might bid just above $60. To simplify the analysis, we assume $60 exactly.

[5] The proof we give here assumes a uniform distribution of values in $[0, 1]$ but the result holds for a larger class of distributions. Suppose that the other $(N - 1)$ bidders all bid their true values times $\frac{N-1}{N}$. A bid b will be higher than another bidder's bid provided the other bidder's value times $\frac{N-1}{N}$ is less than b. The probability that this will occur equals $b\frac{N}{N-1}$. It follows that the probability of being larger than all $(N - 1)$ of the other bids equals this value raised to the $(N-1)$th power. Therefore, if the bidder has a true value of V, the expected payoff from bidding b equals the value minus the bid $(V - b)$ times the probability of b being the highest bid. The expected payoff can then be written as Expected Payoff $= (V - b)\left(b\frac{N}{N-1}\right)^{N-1}$. To maximize this value, we take the derivative with respect to b and set it equal to zero, giving the following condition:

$$V(N-1)\left(\frac{N}{N-1}\right)^{N-1} b^{N-2} - N\left(\frac{N}{N-1}\right)^{N-1} b^{N-1} = 0$$

Simplifying gives: $V(N-1) - Nb = 0$, which can be rewritten as $b = \frac{V(N-1)}{N}$. To show that the bidder with the highest bid pays a price equal to the expected value of the second-highest bidder's bid, note that given N random values drawn from a uniform distribution on the interval $[0, 1]$, the expected value of the highest bid equals $1 \cdot \frac{N}{N+1}$. The expected value of the second-highest bid equals $1 \cdot \frac{N-1}{N+1}$. The expected bid of the highest-value bidder therefore equals

$$\left(\frac{N-1}{N}\right)1\left(\frac{N}{N+1}\right) = 1\left(\frac{N-1}{N+1}\right)$$

which equals the expected value of the second-highest bidder.

[6] Roger Myerson was my PhD advisor and received the Nobel Prize in part for this result.

[7] In an all-pay auction, the optimal strategy when bidders' values lie in the interval $[0, 1]$ can be written as follows: A bidder with a value V bids $V^N \frac{(N-1)}{N}$. So if there are three bidders, a bidder with a value of $\frac{1}{2}$ would bid $\frac{1}{8} \cdot \frac{2}{3} = \frac{1}{12}$.

[8] For experimental evidence, see Lucking-Reiley 1999. For evidence from eBay auction experiments, see Morgan and Hossain 2006. And for the timber auction analysis, see Athey, Levin, and Seira 2011.

[9] Ostrovsky, Edelman, and Schwarz 2007.

[10] See Page 2012 for a brief survey.

Chapter 25: Signaling Models

[1] See Simler and Hanson 2018 for many examples of how status signaling drives behavior and choices.

[2] The cost of a weak type of signaling equals MC. The benefit of signaling equals $\frac{B}{S+1}$ given the assumption that all S strong types signal. Therefore, no weak type will signal if $MC \geq \frac{B}{S+1}$. In contrast, the strong types prefer to send the signal if their benefit from separating $\frac{B}{S} - cM$ exceeds not signaling and all N agents splitting the benefit $\frac{B}{N}$.

From the previous calculation, the minimal signal that will not cause a weak type to signal equals $M = \frac{B}{(S+1)C}$. It follows that if we let $\hat{M} = \frac{B}{SC}M$, then the weak types will not signal. For the strong types to prefer to send signal $\hat{M}$, we must have $\frac{B}{S} - c\frac{B}{SC} \geq \frac{B}{N}$. We can divide both sides by B and multiply by C to obtain $\frac{C}{S} - \frac{c}{S} \geq \frac{C}{N}$, which simplifies to $(C - c)N \geq CS$, which can be rewritten as $C(N-S) \geq cN$.

[3] This possibility was put forth by Michael Spence, who shared the Nobel Memorial Prize in Economic Sciences for constructing a job market model of educational signaling (Spence 1973).

[4] Having huge tail feathers makes a peacock less fit than if the feathers were a more moderate size. See Zahavi 1975.

[5] See Bird and Smith 2005.

[6] See Smith, Bird, and Bird 2003.

Chapter 26: Learning Models

[1] The psychological study of learning encompasses a far broader set of contexts than we cover here. A person can learn a fact, such as what the capital of Arkansas is. A person can acquire tacit knowledge, such as how to bake bread, repair an engine, or program a computer. A person can also learn a corpus of knowledge, such as organic chemistry.

[2] See Thorndike 1911, 244.

[3] See Rescorla and Wagner 1972.

[4] The model that I describe builds from the original Rescorla and Wagner (1972) model as well as the models by Herrnstein (1970), Bush and Mosteller (1955), Cyert and March (1963), Bendor, Diermeier, and Ting (2003), and Epstein (2014).

[5] The parameter γ must be chosen so that the weight on an alternative remains positive. This will be true provided that γ exceeds the inverse of the difference between the highest possible aspiration level and the minimal reward from any alternative.

[6] The presentation here follows Bendor and Swistak 1997.

[7] If we construct a replicator dynamics model with a finite population in which subsequent populations are chosen randomly, then it is possible that the best alternative might not be reproduced. If so, then replicator dynamics would not locate the best alternative because it has no way of reintroducing alternatives into the population.

[8] See Fudenberg and Levine 1998 and Camerer 2003.

[9] The game also has a mixed strategy equilibrium in which two-thirds choose an economy car and one-third choose a gas guzzler, but that equilibrium is not stable under the learning rules that we are using, so we will ignore it.

[10] To show formally: P(Economy, 1) = 0.5, P(Guzzler, 1) = 0.5, Payoff(Economy, 1) = 1.5, Payoff(Guzzler, 1) = 2, Average Payoff = 1.75. Applying the replicator equation gives Prob(Economy, 2) = $0.5 \cdot \frac{1.5}{1.75} = 0.43$ and Prob(Guzzler, 2) = $0.5 \cdot \frac{2}{1.75} = 0.57$.

[11] See Frank 1985 for a more in-depth analysis of these situations.

[12] See Waltz 1979. See also Powell 1991 for a model of relative and absolute gains in international relations.

[13] See Vriend 2000, which analyzes a similar payoff structure and interprets it primarily as competition between firms who produce an identical product and simultaneously choose quantities. Economists call this the *Cournot competition model*.

[14] The *Roth-Erev learning model* updates the weight $W(k, t)$ of an alternative k in period t by the following formula: $W(k, t + 1) = (1 - r) \cdot W(k, t) + \Delta(k, t, e)$. The parameter r denotes the recency parameter, and $\Delta(k, t, e) = (1 - e) \cdot \text{payoff}(k, t)$ if action k was chosen and $\Delta(k, t, e) = e \cdot \text{payoff}(k, t)$ if action k was not chosen. The parameter e, the experimentation parameter, determines the weight on unchosen alternatives.

[15] See Camerer and Ho 1999.

[16] This analysis closely follows the behavioral spillover models of Bednar and Page (2007, 2018) and also borrows from Greif 2006. Bednar and Page stress the importance of initial actions on the equilibrium that emerges, while Grief focuses on the role of beliefs. See Gilboa and Schmeidler 1995 for an introduction to case-based decision theory. See also Akerlof and Kranton 2010 on the role of identity in economic choices.

[17] The formal proof can be written as follows: Let B denote the proportion of players who buy in and choose the innovative action initially. It is then a straightforward exercise to compute the following payoffs for each type of action:

Payoff to cultural action: $(1 - B) \cdot 200 + B \cdot 220.$

Payoff to innovative strategic action: $(1 - B) \cdot 180 + B \cdot 300.$

To prove the result for replicator dynamics, note that the cultural action has a higher payoff if and only if the following holds:

$$(1-B) \cdot 200 + B \cdot 220 > (1-B) \cdot 180 + 300B$$

Rearranging terms gives $20(1 - B) > 80B$. Thus, applying replicator dynamics, the cultural action increases if and only if: $0.2 > B$.

Chapter 27: Multi-Armed Bandit Problems

[1] See Bergemann and Valimaki 2008 for the relevance to economic phenomena.

[2] See Hills et al. 2015 for a survey.

[3] See Scott 2010 for an analysis of the multi-armed bandit problem and various heuristics.

[4] Gittins and Jones (1972) first characterized the optimal rule. The Gittins index can be reformulated as a *Bellman equation,* which applies to any problem that requires a sequence of choices that each produces a payoff. A Bellman equation relies on the construction of a value function that equals the sum of the sequence of payoffs, with future payoffs discounted according to an interest rate.

[5] Roberts 2004.

[6] For an analysis of an experiment with the United States Department of Agriculture's Farm Service Agency (FSA) Microloan program see Bowers et al. 2017.

[7] Data from *Washington Post* 2012 and Dann 2016.

[8] Al Gore, George H. W. Bush, and Hillary Clinton would have received more credit for economic prosperity had they been incumbents. See Markus 1988 for an analysis through the early part of this period, Fair 2012 for more recent evidence, and Campbell, Dettrey, and Yin 2010 for evidence on candidate-versus-party effect size.

Chapter 28: Rugged-Landscape Models

[1] See Page 2007 for a deeper elaboration of the value of diversity.

[2] See Kauffman 1993 for a full treatment of the NK model.

[3] We can derive the expected value of the local peaks and the global peaks when $N = 20, K = 19$. The contribution of each attribute is uniformly distributed on $[0, 1]$. This distribution has a mean of $\frac{1}{2}$ and a variance of $\frac{1}{12}$. The value of each alternative equals the average of the twenty attribute contributions. Therefore, by the central limit theorem, these values are normally distributed with mean $\frac{1}{2}$ and variance $\frac{1}{12N}$, so in the case of $N = 20$, each standard deviation is of size $.0645 = \sqrt{\frac{1}{240}}$. We can then estimate the average value of a local peak: 0.609. A local peak can be thought of as the best from among 21 random draws from this distribution. Therefore, its expected value will be approximately equal to that of a value drawn from the normal distribution such that $\frac{21}{22}$ of the alternatives have lower values. This will be a little less than two standard deviations above the mean. Using a normal distribution calculator, the expected average value equals 0.609. To estimate the expected value of the global peak, 0.759, we note that the global peak has the highest value of all 2^{20} alternatives. Each alternative can be thought of as a value drawn from the distribution, so the expected value will be approximately equal to that of value drawn from the normal distribution such that $\frac{2^{20}}{2^{20}+1}$ of the alternatives have lower value. Using a normal distribution, the expected average value equals 0.759. The global peak has a higher value than would be expected in a million draws.

[4] See Wright 2001, which argues that these positive-sum recombinations helped produce humans, societies, and the technical and scientific advancements of our age.

[5] See Kauffman 1993 and Miller and Page 2007.

[6] The United States allows for seventeen years from the issue date if issuing the patent takes more than three years.

[7] Boldrin and Levine 2010.

Chapter 29: Opioids, Inequality, and Humility

[1] Given the probabilities, the exact statistical equilibrium is 70.7% in no pain, 19.5% on opioids, and 9.8% addicted. For the first case, the percentages are 76.3%, 21.5%, and 2.2% respectively.

[2] See Wakeland, Nielsen, and Geissert 2015.

[3] I thank Abbie Jacobs for her commentary and insights on this section of the book.

[4] See Wilkinson and Pickett 2009.

[5] Comments made at the Becker Friedman Institute at the University of Chicago, "Understanding Inequality and What to Do About It," November 6, 2015.

[6] See Goldin and Katz 2008, Acemoglu and Autor 2011, and Murphy and Topel 2016.

[7] See Mas-Colell, Whinston, and Green 1995 for how to prove such a result.

[8] See Kaplan and Rauh 2013a, and a model by Jones and Kim (2018) that uses ability as a proxy for the scalability of an entrepreneurial idea. See Frank 1996 for early research on how inequality has occurred within every profession. For a more recent study, see Xie, Killewald, and Near 2016.

[9] See Ormerod 2012.

[10] Ormerod 2012 describes in detail how our increased connectedness contributes to inequality.

[11] See Cancian and Reed 1999 and Schwartz and Mare 2005.

[12] See Greenwood et al. 2014 for the full model.

[13] The estimate is that it would have been 0.34 rather than 0.43. See Greenwood et al. 2014. The *Gini coefficient* measures the distance between the income distribution and an equal distribution. Let $S(P)$ denote the total share of income (or wealth) earned (held) by the lowest P% of the population (i.e., if the bottom 30% of the population earns 2% of the income, then $S(30) = 2$):

$$\text{GINI} = \frac{2}{99} \times \sum_{P=1}^{100} \left[\frac{P}{100} - S(P) \right]$$

If income is evenly distributed, $S(P) = \frac{P}{100}$, and GINI = 0. If all of the income goes to the top 1%, $S(P) = 0$ for $P < 100$, and $S(100) = 1$, so GINI = 1.

[14] The calculation can be made as follows: Children belong to the four categories with probabilities $(0.6, 0.25, 0.1, 0.05)$. That is, 60% of the offspring of high-income earners, 20% of the offspring of upper-middle-income earners, 15% of the offspring of lower-middle-income earners, and 5% of the offspring of low-income earners have high income. The percentage of high-income grandchildren equals $(0.6)(0.6) + (0.25)(0.2) + (0.1)(0.15) + (0.05)(0.05) = 0.4275$ and the percentage of low-income grandchildren equals $(0.6)(0.05) + (0.25)(0.1) + (0.1)(0.15) + (0.05)(0.7) = 0.105$.

[15] Pfeffer and Killewald 2017.

[16] Kaplan and Rao 2013b.

[17] See Farmer 2018.

Bibliography

Acemoglu, Daron, and David Autor. 2011. "Skills, Tasks and Technologies: Implications for Employment and Earnings." In Orley Ashenfelter and David Card, eds., *Handbook of Labor Economics,* 4: 1043–1171. Amsterdam: Elsevier-North Holland.

Acemoglu, Daron, and James Robinson. 2012. *Why Nations Fail: The Origins of Power, Prosperity, and Poverty.* Cambridge, MA: Harvard University Press.

Adler, Mortimer Jerome. 1970. *The Time of Our Lives: The Ethics of Common Sense.* New York: Holt, Rinehart and Winston.

Akerlof, G., and R. Kranton. 2010. *Identity Economics.* Princeton, NJ: Princeton University Press.

Albert, Rika, Istvan Albert, and Gary L. Nakarado. 2004. "Structural Vulnerability of the North American Power Grid." *Physical Review E* 69: 025103.

Allesina, Stefano, and Mercedes Pascual. 2009. "Googling Food Webs: Can an Eigenvector Measure Species' Importance for Coextinctions?" *PLOS: Computational Biology* 9, no. 4.

Allison, Graham. 1971. *Essence of Decision: Explaining the Cuban Missile Crisis.* New York: Little, Brown.

Alvaredo, Facundo, Anthony B. Atkinson, Thomas Piketty, and Emmanuel Saez. 2013. "The World Top Incomes Database." https://www.inet.ox.ac.uk/projects/view/149.

Anderson, Chris. 2008a. "The End of Theory: The Data Deluge Makes the Scientific Method Obsolete." *Wired* 16, no. 7.

Anderson, Chris. 2008b. *The Long Tail: Why the Future of Business Is Selling Less of More.* New York: Hachette.

Anderson, Phillip. 1972. "More Is Different." *Science* 177, no. 4047: 393–396.

Arrow, Kenneth. 1963. *Social Choice and Individual Values*. New Haven, CT: Yale University Press.

Arthur, W. B. 1994. "Inductive Reasoning and Bounded Rationality (The El Farol Problem)." *American Economic Review Papers and Proceedings* 84: 406–411.

Arthur, W. B. 2011. *The Nature of Technology: What It Is and How It Evolves.* New York: Free Press.

Ashenfelter, Orley. 2010. "Predicting the Quality and Prices of Bordeaux Wine." *Journal of Wine Economics* 5, no. 1: 40–52.

Athey, Susan, Jonathan Levin, and Enrique Seira. 2011. "Comparing Open and Sealed Bid Auctions: Evidence from Timber Auctions." *Quarterly Journal of Economics* 126, no. 1: 207–257.

Austin, David. 2008. "Percolation: Slipping Through the Cracks." American Mathematical Society. www.ams.org/publicoutreach/feature-column/fcarc-percolation.

Axelrod, Robert. 1984. *The Evolution of Cooperation.* New York: Basic Books.

Axelrod, David, Robert Axelrod, and Kenneth J. Pienta. 2006. "Evolution of Cooperation Among Tumor Cells." *Proceedings of the National Academy of Sciences* 103, no. 36: 13474–13479.

Axtell, Robert L. 2001. "Zipf Distribution of U.S. Firm Sizes." *Science* 293: 1818–1820.

Bajari, Patrick, and Matthew E. Kahn. 2008. "Estimating Hedonic Models of Consumer Demand with an Application to Urban Sprawl." In *Hedonic Methods in Housing Markets,* 129–155. New York: Springer.

Bak, Per. 1996. *How Nature Works: The Science of Self-Organized Criticality.* New York: Springer.

Baldwin, Carliss Y., and Kim B. Clark. 2000. *Design Rules. Vol. 1, The Power of Modularity*. Cambridge, MA: MIT Press.

Ball, Eric, and Joseph LiPuma. 2012. *Unlocking the Ivory Tower: How Management Research Can Transform Your Business.* Palo Alto, CA: Kauffman Fellow Press.

Banzhaf, John F. 1965. "Weighted Voting Doesn't Work: A Mathematical Analysis." *Rutgers Law Review* 19, no. 2: 317–343.

Barber, Gerald M. 1997. "Sequencing Highway Network Improvements: A Case Study of South Sulawesi." *Economic Geography* 53, no. 1: 55–69.

Bass, Frank. 1969. "A New Product Growth Model for Consumer Durables." *Management Science* 15, no. 5: 215–227.

Baxter, G. William. 2009. "The Dynamics of Foraging Ants." Paper presented at the annual meeting of the American Physical Society, March 16–20, abstract H40.00011.

Bednar, Jenna. 2007. "Credit Assignment and Federal Encroachment." *Supreme Court Economic Review* 15: 285–308.

Bednar, Jenna. 2008. *The Robust Federation: Principle of Design.* Cambridge: Cambridge University Press.

Bednar, Jenna, Aaron Bramson, Andrea Jones-Rooy, and Scott E. Page. 2010. "Emergent Cultural Signatures and Persistent Diversity: A Model of Conformity and Consistency." *Rationality and Society* 22, no. 4: 407–444.

Bednar, Jenna, and Scott E. Page. 2007. "Can Game(s) Theory Explain Culture? The Emergence of Cultural Behavior Within Multiple Games." *Rationality and Society* 19, no. 1: 65–97.

Bednar, Jenna, and Scott E. Page. 2018. "When Order Affects Performance: Culture, Behavioral Spillovers and Institutional Path Dependence." *American Political Science Review* 112, no. 1: 82–98.

Bell, Alex, Raj Chetty, Xavier Jaravel, Neviana Petkova, and John Van Reenen. 2018 "Who Becomes an Inventor in America? The Importance of Exposure to Innovation: Executive Summary." www.equality-of-opportunity.org.

Bendor, Jonathan, Daniel Diermeier, and Michael Ting. 2003. "A Behavioral Model of Turnout." *American Political Science Review* 97, no. 2: 261–280.

Bendor, Jonathan, and Piotr Swistak. 1997. "The Evolutionary Stability of Cooperation." *American Political Science Review* 91: 290–307.

Bendor, Jonathan, and Scott E. Page. 2018. "A Model of Team Problem Solving." Unpublished manuscript.

Berg, Nathan, and Gerd Gigerenzer. 2010. "As-If Behavioral Economics: Neoclassical Economics in Disguise?" *History of Economic Ideas* 18, no. 1: 133–166.

Bergemann, Dirk, and Juuso Välimäki. 2008. "Bandit Problems." In *The New Palgrave Dictionary of Economics,* 2nd ed., ed. Steven N. Durlauf and Lawrence E. Blume. London: Palgrave Macmillan.

Berlekamp, Elwyn R., John H. Conway, and Richard K. Guy. 1982. "What Is Life?" In *Winning Ways for Your Mathematical Plays*. Vol. 2, *Games in Particular*. London: Academic Press.

Bertrand, Marianne, and Sendhil Mullainathan. 2001. "Are CEOs Rewarded for Luck? The Ones Without Principles Are." *Quarterly Journal of Economics* 116: 901–932.

Bickel, P. J., E. A. Hammel, and J. W. O'Connell. 1974. "Sex Bias in Graduate Admissions: Data from Berkeley." *Science* 187 (4175): 398–404.

Biernaskie, Jay, M. 2011. "Evidence for Competition and Cooperation Among Climbing Plants." *Proceedings of the Royal Society B* 278: 1989–1996.

Bird, Rebecca, and Eric Smith. 2004. "Signaling Theory, Strategic Interaction, and Symbolic Capital." *Current Anthropology* 46, no. 2: 222–248.

Boldrin, Michele, and David Levine. 2010. *Against Intellectual Monopoly*. Cambridge: Cambridge University Press.

Borges, Jorge Luis. 1974. *A Universal History of Infamy*. Trans. Norman Thomas de Giovanni. London: Penguin.

Bowers, Jake, Nathaniel Higgins, Dean Karlan, Sarah Tulman, and Jonathan Zinman. 2017. "Challenges to Replication and Iteration in Field Experiments: Evidence from Two Direct Mail Shots." *American Economic Review Papers & Proceedings* 107, no. 5: 1–3.

Bowles, Samuel, and Herbert Gintis. 2002. "The Inheritance of Inequality." *Journal of Economic Perspectives* 16, no. 3: 3–30.

Box, George E. P., and Norman Draper. 1987. *Empirical Model-Building and Response Surfaces*. New York: Wiley.

Boyd, Robert. 2006. "Reciprocity: You Have to Think Different." *Journal of Evolutionary Biology* 19: 1380–1382.

Breiman, Leo. 1996. "Bagging Predictors." *Machine Learning* 24, no. 2: 123–140.

Briggs, Andrew, and Mark Sculpher. 1998. "An Introduction to Markov Modeling for Economic Evaluation." *Pharmaco Economics* 13, no. 4: 397–409.

Brock, William, and Steven Durlauf. 2001. "Discrete Choice with Social Interactions." *Review of Economic Studies* 68: 235–260.

Broido, A. D., and A. Clauset. 2018. "Scale-Free Networks Are Rare." Working paper.

Bshary, R., and A. S. Grutter. 2006. "Image Scoring and Cooperation in a Cleaner Fish Mutualism." *Nature* 441, no. 7096: 975–978.

Burt, Ronald. 1995. *Structural Holes: The Social Structure of Competition.* Cambridge, MA: Harvard University Press.

Bush, Robert, and Frederick Mosteller. 1954. *Stochastic Models for Learning.* New York: John Wiley and Sons.

Camerer, Colin F. 2003. *Behavioral Game Theory: Experiments in Strategic Interaction.* Princeton, NJ: Princeton University Press.

Camerer, Colin, Linda Babcock, George Loewenstein, and Richard Thaler. 1997. "Labor Supply of New York City Cabdrivers: One Day at a Time." *Quarterly Journal of Economics* 112, no. 2: 407–441.

Camerer, Colin, and Tek Ho. 1999. "Experience-Weighted Attraction Learning in Normal Form Games." *Econometrica* 67, no. 4: 827–874.

Camerer, Colin, George Loewenstein, and Drazen Prelec. 2005. "Neuroeconomics: How Neuroscience Can Inform Economics." *Journal of Economic Literature* 43: 9–64.

Campbell, Donald T. 1976. "Assessing the Impact of Planned Social Change." Public Affairs Center, Dartmouth College.

Campbell, James E., Bryan J. Dettrey, and Hongxing Yin. 2010. "The Theory of Conditional Retrospective Voting: Does the Presidential Record Matter Less in Open-Seat Elections?" *Journal of Politics* 72, no. 4: 1083–1095.

Cancian, Maria, and Deborah Reed. 1999. "The Impact of Wives' Earnings on Income Inequality: Issues and Estimates." *Demography* 36, no. 2: 173–184.

Carvalho, Vasco, and Xavier Gabaix. 2013 "The Great Diversification and Its Undoing," *American Economic Review* 103, no. 5: 1697–1727.

Castellano, Claudio, Santo Fortunato, and Vittorio Loreto. 2009. "Statistical Physics of Social Dynamics." *Review of Modern Physics* 81: 591–646.

Cederman, Lars Erik. 2003. "Modeling the Size of Wars: From Billiard Balls to Sandpiles." *American Political Science Review* 97: 135–150.

Centola, Damon, and Michael Macy. 2007. "Complex Contagions and the Weakness of Long Ties." *American Journal of Sociology* 113: 702–734.

Chance, Donald. 2009. "What Are the Odds? Another Look at DiMaggio's Streak." *Chance* 22, no. 2: 33–42.

Christakis, N. A., and J. Fowler. 2009. *Connected: The Surprising Power of Our Social Networks and How They Shape Our Lives*. New York: Little, Brown.

Churchland, Patricia, and Terry J. Sejnowski. 1992. *The Computational Brain*. Cambridge, MA: MIT Press.

Chwe, Michael. 2013. *Jane Austen: Game Theorist*. Princeton, NJ: Princeton University Press.

Clarida, Richard, Jordi Galí, and Mark Gertler. 2000. "Monetary Policy Rules and Macroeconomic Stability: Evidence and Some Theory." *Quarterly Journal of Economics* 115, no. 1: 147–180.

Clark, Gregory. 2014. *The Son Also Rises: Surnames and the History of Social Mobility*. Princeton, NJ: Princeton University Press.

Clark, William, Matt Golder, and Sona Nadenicheck Golder. 2008. *Principles of Comparative Politics*. Washington, DC: Congressional Quarterly Press.

Clauset, Aaron, M. Young, and K. S. Gleditsch. 2007. "On the Frequency of Severe Terrorist Attacks." *Journal of Conflict Resolution* 51, no. 1: 58–88.

Cohen, Tyler. 2013. *Average Is Over: Powering America Beyond the Age of the Great Stagnation*. New York: Dutton.

Cooke, Nancy J., and Margaret L. Hilton, eds. 2014. *Enhancing the Effectiveness of Team Science*. Washington, DC: National Academies Press.

Cornes, Richard, and Todd Sandler. 1996. *The Theory of Externalities, Public Goods, and Club Goods*. 2nd ed. Cambridge: Cambridge University Press.

Craine, Joseph, and Ray Dybzinski. 2013. "Mechanisms of Plant Competition for Nutrients, Water and Light." *Functional Ecology* 27: 833–840.

Cyert, Richard M., and James G. March. 1963. *A Behavioral Theory of the Firm*. Englewood Cliffs, NJ: Prentice-Hall.

Dan, Avi. 2018. "How Michelle Peluso Is Redefining Marketing at IBM." *Forbes,* January 18.

Dann, Carrie. 2016. "Pro-Clinton Battleground Ad Spending Outstrips Trump Team by 2." NBC News, November 4.

Dawes, Robyn. 1979. "The Robust Beauty of Improper Linear Models in Decision Making." *American Psychologist* 34: 571–582.

de Marchi, Scott. 2005. *Computational and Mathematical Modeling in the Social Sciences*. Cambridge: Cambridge University Press.

DeMiguel, Victor, Lorenzo Garlappi, and Raman Uppal. 2009. "Optimal Versus Naive Diversification: How Inefficient Is the $\frac{1}{N}$ Portfolio Strategy?" *Review of Financial Studies* 22, no. 5: 1915–1953.

Dennett, Daniel C. 1991. *Consciousness Explained*. Boston: Back Bay Books.

Dennett, Daniel C. 1994. *Darwin's Dangerous Idea: Evolution and the Meanings of Life*. New York: Simon & Schuster.

Denrell, Jerker, and Chengwei Liu. 2012. "Top Performers Are Not the Most Impressive When Extreme Performance Indicates Unreliability." *Proceedings of the National Academy of Sciences* 109, no. 24: 9331–9336.

Diamond, Jared. 2005. *Collapse: How Societies Choose to Fail or Succeed*. New York: Viking Penguin.

Dodds, Peter, Robby Muhamad, and Duncan Watts. 2003. "An Experimental Study of Search in Global Social Networks." *Science* 301: 827–829.

Downing, John A., et al. 2006. "The Global Abundance and Size Distribution of Lakes, Ponds, and Impoundments." *Limnology and Oceanography* 51, no. 5: 2388–2397.

Dragulescu, Adrian, and Victor M. Yakovenko. 2001. "Exponential and Power-Law Probability Distributions of Wealth and Income in the United Kingdom and the United States." *Physica A* 299: 213–221.

Drucker, Peter. 1969. *The Age of Discontinuity: Guidelines to Our Changing Society*. New York: Harper and Row.

Dubos, Jean. 1987. *The White Plague: Tuberculosis, Man and Society*. New Brunswick, NJ: Rutgers University Press.

Dunne, Anthony. 1999. *Hertzian Tales: Electronic Products, Aesthetic Experience and Critical Design*. London: Royal College of Art.

Dyson, Freeman. 2004. "A Meeting with Enrico Fermi." *Nature* 427: 297.

Easley, David, and Jon Kleinberg. 2010. *Networks, Crowds, and Markets: Reasoning About a Highly Connected World*. Cambridge: Cambridge University Press.

Easley, David, Marcos Lopez de Prado, and Maureen O'Hara. 2012. "Flow Toxicity and Liquidity in a High Frequency World." *Review of Financial Studies* 24, no. 5: 1457–1493.

Easterly, William, and Stanley Fischer. 1995. "The Soviet Economic Decline." *World Bank Economic Review* 9, no. 3: 341–371.

Ebbinghaus, Herman. 1885. *Memory: A Contribution to Experimental Psychology*. Online in *Classics in the History of Psychology*. http://psychclassics.yorku.ca/Ebbinghaus/index.htm.

Ehrenberg, Andrew. 1969. "Towards an Integrated Theory of Consumer Behaviour." *Journal of the Market Research Society* 11, no. 4: 305–337.

Einstein, Albert. 1934. "On the Method of Theoretical Physics." *Philosophy of Science* 1, no. 2: 163–169.

Eliot, Matt, Ben Golub, and Matthew Jackson. 2014. "Financial Networks and Contagion." *American Economic Review* 104, no. 10: 3115–3153.

Eom, Young-Ho, and Hang-Hyun Jo. 2014. "Generalized Friendship Paradox in Complex Networks: The Case of Scientific Collaboration." *Scientific Reports* 4: 4603.

Epstein, Josh. 2006. *Generative Social Science: Studies in Agent-Based Computational Modeling*. Princeton, NJ: Princeton University Press.

Epstein, Joshua. 2008. "Why Model?" *Journal of Artificial Societies and Social Simulation* 11, no. 4: 12.

Epstein, Joshua. 2014. *Agent Zero: Toward Neurocognitive Foundations for Generative Social Science*. Princeton, NJ: Princeton University Press.

Ericsson, K. A. 1996. "The Acquisition of Expert Performance: An Introduction to Some of the Issues." In *The Road to Excellence: The Acquisition of Expert Performance in the Arts and Sciences, Sports, and Games,* ed. K. A. Ericsson, 1–50. Mahwah, NJ: Erlbaum.

Fair, Raymond. 2012. *Predicting Presidential Elections and Other Things*. 2nd ed. Stanford, CA: Stanford University Press.

Farmer, J. Doyne 2018. "Collective Awareness: A Conversation with J. Doyne Farmer." *The Edge*. https://www.edge.org/conversation/j doyne farmer-collective-awareness.

Feld, Scott L. 1991. "Why Your Friends Have More Friends than You Do." *American Journal of Sociology* 96, no. 6: 1464–1477.

Flegal, Katherine M., Brian K. Kit, Heather Orpana, and Barry I. Graubard. 2012. "Association of All-Cause Mortality with Overweight and Obesity Using Standard Body Mass Index Categories: A Systematic Review and Meta-analysis." *Journal of the American Medical Association* 309, no. 1: 71–82.

Flores, Thomas, and Irfan Nooruddin. 2016. *Elections in Hard Times: Building Stronger Democracies in the 21st Century.* Cambridge: Cambridge University Press.

Florida, Richard. 2005. *Cities and the Creative Class.* New York: Routledge.

Foster, Dean, and H. Peyton Young. 2001. "On the Impossibility of Predicting the Behavior of Rational Agents." *Proceedings of the National Academy of Sciences* 98, no. 22: 12848–12853.

Frank, Kenneth, et al. 2018. "Teacher Networks and Educational Opportunity." In *Handbook on the Sociology of Education,* ed. Barbara Schneider and Guan Saw. New York: Oxford University Press.

Frank, Robert. 1984. *Choosing the Right Pond.* Oxford: Oxford University Press.

Frank, Robert. 1996. *The Winner-Take-All Society: Why the Few at the Top Get So Much More than the Rest of Us.* New York: Penguin.

Freeman, Richard, and Wei Huang. 2015. "Collaborating with People Like Me: Ethnic Co-authorship Within the U.S." *Journal of Labor Economics* 33 no. S1: S289-S318.

Fudenberg, Drew, and David Levine. 1998. *Theory of Learning in Games.* Cambridge, MA: MIT Press.

Fudenberg, Drew, and David Levine. 2006. "A Dual-Self Model of Impulse Control." *American Economic Review* 96: 1449–1476.

Gammill, James F. , Jr., and Terry A. Marsh. 1988. "Trading Activity and Price Behavior in the Stock and Stock Index Futures Markets in October 1987." *Journal of Economic Perspectives* 2, no. 3: 25–44.

Gawande, Atul. 2009. *The Checklist Manifesto: How to Get Things Right.* New York: Henry Holt.

Geithner, Timothy. 2014. *Stress Test: Reflections on Financial Crises.* New York: Crown.

Gerschenkron, Alexander. 1952. "Economic Backwardness in Historical Perspective." In *The Progress of Underdeveloped Areas,* ed. B. F. Hoselitz. Chicago: University of Chicago Press.

Gertner, Jon. 2012. *The Idea Factory: Bell Labs and the Great Age of American Innovation.* New York: Penguin.

Gibrat, Robert. 1931. *Les inégalités economique.* Paris: Sirely.

Gigerenzer, Gerd, and Reinhard Selten. 2002. *Bounded Rationality: The Adaptive Toolbox.* Cambridge, MA: MIT Press.

Gigerenzer, Gerd, and Peter Todd. 2000. *Simple Heuristics That Make Us Smart*. New York: Oxford University Press.

Gilboa, Itzhak, and David Schmeidler. 1994. "Case-Based Decision Theory." *Quarterly Journal of Economics* 110: 605–639.

Gilovich, Thomas, Amos Tversky, and R. Vallone. 1984. "The Hot Hand in Basketball: On the Misperception of Random Sequences." *Cognitive Psychology* 17, no. 3: 295–314.

Glaeser, Edward, Bruce Sacerdote, and Jose Scheinkman. 1996. "Crime and Social Interactions." *Quarterly Journal of Economics* 111, no. 2: 507–548.

Glantz, Andrew. 2008. "A Tax on Light and Air: Impact of the Window Duty on Tax Administration and Architecture, 1696–1851." *Penn History Review* 15, no. 2: 18–40.

Glasserman, Paul, and H. Peyton Young. 2014. "Contagion in Financial Networks." Office of Financial Research Working Paper.

Godard, Renee. 1993. "Tit for Tat Among Neighboring Hooded Warblers." *Behavioral Ecology and Sociobiology* 33, no. 1: 45–50.

Gode, Dhananjay K., and Shyam Sunder. 1993. "Allocative Efficiency of Markets with Zero-Intelligence Traders: Market as a Partial Substitute for Individual Rationality." *Journal of Political Economy* 101, no. 1: 119–137.

Goldin, Claudia, and Lawrence F. Katz. 2008. *The Race Between Education and Technology*. Cambridge, MA: Harvard University Press.

Gordon, Robert J. 2016. *The Rise and Fall of American Growth: The U.S. Standard of Living Since the Civil War*. Princeton, NJ: Princeton University Press.

Granovetter, Mark. 1973. "The Strength of Weak Ties." *American Journal of Sociology* 78, no. 6: 1360–1380.

Granovetter, Mark. 1978. "Threshold Models of Collective Behavior." *American Journal of Sociology* 83, no. 6: 1360–1443.

Greenwood, Jeremy, Nezih Guner, Georgi Kocharkov, and Cezar Santos. 2014. "Marry Your Like: Assortative Mating and Income Inequality." *American Economic Review: Papers & Proceedings* 104, no 5: 348-353.

Greif, Avner. 2006. *Institutions and the Path to the Modern Economy: Lessons from Medieval Trade*. Cambridge: Cambridge University Press.

Griliches, Zvi. 1957, 1988. "Hybrid Corn: An Exploration of the Economics of Technological Change." In *Technology, Education and Productivity: Early Papers with Notes to Subsequent Literature*. New York: Basil Blackwell.

Groseclose, Tim, and James Snyder. 1996. "Buying Supermajorities." *American Political Science Review* 90: 303–315.

Grossman, S., and J. Stiglitz. 1980. "On the Impossibility of Informationally Efficient Markets." *American Economic Review* 70, no. 3: 393–408.

Groysberg, Boris. 2012. *Chasing Stars: The Myth of Talent and the Portability of Performance*. Princeton, NJ: Princeton University Press.

Guy, Richard. 1983. "Don't Try to Solve These Problems." *American Mathematical Monthly* 90: 35–41.

Haidt, Jonathan. 2006. *The Happiness Hypothesis: Finding Modern Truth in Ancient Wisdom*. Basic Books. New York: NY.

Haldene, Andrew. 2012. "The Dog and the Frisbee." Speech given at the Federal Reserve Bank of Kansas City's 36th Economic Policy Symposium, Jackson Hole, WY.

Haldene, Andrew. 2014. "The Dappled World." Speech given at the University of Michigan Law School, Ann Arbor, October 23.

Haldane, John B. S. 1928. "On Being the Right Size." Online version available at http://irl.cs.ucla.edu/papers/right-size.html.

Hardin, Garret. 1968. "The Tragedy of the Commons." *Science* 162, no. 3859: 1243–1248.

Harrell, Frank E. 2001. *Regression Modeling Strategies with Applications to Linear Models, Logistic Regression, and Survival Analysis*. New York: Springer.

Harstad, Ronald M., and Reinhard Selten. 2013. "Bounded Rationality Models: Tasks to Become Intellectually Competitive." *Journal of Economic Literature* 51, no. 2: 496–511.

Harte, John. 1988. *Consider a Spherical Cow*. Mill Valley, CA: University Science Books.

Hathaway, Oona. 2001. "Path Dependence in the Law: The Course and Pattern of Change in a Common Law Legal System." *Iowa Law Review* 86.

Havel, Václav. 1985. *The Power of the Powerless: Citizens Against the State in Central-Eastern Europe*. Ed. John Keane. Armonk, NY: M. E. Sharpe.

Hawking, Stephen, and Leonard Mlodinow. 2011. *The Grand Design*. New York: Bantam.

Hecht, Jeff. 2008. "Prophecy of Economic Collapse 'Coming True.'" *New Scientist*. November 17.

Herrnstein, Richard J. 1970. "On the Law of Effect." *Journal of the Experimental Analysis of Behavior* 13: 243–266.

Hills, Thomas, Peter M. Todd, David Lazer, A. David Redish, Iain D. Couzin, and the Cognitive Search Research Group. 2015. "Exploration Versus Exploitation in Space, Mind, and Society." *Trends in Cognitive Science* 19, no. 1: 46–54.

Hofstadter, Douglas, and Emmanuel Sander. 2013. *Surfaces and Essences: Analogy as the Fuel and Fire of Thinking*. New York: Basic Books.

Holland, John. 1975. *Adaptation in Natural and Artificial Systems*. Ann Arbor: University of Michigan Press.

Hong, Lu, and Scott E. Page. 2009. "Interpreted and Generated Signals." *Journal of Economic Theory* 144: 2174–2196.

Hotelling, Harold. 1929. "Stability in Competition." *Economic Journal* 39, no. 153: 41–57.

Huffaker, Carl Burton. 1958. "Experimental Studies on Predation: Dispersion Factors and Predator-Prey Oscillations." *Hilgardia* 27, no. 14: 343–383.

Hurwicz, Leo, and David Schmeidler. 1978. "Outcome Functions Which Guarantee the Existence and Pareto Optimality of Nash Equilibria." *Econometrica* 46: 144–174.

Inman, Mason. 2011. "Sending Out an SOS." *Nature Climate Change* 1: 180–183.

International Monetary Fund. 2009. *Global Financial Stability Report*.

Jackson, Matthew. 2008. *Social and Economic Networks*. Princeton, NJ: Princeton University Press.

Jackson, Matthew and Asher Wolinsky. 1996. "A Strategic Model of Social and Economic Networks." *Journal of Economic Theory* 71: 44–74.

Jacob, Francois. 1977. "Evolution and Tinkering." *Science* 196: 1161–1166.

Jacobs, Jane. 1989. *Revolving Doors: Sex Segregation and Women's Careers*. Stanford, CA: Stanford University Press.

Johnson, James. 2014. "Models Among the Political Theorists." *American Journal of Political Science* 58, no. 33: 547–560.

Johnson-Laird, Philip. 2009. *How We Reason*. New York: Oxford University Press.

Jones, Benjamin F., Brian Uzzi, and Stefan Wuchty. 2008. "Multi-University Research Teams: Shifting Impact, Geography and Social Stratification in Science." *Science* 322: 1259–1262.

Jones, Charles, and Jihee Kim. 2018 "A Schumpeterian Model of Top Income Inequality." *Journal of Political Economy*. Forthcoming.

Kahneman, Daniel. 2011. *Thinking Fast and Slow*. New York: Farrar, Straus and Giroux.

Kahneman, Daniel, and Amos Tversky. 1979. "Prospect Theory: An Analysis of Decisions Under Risk." *Econometrica* 47, no. 2: 263–291.

Kalyvas, Stathis. 1999. "The Decay and Breakdown of Communist One-Party Systems." *Annual Review of Political Science* 2: 323–343.

Kamin, Leon J. 1969. "Predictability, Surprise, Attention and Conditioning." In *Punishment and Aversive Behavior,* ed. B. A. Campbell and R. M. Church, 279–296. New York: Appleton-Century-Crofts.

Kaplan, Steven, and Joshua D. Rauh. 2013a. "Family, Education, and Sources of Wealth Among the Richest Americans, 1982–2012." *American Economic Review Papers and Proceedings* 103, no. 3: 158–162.

Kaplan, Steven, and Joshua D. Rauh. 2013b. "It's the Market: The Broad-Based Rise in the Return to Top Talent." *Journal of Economic Perspectives* 27, no. 3: 35–56.

Karlsson, Bengt. 2016. "The Forest of Our Lives: In and Out of Political Ecology." *Conservation and Society* 14, no. 4: 380–390.

Kauffman, Stuart. 1993. *The Origins of Order: Self-Organization and Selection in Evolution*. Oxford: Oxford University Press.

Kennedy, John F. 1956. *Profiles in Courage*. New York: Harper & Brothers.

Khmelev, Dmitri, and F. J. Tweedie. 2001. "Using Markov Chains for Identification of Writers." *Literary and Linguistic Computing* 16, no. 4: 299–307.

Kleinberg, Jon, and M. Raghu. 2015. "Team Performance with Test Scores." Working paper, Cornell University School of Information.

Knox, Grahame. n.d. "Lost at Sea." *Insight*, http://insight.typepad.co.uk/lost_at_sea.pdf.

Kollman, Ken, J. Miller, and S. Page. 1992. "Adaptive Parties in Spatial Elections." *American Political Science Review* 86: 929–937.

Kooti, Farshad, Nathan O. Hodas, and Kristina Lerman. 2014. "Network Weirdness: Exploring the Origins of Network Paradoxes." Paper presented at the International Conference on Weblogs and Social Media (ICWSM), March.

Kurlansky, Mark. 1998. *Cod: A Biography of the Fish That Changed the World.* New York: Penguin.

Kydland, Finn E., and Edward C. Prescott. 1977. "Rules Rather than Discretion: The Inconsistency of Optimal Plans." *Journal of Political Economy* 85, no. 3: 473–491.

Lai, T. L., and Herbert Robbins. 1985. "Asymptotically Efficient Adaptive Allocation Rules." *Advances in Applied Mathematics* 6, no. 1: 4–22.

Laibson, David. 1997. "Golden Eggs and Hyperbolic Discounting." *Quarterly Journal of Economics* 112, no. 2: 443–477.

Lamberson, P. J., and Scott E. Page. 2012a. "The Effect of Feedback Variability on Success in Markets with Positive Feedbacks." *Economics Letters* 114: 259–261.

Lamberson, P. J., and Scott E. Page. 2012b. "Tipping Points." *Quarterly Journal of Political Science* 7, no. 2: 175–208.

Lancaster, Kelvin J. 1966. "A New Approach to Consumer Theory." *Journal of Political Economy* 74: 132–157.

Landemore, Helene. 2013. *Democratic Reason: Politics, Collective Intelligence, and the Rule of the Many.* Princeton, NJ: Princeton University Press.

Lango, Allen H., et al. 2010. "Hundreds of Variants Clustered in Genomic Loci and Biological Pathways Affect Human Height." *Nature* 467, no. 7317: 832–838.

Langville, Amy N., and Carl D. Meyer. 2012. *Who's #1?: The Science of Rating and Ranking.* Princeton, NJ: Princeton University Press.

Lave, Charles, and James G. March. 1975. *An Introduction to Models in the Social Sciences.* Lanham, MD: University Press of America.

Ledyard, John, David Porter, and Antonio Rangle. 1997. "Experiments Testing Multiobject Allocation Mechanisms." *Journal of Economics and Management Strategy* 6, no. 3: 639–675.

Ledyard, John, David Porter, and Randii Wessen. 2000. "A Market-Based Mechanism for Allocating Space Shuttle Secondary Payload Priority." *Experimental Economics* 2, no. 3: 173–195.

Levins, Richard. 1966. "The Strategy of Model Building in Population Biology." *American Scientist* 54: 421–431.

Levinthal, Daniel A. 1997. "Adaptation on Rugged Landscapes." *Management Science* 43: 934–950.

Levinthal, Daniel. 1991. "Random Walks and Organizational Mortality." *Administrative Science Quarterly* 36, no. 3: 397–420.

Levitt, Steven, and Stephen Dubner. 2009. *SuperFreakonomics: Global Cooling, Patriotic Prostitutes, and Why Suicide Bombers Should Buy Life Insurance.* New York: William Morrow.

Lewis, Michael. 2014. *Flash Boys: A Wall Street Revolt.* New York: W. W. Norton.

Limpert, Eckhard, Werner A. Stahel, and Markus Abbt. 2001. "Log-normal Distributions Across the Sciences: Keys and Clues." *BioScience* 51, no. 5: 341–352.

Little, Daniel. 1998. *Microfoundations, Method, and Causation: On the Philosophy of the Social Sciences.* Piscataway, NJ: Transaction Publishers.

Lo, Andrew W., and A. Craig MacKinlay. 2007. *A Non-Random Walk Down Wall Street.* Princeton, NJ: Princeton University Press.

Lo, Andrew W. 2012. "Reading About the Financial Crisis: A Twenty-One-Book Review." *Journal of Economic Literature* 50, no. 1: 151–178.

Lucas, Robert. 1976. "Econometric Policy Evaluation: A Critique." In *The Phillips Curve and Labor Markets,* ed. K. Brunner and A. Meltzer, 19–46. Carnegie-Rochester Conference Series on Public Policy 1. New York: Elsevier.

Lucking-Reiley, David. 1999. "Using Field Experiments to Test Equivalence Between Auction Formats: Magic on the Internet." *American Economic Review* 89, no. 5: 1063–1080.

MacKenzie, Debora. 2012. "Boom and Doom: Revisiting Prophecies of Collapse." *New Scientist,* January.

Mannes, Albert E., Jack B. Soll, and Richard P. Larrick. 2014. "The Wisdom of Select Crowds." *Journal of Personality and Social Psychology* 107: 276–299.

Markowitz, Harold M. 1952. "Portfolio Selection." *Journal of Finance* 7, no. 1: 77–91.

Markus, Greg B. 1988. "The Impact of Personal and National Economic Conditions on the Presidential Vote: A Pooled Cross-Sectional Analysis." *American Journal of Political Science* 32: 137–154.

Martin, Andrew D., and Kevin M. Quinn. 2002. "Dynamic Ideal Point Estimation via Markov Chain Monte Carlo for the U.S. Supreme Court, 1953–1999." *Political Analysis* 10: 134–153.

Martin, Francis, et al. 2008. "The Genome of *Laccaria bicolor* Provides Insights into Mycorrhizal Symbiosis." *Nature* 452: 88–92.

Martinez Peria, Maria Soledad, Giovanni Majnoni, Matthew T. Jones, and Winfrid Blaschke. 2001. "Stress Testing of Financial Systems: An Overview of Issues, Methodologies, and FSAP Experiences." IMF Working Paper no. 01/88.

Mas-Colell, Andreu, Michael D. Whinston, and Jerry R. Green. 1994. *Microeconomic Theory.* New York: Oxford University Press.

Mauboussin, Michael. 2012. *The Success Equation: Untangling Skill and Luck in Business.* Cambridge, MA: Harvard University Press.

May, Robert M., Simon A. Levin, and George Sugihara. 2008. "Ecology for Bankers." *Nature* 451: 893–895.

McCarty, Nolan. 2011. "Measuring Legislative Preferences." In *Oxford Handbook of Congress,* ed. Eric Schickler and Frances Lee. New York: Oxford University Press.

McCarty, Nolan, and Adam Meirowitz. 2014. *Political Game Theory: An Introduction.* Cambridge: Cambridge University Press.

McKelvey, Richard. 1979. "General Conditions for Global Intransitivities in Formal Voting Models." *Econometrica* 47: 1085–1112.

McPhee, William N. 1963. *Formal Theories of Mass Behaviour.* New York: Free Press of Glencoe.

Meadows, D., G. Meadows, J. Randers, and W. W. Behrens III. 1972. *The Limits to Growth.* New York: Universe Books.

Medin, Douglas, Will Bennis, and Michel Chandler. 2010. "The Home-Field Disadvantage." *Perspectives on Psychological Science* 5, no. 6: 708–713.

Merriam, Daniel F., and John C. Davis. 2009. "Using Zipf's Law to Predict Future Earthquakes in Kansas." *Transactions of the Kansas Academy of Science* 112, nos. 1&2: 127–129.

Merton, Robert C. 1969. "Lifetime Portfolio Selection Under Uncertainty: The Continuous-Time Case." *Review of Economics and Statistics* 51, no. 3: 247–257.

Merton, Robert K. 1963. "Resistance to the Systematic Study of Multiple Discoveries in Science." *European Journal of Sociology* 4, no. 2: 237–282.

Milgrom, Paul, and John Roberts. 1986. "Pricing and Advertising Signals of Product Quality." *Journal of Political Economy* 94, no. 4: 796–821.

Miller, John H. 1998. "Active Nonlinear Tests (ANTs) of Complex Simulation Models." *Management Science* 44, no. 6: 820–830.

Miller, John H. 2015. *A Crude Look at the Whole*. New York: Basic Books.

Miller, John H., and Scott E. Page. 2004. "The Standing Ovation Problem." *Complexity* 9, no. 5: 8–16.

Miller, John H., and Scott E. Page. 2007. *Complex Adaptive Systems: An Introduction to Computational Models of Social Life*. Princeton, NJ: Princeton University Press.

Miller, Joshua B., and Adam Sanjurjo. 2015. "Surprised by the Gambler's and Hot Hand Fallacies: A Truth in the Law of Small Numbers." IGIER Working Paper no. 552.

Mitchell, Melanie. 1996. *An Introduction to Genetic Algorithms*. Cambridge, MA: MIT Press.

Mitchell, Melanie. 2009. *Complexity: A Guided Tour*. Oxford: Oxford University Press.

Mlodinow, Leonard. 2009. *The Drunkard's Walk: How Randomness Rules Our Lives*. New York: Penguin.

Mokyr, Joel. 2002. *The Gifts of Athena: Historical Origins of the Knowledge Economy*. Princeton, NJ: Princeton University Press.

Morgan, John, and Tanjim Hossain. 2006. ". . . Plus Shipping and Handling: Revenue (Non)Equivalence in Field Experiments on eBay." *Advances in Economic Analysis & Policy* 6, no. 2: 3.

Moss-Racusin, Corinne, John F. Dovidio, Victoria L. Brescoll, Mark J. Graham, and Jo Handelsman. 2012. "Science Faculty's Subtle Gender Biases Favor Male Students." *Proceedings of the National Academy of Sciences*. 1647–1649.

Munger, Charles. 1994. "A Lesson on Elementary, Worldly Wisdom as It Relates to Investment Management & Business." University of Southern California Business School.

Murphy, Kevin M., and Robert H. Topel. 2016. "Human Capital Investment, Inequality and Growth." *Journal of Labor Economics* 34: 99–127.

Murray, J. D. 1988. "Mammalian Coat Patterns: How the Leopard Gets Its Spots." *Scientific American* 256: 80–87.

Myerson, Roger B. 1999. "On the Value of Game Theory in Social Science." *Rationality and Society* 4: 62–73.

Myerson, Roger B. 1999. "Nash Equilibrium and the History of Economic Theory." *Journal of Economic Literature* 37, no. 3: 1067–1082.

Nagel, Rosemarie. 1995. "Unraveling in Guessing Games: An Experimental Study." *American Economic Review* 85, no. 5: 1313–1326.

Newman, Mark E. 2005. "Power Laws, Pareto Distributions and Zipf's Law." *Contemporary Physics* 46: 323–351.

Newman, Mark E. 2010. *Networks: An Introduction*. Oxford: Oxford University Press.

Nowak, Martin. 2006. "Five Rules for the Evolution of Cooperation." *Science* 314, no. 5805: 1560–1563.

Nowak, Martin A., and Karl Sigmund. 1998. "Evolution of Indirect Reciprocity by Image Scoring." *Nature* 393: 573–577.

Olson, Mancur. 1965. *The Logic of Collective Action: Public Goods and the Theory of Groups*. Cambridge, MA: Harvard University Press.

O'Neil, Cathy 2016. *Weapons of Math Destruction: How Big Data Increases Inequality and Threatens Democracy*. New York, NY: Crown.

Open Science Collaboration. 2015. "Estimating the Reproducibility of Psychological Science." *Science* 349: 6251.

Organization for Economic Co-operation and Development. 1996. *The Knowledge Based Economy*. Paris: OECD.

Ormerod, Paul. 2012. *Positive Linking: How Networks Can Revolutionise the World*. London: Faber and Faber.

Ostrom, Elinor. 2004. *Understanding Institutional Diversity*. Princeton, NJ: Princeton University Press.

Ostrom, Elinor. 2010. "Beyond Markets and States: Polycentric Governance of Complex Economic Systems." *Transnational Corporations Review* 2, no. 2: 1–12.

Ostrom, Elinor, Marco A. Janssen, and John M. Anderies. 2007. "Going Beyond Panaceas." *Proceedings of the National Academy of Sciences* 104: 15176–15178.

Ostrovsky, Michael, Benjamin Edelman, and Michael Schwarz. 2007. "Internet Advertising and the Generalized Second Price Auction: Selling Billions of Dollars Worth of Keywords." *American Economic Review* 97, no. 1: 242–259.

Paarsch, Harry J., and Bruce S. Shearer. 1999. "The Response of Worker Effort to Piece Rates: Evidence from the British Columbia Tree-Planting Industry." *Journal of Human Resources* 34, no. 4: 643–667.

Packer, Craig, and Anne E. Pusey. 1997. "Divided We Fall: Cooperation Among Lions." *Scientific American,* May, 52–59.

Paczuski, Maya, and Kai Nagel. 1996. "Self-Organized Criticality and $1/f$ Noise in Traffic." arXiv: cond-mat/9602011.

Page, Scott E. 1997. "An Appending Efficient Algorithm for Allocating Public Projects with Complementarities," *Journal of Public Economics* 64, no 3: 291–322.

Page, Scott E. 2001. "Self Organization and Coordination." *Computational Economics* 18: 25–48.

Page, Scott E. 2006. "Essay: Path Dependence." *Quarterly Journal of Political Science* 1: 87–115.

Page, Scott E. 2007. *The Difference: How the Power of Diversity Creates Better Groups, Teams, Schools, and Societies.* Princeton, NJ: Princeton University Press.

Page, Scott E. 2010a. *Diversity and Complexity.* Princeton, NJ: Princeton University Press.

Page, Scott E. 2010b. "Building a Science of Economics for the Real World." Presentation to the House Committee on Science and Technology Subcommittee on Investigations and Oversight, July 20.

Page, Scott E. 2012. "A Complexity Perspective on Institutional Design." *Politics, Philosophy and Economics* 11: 5–25.

Page, Scott E. 2017. *The Diversity Bonus.* Princeton, NJ: Princeton University Press.

Pan, Jessica. 2014. " Gender Segregation in Occupations: The Role of Tipping and Social Interactions." *Journal of Labor Economics* 33, no. 2: 365–408.

Parrish, Susan Scott. 2017. *The Flood Year 1927: A Cultural History*. Princeton, NJ: Princeton University Press.

Parsa, H. G., John T. Self, David Njite, and Tiffany King. 2005. "Why Restaurants Fail." *Cornell Hospitality Quarterly* 46, no. 3: 304–322.

Patel, Kayur, Steven Drucker, James Fogarty, Ashish Kapoor, and Desney Tan. 2011. "Using Multiple Models to Understand Data." *Proceedings of the International Joint Conference on Artificial Intelligence*, 1723-1728.

Peel, L., and A. Clauset. 2014. "Predicting Sports Scoring Dynamics with Restoration and Anti-Persistence." *Proceedings of the International Conference on Data Mining*. Philadelphia: SIAM.

Pfeffer, Fabian T., and Alexandra Killewald. 2017. "Generations of Advantage: Multigenerational Correlations in Family Wealth." *Social Forces*, 1–31.

Piantadosi, Steven. 2014. "Zipf's Word Frequency Law in Natural Language: A Critical Review and Future Directions." *Psychonomic Bulletin & Review* 21, no. 5: 1112–1130.

Pierson, Paul. 2004. *Politics in Time: History, Institutions, and Social Analysis*. Princeton, NJ: Princeton University Press.

Piketty, Thomas. 2014. *Capital in the 21st Century*. Trans. Arthur Goldhammer. Cambridge, MA: Belknap Press.

Pollack, John. 2014. *Shortcut: How Analogies Reveal Connections, Spark Innovation, and Sell Our Greatest Ideas*. New York: Gotham.

Poole, Keith T., and Howard Rosenthal. 1984. "A Spatial Model for Legislative Roll Call Analysis." *American Journal of Political Science* 29, no. 2: 357–384.

Porter, David, and Vernon Smith. 2007. "FCC Spectrum Auction Design: A 12-Year Experiment." *Journal of Law, Economics, and Policy* 3, no. 1: 63–80.

Powell, Robert. 1991. "Absolute and Relative Gains in International Relations Theory." *American Political Science Review* 85, no. 4: 1303–1320.

Przeworski, Adam, Jose Antonio Cheibub, Michael E. Alvarez, and Fernando Limongi. 2000. *Democracy and Development: Political Institutions and Material Well-Being in the World, 1950–1990*. Cambridge: Cambridge University Press.

Raby, Fiona. 2001. *Design Noir: The Secret Life of Electronic Objects*. Basel: Birkhauser.

Ramo, Joshua Cooper. 2016. *The Seventh Sense: Power, Fortune, and Success, in the Age of Networks*. New York: Little, Brown and Company.

Rand, David G., Hisashi Ohtsukia, and Martin A. Nowak. 2009. "Direct Reciprocity with Costly Punishment: Generous Tit-for-Tat Prevails." *Journal of Theoretical Biology* 256, no. 1: 45–57.

Rapoport, Anatol. 1978. "Reality-Simulation: A Feedback Loop." *Sociocybernetics*, 123–141.

Rauch, Jeffrey. 2012. *Hyperbolic Partial Differential Equations and Geometric Optics*. Graduate Studies in Mathematics. Providence, RI: American Mathematical Society.

Rawls, John. 1971. *A Theory of Justice*. Cambridge, MA: Harvard University Press.

Rescorla, Robert, and Allan Wagner. 1972. "A Theory of Pavlovian Conditioning: Variations in the Effectiveness of Reinforcement and Nonreinforcement." In *Classical Conditioning II,* ed. A. H. Black and W. F. Prokasy, 64–99. New York: Appleton-Century-Crofts.

Reynolds, Noel B., and Arlene Saxonhouse. 1994. *Three Discourses*. Chicago: University of Chicago Press.

Roberts, D. C., and D. L. Turcotte. 1998. "Fractality and Self-Organized Criticality of Wars." *Fractals* 6: 351–357.

Roberts, Seth. 2004. "Self-Experimentation as a Source of New Ideas: Ten Examples About Sleep, Mood, Health, and Weight." *Behavioral and Brain Sciences* 27, no. 2: 227–262

Romer, Paul. 1986. "Increasing Returns and Long-Run Growth." *Journal of Political Economy* 94: 1002–1037.

Rosen, Sherwin. 1981. "The Economics of Superstars." *American Economic Review* 71: 845–858.

Roth, Alvin, and Ido Erev. 1995. "Learning in Extensive Form Games: Experimental Data and Simple Dynamic Models in the Intermediate Term." *Games and Economics Behavior* 8: 164–212.

Russakoff, Dale. 2015. *The Prize: Who's in Charge of America's Schools?* Boston: Houghton Mifflin Harcourt.

Rust, Jon. 1987. "Optimal Replacement of GMC Bus Engines: An Empirical Model of Harold Zurcher." *Econometrica* 55, no. 5: 999–1033.

Ryall, Michael D., and Aaron Bramson *Inference and Intervention: Causal Models for Business Analysis*. New York: Routledge.

Salganik, Matthew, Peter Dodds, and Duncan J. Watts. 2006. "Experimental Study of Inequality and Unpredictability in an Artificial Cultural Market." *Science* 311: 854–856.

Samuelson, Paul. 1964. "Proof That Properly Anticipated Prices Fluctuate Randomly." *Industrial Management Review* 6: 41–49.

Schiller, Robert. 2004. *Irrational Exuberance.* 2nd ed. Princeton, NJ: Princeton University Press.

Schrodt, Philip. 1998. "Pattern Recognition of International Crises Using Hidden Markov Models." In *Non-Linear Models and Methods in Political Science,* ed. Diana Richards. Ann Arbor: University of Michigan Press.

Schwartz, Christine R., and Robert D. Mare. 2004. "Trends in Educational Assortative Marriage from 1940 to 2003." *Demography* 42, no. 4: 621–646.

Schelling, Thomas. 1978. *Micromotives and Macrobehavior.* New York: W. W. Norton.

Scott, Steven L. 2010. "A Modern Bayesian Look at the Multi-Armed Bandit." *Applied Stochastic Models in Business and Industry* 26: 639–658.

Shalizi, Cosma, and Andrew C. Thomas. 2011. "Homophily and Contagion Are Generically Confounded in Observational Social Network Studies." *Sociological Methods and Research* 40: 211–239.

Shapiro, Thomas, Tatjana Meschede, and Sam Osoro. 2013. "The Roots of the Widening Racial Wealth Gap: Explaining the Black-White Economic Divide." Research and Policy Brief, Institute on Assets and Social Policy, Brandeis University, Waltham, MA.

Shi, Xiaolin, Lada A. Adamic, Belle L. Tseng, and Gavin S. Clarkson. 2009. "The Impact of Boundary Spanning Scholarly Publications and Patents." *PLoS ONE* 4, no. 8: e6547.

Silver, Nate. 2012. *The Signal and the Noise: Why So Many Predictions Fail—but Some Don't.* New York: Penguin.

Simler, Kevin, and Robin Hanson. 2018. *The Elephant in the Brain: Hidden Motives in Everyday Life.* Oxford: Oxford University Press.

Simmons, Matthew, Lada Adamic, and Eytan Adar. 2011. "Memes Online: Extracted, Subtracted, Injected, and Recollected." Paper presented at the International Conference on Web and Social Media.

Slaughter, Ann Marie. 2017. *The Chessboard and the Web: Strategies of Connection in a Networked World.* New Haven, CT: Yale University Press.

Smaldino, Paul. 2013. "Measures of Individual Uncertainty for Ecological Models: Variance and Entropy." *Ecological Modelling* 254: 50–53.

Small, Dana M., Robert J. Zatorre, Alain Dagher, Alan C. Evans, and Marilyn Jones-Gotman. 2001. "Changes in Brain Activity Related to Eating Chocolate: From Pleasure to Aversion." *Brain* 124, no. 9: 1720–1733.

Smith, Eric, Rebecca Bliege Bird, and D. Bird. 2003. "The Benefits of Costly Signaling: Meriam Turtle Hunters." *Behavioral Ecology* 14: 116–126.

Smith, Vernon. 2002. "Constructivist and Ecological Rationality." Nobel Prize lecture.

Sneppen, Kim, Per Bak, Henrik Flyvbjerg, and Mogens Jensen. 1994. "Evolution as a Self-Organized Critical Phenomenon." *Proceedings of the National Academy of Sciences* 92: 5209–5213.

Solow, Robert M. 1956. "A Contribution to the Theory of Economic Growth." *Quarterly Journal of Economics* 70, no. 1: 65–94.

Spence, A. Michael. 1973. "Job Market Signaling." *Quarterly Journal of Economics* 87, no. 3: 355–374.

Squicciarini, Mara, and Nico Voigtländer. 2015. "Human Capital and Industrialization: Evidence from the Age of Enlightenment." *Quarterly Journal of Economics* 30, no. 4: 1825–1883.

Starfield, Anthony, Karl Smith, and Andrew Bleloch. 1994. *How to Model It: Problem Solving for the Computer Age.* Minneapolis, MN: Burgess International.

Stein, Richard A. 2011. "Superspreaders in Infectious Diseases." *International Journal of Infectious Diseases* 15, no. 8: e510–e513.

Sterman, John D. 2000. *Business Dynamics: Systems Thinking and Modeling for a Complex World.* New York: McGraw-Hill.

Sterman, John. 2006. "Learning from Evidence in a Complex World." *American Journal of Public Health* 96, no. 3: 505–515.

Stiglitz, Joseph. 2013. *The Price of Inequality: How Today's Divided Society Endangers Our Future.* New York: W. W. Norton.

Stock, James H., and Mark W. Watson. 2003. "Has the Business Cycle Changed and Why?" In *National Bureau of Economic Research Macroeconomics Annual 2002,* vol. 17, ed. Mark Gertler and Kenneth Rogoff, 159–218. Cambridge, MA: MIT Press.

Stone, Lawrence D., Colleen M. Keller, Thomas M. Kratzke, and Johan P. Strumpfer. 2014. "Search for the Wreckage of Air France Flight AF 447." *Statistical Science* 29, no. 1: 69–80.

Storchmann, Karl. 2011. "Wine Economics: Emergence, Developments, Topics." *Agrekon* 50, no. 3: 1–28.

Suki, Bela, and Urs Frey. 2017. "A Time Varying Biased Random Walk Model of Growth: Application to Height from Birth to Childhood." *Journal of Critical Care* 38: 362–370.

Suroweicki, James. 2006. *The Wisdom of Crowds*. New York: Anchor Press.

Syverson, Chad. 2007. "Prices, Spatial Competition, and Heterogeneous Producers: An Empirical Test." *Journal of Industrial Economics* 55, no. 2: 197–222.

Taleb, Nassim. 2001. *Fooled by Randomness*. New York: Random House.

Taleb, Nassim. 2007. *The Black Swan: The Impact of the Highly Improbable*. New York: Random House.

Taleb, Nassim. 2012. *Antifragile: Things That Gain from Disorder*. New York: Random House.

Tassier, Troy. 2013. *The Economics of Epidemiology*. Amsterdam: Springer.

Tetlock, Phillip. 2005. *Expert Political Judgment: How Good Is It? How Can We Know?* Princeton, NJ: Princeton University Press.

Thaler, R. H. 1981. "Some Empirical Evidence on Dynamic Inconsistency." *Economic Letters* 8, no. 3: 201–207.

Thompson, Derek. 2014. "How You, I, and Everyone Got the Top 1 Percent All Wrong: Unveiling the Real Story Behind the Richest of the Rich." *Atlantic,* March 30.

Thorndike, Edward L. 1911. *Animal Intelligence*. New York: Macmillan.

Tilly, Charles. 1998. *Durable Inequality*. Berkeley: University of California Press.

Tsebelis, George. 2002. *Veto Players: How Political Institutions Work*. Princeton, NJ: Princeton University Press.

Turchin, Peter. 1998. *Quantitative Analysis of Movement: Measuring and Modeling Population Redistribution in Animals and Plants*. Sunderland, MA: Sinauer Associates.

Tweedle, Valerie, and Robert J. Smith. 2012. "A Mathematical Model of Bieber Fever: The Most Infectious Disease of Our Time?" In *Understanding the Dynamics of Emerging and Re-Emerging Infectious Diseases Using*

Mathematical Models, ed. Steady Mushayabasa and Claver P. Bhunu. Cham, Switzerland: Springer.

Ugander, Johan, Brian Karrer, Lars Backstrom, and Cameron Marlow. 2011. "The Anatomy of the Facebook Social Graph." arXiv:1111.4503.

Updike, John. 1960. "Hub Fans Bid Adieu." *New Yorker,* October 22.

US Bureau of Labor Statistics. 2013. *Consumer Expenditures in 2011.* Report 1042, April. Washington, DC: BLS.

Uzzi, Brian, Satyam Mukherjee, Michael Stringer, and Ben Jones. 2013. "Atypical Combinations and Scientific Impact." *Science* 342: 468–471.

Van Noorden, Richard. 2015. "Interdisciplinary Research by the Numbers." *Nature,* September 16.

von Neumann, John, and Morgenstern, Oskar. 1953. *Theory of Games and Economic Behavior.* Princeton, NJ: Princeton University Press.

Vriend, Nicolaas J. 2000. "An Illustration of the Essential Difference Between Individual and Social Learning, and Its Consequences for Computational Analyses." *Journal of Economic Dynamics and Control* 24: 1–19.

Wainer, Howard. 2009. *Picturing the Uncertain World.* Princeton, NJ: Princeton University Press.

Wakeland, W., A. Nielsen, and P. Geissert. 2015. "Dynamic Model of Nonmedical Opioid Use Trajectories and Potential Policy Interventions." *American Journal of Drug and Alcohol Abuse* 41, no. 6: 508–518.

Waltz, Kenneth. 1979. *Theory of International Politics.* New York: McGraw-Hill.

Washington Post. 2012. "Mad Money: TV Ads in the 2012 Presidential Campaign." http://www.washingtonpost.com/wp-srv/special/politics/track-presidential-campaign-ads-2012.

Watts, Duncan. 2011. *Everything Is Obvious Once You Know the Answer.* New York: Crown Business.

Watts, Duncan, and Steven Strogatz. 1998. "Collective Dynamics of 'Small-World' Networks." *Nature* 393, no. 6684: 440–442.

Weisberg, Michael. 2007. "Three Kinds of Idealization." *Journal of Philosophy* 104, no. 12: 639–659.

Weisberg, Michael. 2012. *Simulation and Similarity: Using Models to Understand the World.* Oxford: Oxford University Press.

Weisberg, Michael, and Muldoon, Ryan. 2009. "Epistemic Landscapes and the Division of Cognitive Labor." *Philosophy of Science* 76, no. 2: 225–252.

Weitzman, Martin L. 1979. "Optimal Search for the Best Alternative." *Econometrica* 77: 641–654.

Weitzman, Martin L. 1998. "Recombinant Growth." *Quarterly Journal of Economics* 2: 331–361.

Wellman, Michael. 1990. "Fundamental Concepts of Qualitative Probabilistic Networks." *Artificial Intelligence* 44: 257–303.

Wellman, Michael. 2013. "Head to Head: Does US High-Frequency Trading Need Stricter Regulatory Oversight? (YES)." *International Financial Law Review,* September.

West, Geoffrey. 2017. *Scale: The Universal Laws of Growth, Innovation, Sustainability, and the Pace of Life in Organisms, Cities, Economies, and Companies.* New York: Penguin.

Whittle, Peter. 1979. "Discussion of Dr Gittins' Paper." *Journal of the Royal Statistical Society, Series B* 41, no. 2: 148–177.

Whitty, Robin W. 2017. "Some Comments on Multiple Discovery in Mathematics." *Journal of Humanistic Mathematics* 7, no. 1: 172–188.

Wigner, Eugene. 1960. "The Unreasonable Effectiveness of Mathematics in the Natural Sciences." *Communications in Pure and Applied Mathematics* 13, no. 1.

Wilkinson, Richard, and Kate Pickett. 2009. *The Spirit Level: Why Greater Equality Makes Societies Stronger.* London: Bloomsbury.

Wilson, David Sloan. 1975. "A Theory of Group Selection." *Proceedings of the National Academy of Sciences* 72, no. 1: 143–146.

Wolfram, Stephen. 2001. *A New Kind of Science.* Champaign, IL: Wolfram Media.

Wright, Robert. 2001. *Nonzero: The Logic of Human Destiny.* New York: Vintage.

Wu, Jianzhong, and Robert Axelrod. 1995. "How to Cope with Noise in the Iterated Prisoner's Dilemma." *Journal of Conflict Resolution* 39, no. 1: 183–189.

Wuchty, Stefan, Benjamin F. Jones, and Brian Uzzi. 2007. "The Increasing Dominance of Teams in the Production of Knowledge." *Science* 316, no. 5827: 1036–1039.

Xie, Yu. 2007. "Otis Dudley Duncan's Legacy: The Demographic Approach to Quantitative Reasoning in Social Science." *Research in Social Stratification and Mobility* 25: 141–156.

Xie, Yu, Alexandra Killewald, and Christopher Near. 2016. "Between- and Within-Occupation Inequality: The Case of High Status Professions." *Annals of the American Academy of Political and Social Science* 663, no. 1: 53–79.

Youn, Hyejin, Deborah Strumsky, Luis Bettencourt, and José Lobo. 2015. "Inventions as a Combinatorial Process: Evidence From US Patents." *Journal of the Royal Society Interfaces* 12: 0272.

Zagorsky, Jay. 2007. "Do You Have to Be Smart to Be Rich? The Impact of IQ on Wealth, Income and Financial Distress." *Intelligence* 35: 489–501.

Zahavi, Amotz. 1974. "Mate Selection: A Selection for a Handicap." *Journal of Theoretical Biology* 53, no. 1: 205–214.

Zak, Paul, and Stephen Knack. 2001. "Trust and Growth." *Economic Journal* 111, no. 470: 295–321.

Zaretsky, Adam. 1998. "Have Computers Made Us More Productive? A Puzzle." *Regional Economist,* Federal Reserve Bank of St. Louis.

Ziliak, Stephen T., and Deirdre N. McCloskey. 2008. *The Cult of Statistical Significance: How the Standard Error Costs Us Jobs, Justice, and Lives.* Ann Arbor: University of Michigan Press.

Index

SCOTT E. PAGE is the Leonid Hurwicz Collegiate Professor of Complex Systems, Political Science, and Economics at the University of Michigan and an external faculty member of the Santa Fe Institute.

Photo Credit: Cooper Page